HOW BIOLOGY WORKS

HOW BIOLOGY WORKS

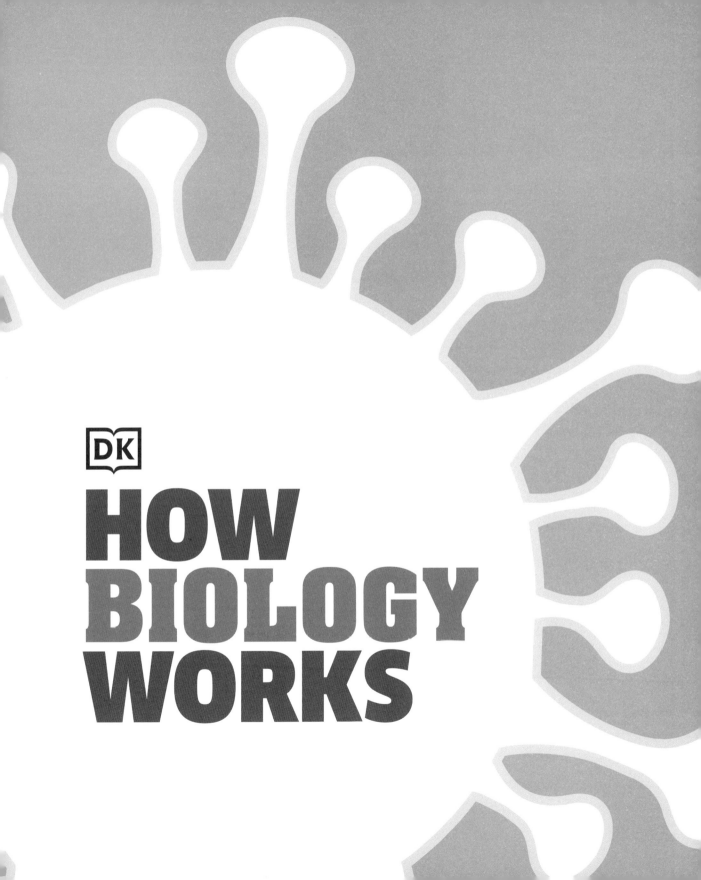

DK | Penguin Random House

Editorial Consultants
Chris Clennett, Jo Locke, Tom Jackson

Project Art Editors
Francis Wong, Steve Woosnam-Savage

Art Editors
Stephen Bere, Amy Child, Mik Gates

Illustrators
Edwood Burn, Victoria Clark, Mark Clifton, Dan Crisp

Managing Art Editor
Michael Duffy

Jacket Designer
Tanya Mehrotra

Production Editor
Rob Dunn

Senior Production Controller
Meskerem Berhane

Art Director
Karen Self

Contributors
Olivia Drake, Jack Challoner, Tim Harris, Alina Ivan, Tom Jackson, Nicola Temple

Senior Editors
Peter Frances, Miezan van Zyl

Project Editors
Michael Clark, Sarah MacLeod, Martyn Page

Editors
Jemima Dunne, Annie Moss

Editorial Assistant
Emily Kho

US Editor
Jill Hamilton

Managing Editor
Angeles Gavira Guerrero

Publisher
Liz Wheeler

Publishing Director
Jonathan Metcalf

First American Edition, 2023
Published in the United States by DK Publishing
1745 Broadway, 20th Floor, New York, NY 10019

23 24 25 26 27 10 9 8 7 6 5 4 3 2

004–333998–Jun/23

Published in Great Britain by Dorling Kindersley Limited

A catalog record for this book is available from the Library of Congress.

ISBN 978-0-7440-8074-2

Printed and bound in China

For the curious

www.dk.com

MIX
Paper from responsible sources
FSC™ C018179
FSC www.fsc.org

This book was made with Forest Stewardship Council™ certified paper—one small step in DK's commitment to a sustainable future. For more information go to www.dk.com/our-green-pledge

CONTENTS

WHAT IS LIFE?

THE CHEMISTRY OF LIFE

HOW CELLS WORK

REPRODUCTION AND GENETICS

EVOLUTION

THE TREE OF LIFE

HOW PLANTS WORK

HOW ANIMALS WORK

ECOLOGY

BIOTECHNOLOGY

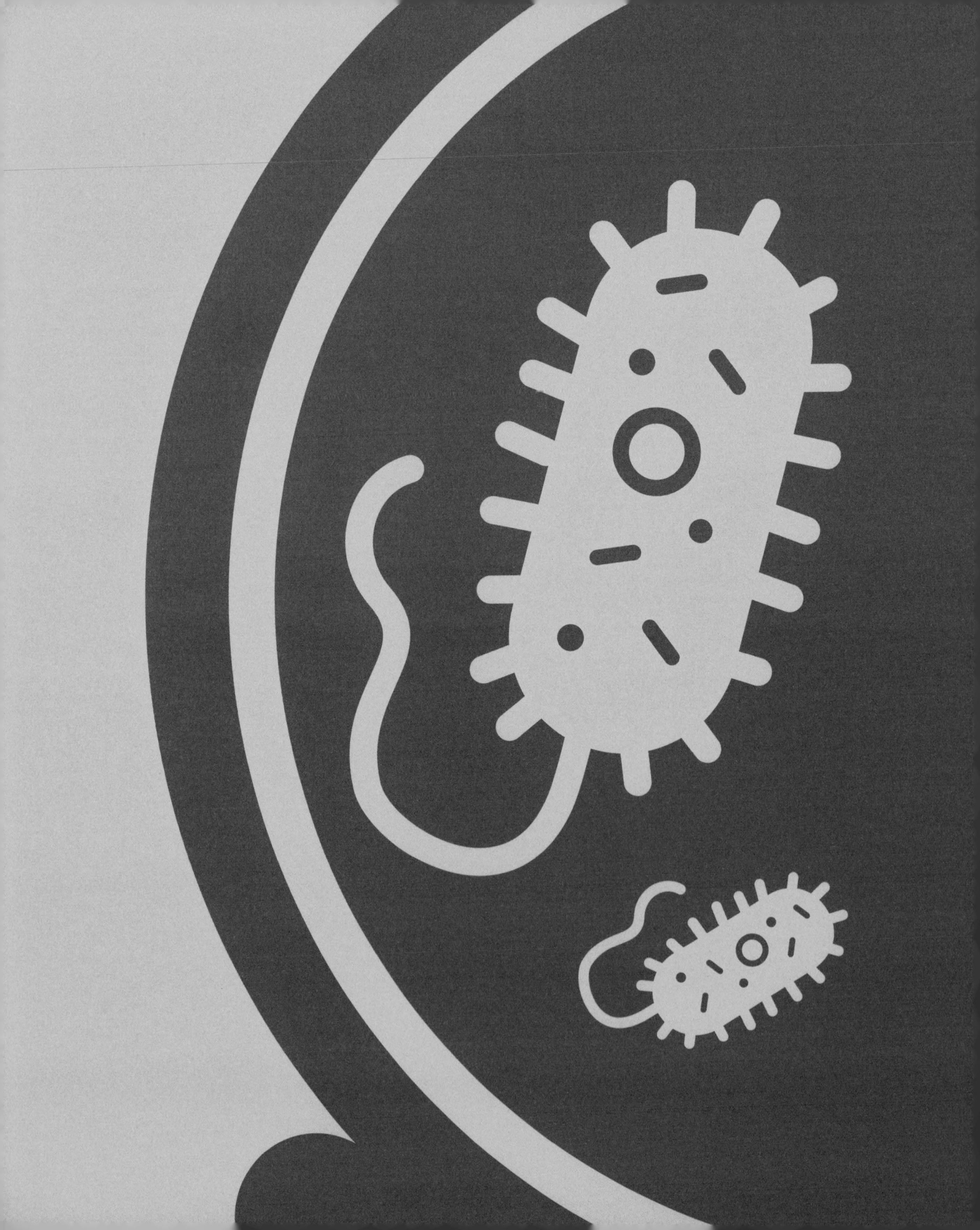

WHAT IS LIFE?

Features of life

What does it mean to be a living thing? There are many varieties and forms of living organisms, and they use different strategies to survive and thrive, but they all share certain basic features that set them apart from inanimate objects.

REPRODUCTION

A living body is produced under the guidance of a set of DNA instructions (genes). The "purpose" of the body is to generate fresh copies of that DNA and create new bodies that will carry the DNA copies and pass them on to the next generation.

GROWTH

All living things increase in size. Complex organisms, such as mammals and trees, develop from a single cell into an intricate, multicellular body with many distinct parts. Single-celled organisms also grow, by enlarging their body prior to it dividing into two.

LIVING ORGANISM

All living things use all of the fundamental life processes at some point in their life. *Dinobryon*—a type of microscopic freshwater alga (shown above)—is unusual because it has two methods of nutrition, one by photosynthesis, the other by eating bacteria.

NUTRITION

Living things need a supply of nutrients to provide the raw materials for building and maintaining their bodies and as a source of energy. Plants obtain these nutrients from the soil, water, and air; animals and fungi do it by consuming the bodies or wastes of other organisms.

The fundamental processes of life

Living organisms are among the most complex things known. However, underlying this complexity are seven fundamental processes that come together to produce life. Nonliving things may be able to carry out some of these functions; for example, a crystal can grow, while an engine can release energy from fuel, move around, and remove its own waste. However, only living things use all of these processes at once.

SENSITIVITY

Organisms are sensitive to changes in their surroundings and can respond to these changes. The responses may vary from simple, such as a bacterium forming a protective cyst in dry conditions, to complex, such as the "fight or flight" response in mammals.

RESPIRATION

The energy needed to sustain life comes from respiration, a chemical process that takes place inside every living cell. This process breaks down chemical fuels, like sugars, into simpler substances, releasing energy that can be used by the organism's cells.

THERE ARE AN ESTIMATED **8.7 MILLION SPECIES** OF LIFE FORM **LIVING** ON EARTH **TODAY**

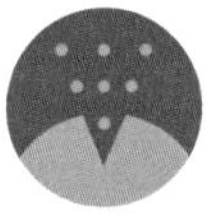

EXCRETION

Life processes create waste products that must be removed from the body. For example, respiration in animals and plants produces carbon dioxide as waste. Animals also produce urine, which contains toxic chemicals, and expel it from their bodies.

MOVEMENT

To some degree, all forms of life are able to move. Animals are the most mobile, while plants can open and close pores in their leaves and orient themselves toward light. Many single-celled organisms move by beating hairlike cilia or flagella.

Vital functions
Life is a set of vital processes that sustain the form and function of a body, at least for a short time and long enough for it to produce offspring.

Cells and life

Cells are the basic structural and functional units of all living things, including animals, plants, and microorganisms. Each cell contains DNA (genetic information that controls cell function and enables new cells to be made). One of the basic principles of biology is cell theory. This was formulated in the 1830s after scientists were able to observe the detailed structure of cells clearly through microscopes for the first time. The theory states that life is based on cells, every living body is made from at least one cell, and every new cell can only ever arise from an older one.

WHY IS LIFE BASED ON CARBON?

Carbon atoms can combine with themselves and many other types of atoms, enabling them to form a huge number of complex compounds that can be used by life.

A body of cells
A living body starts with a single cell, which itself was made by an older organism. That cell divides and specializes into different types of cells to build the body.

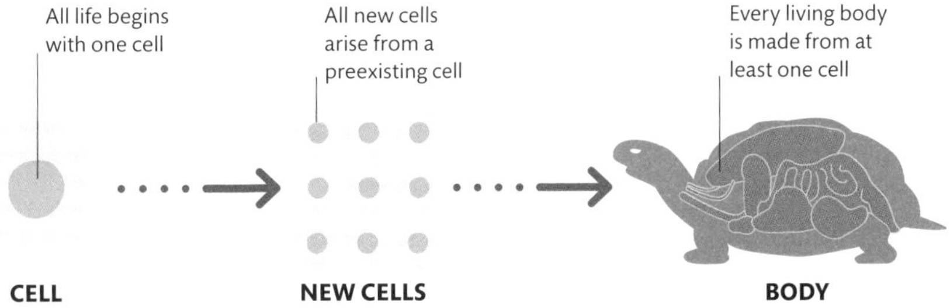
All life begins with one cell
All new cells arise from a preexisting cell
Every living body is made from at least one cell
CELL
NEW CELLS
BODY

ENTROPY AND LIFE

Life processes create ordered structures, such as cells and bodies, from disordered raw materials. This creation of order would seem to contravene the laws of thermodynamics, which say that natural systems always increase in entropy (become more disordered). However, the metabolic processes of life also break down some large molecules and produce heat, which increases the entropy of the system as a whole.

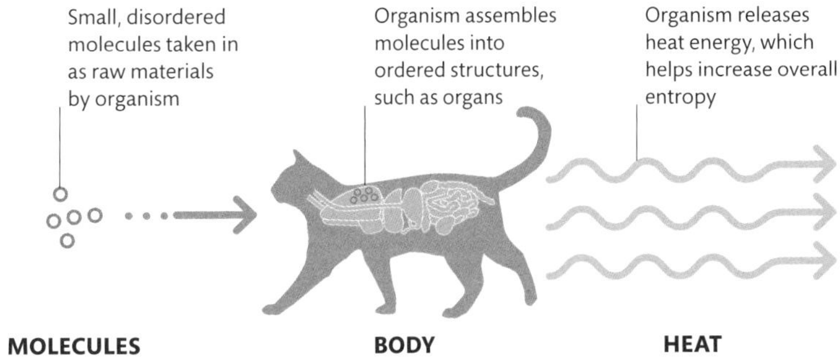
Small, disordered molecules taken in as raw materials by organism
Organism assembles molecules into ordered structures, such as organs
Organism releases heat energy, which helps increase overall entropy
MOLECULES
BODY
HEAT

Homeostasis

To function optimally, living things maintain a constant internal environment. They achieve this through a process called homeostasis. This involves various interconnected mechanisms that regulate internal factors such as body temperature, water level, and chemical balance of body fluids.

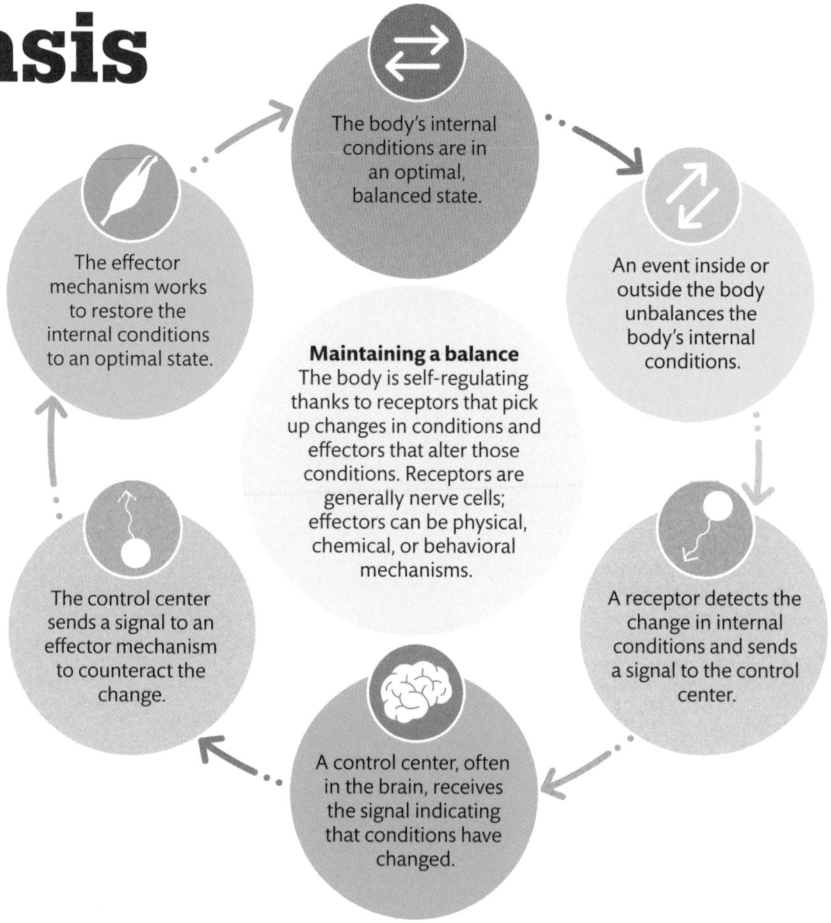

The body's internal conditions are in an optimal, balanced state.
An event inside or outside the body unbalances the body's internal conditions.
A receptor detects the change in internal conditions and sends a signal to the control center.
A control center, often in the brain, receives the signal indicating that conditions have changed.
The control center sends a signal to an effector mechanism to counteract the change.
The effector mechanism works to restore the internal conditions to an optimal state.
Maintaining a balance
The body is self-regulating thanks to receptors that pick up changes in conditions and effectors that alter those conditions. Receptors are generally nerve cells; effectors can be physical, chemical, or behavioral mechanisms.

Feedback loops

Homeostasis relies on a system known as a negative feedback loop to respond to any changes in internal conditions in order to keep them as stable as possible. A feedback loop involves several steps that work to restore the internal conditions to an optimal state by counteracting any changes to that state that are detected by internal receptors.

Balancing water

Water is the primary constituent of all living bodies, and the control of water levels (called osmoregulation) is vital. Water moves by osmosis from an area of high water concentration (low salt concentration) to an area of low water concentration (high salt concentration). The goal of osmoregulation is to keep concentrations of chemicals in the cells' cytoplasm and other body fluids at optimal levels for metabolic processes to work efficiently. This involves removing excess water and taking in more when supplies run low. Water supplies are depleted by various body processes, such as the production of liquid urine to remove toxins and, in land animals, sweating to regulate body temperature.

Water balance in different environments
The way in which an organism's water balance is managed depends on how water enters and leaves the organism's body, which is greatly influenced by its habitat.

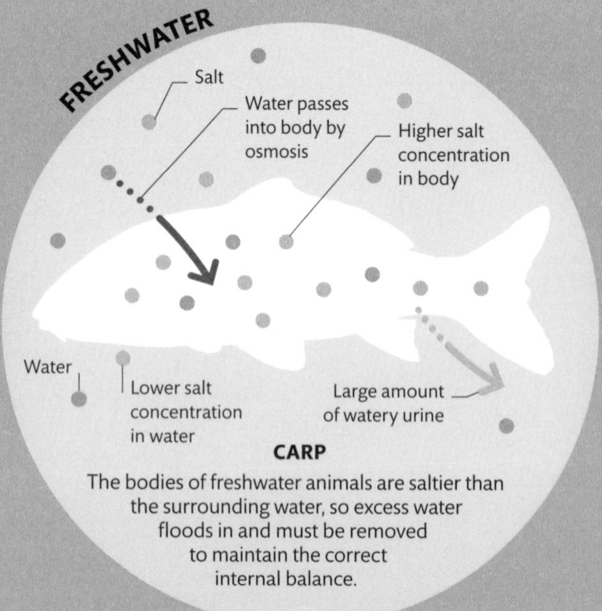

FRESHWATER
Salt
Water passes into body by osmosis
Higher salt concentration in body
Water
Lower salt concentration in water
Large amount of watery urine
CARP
The bodies of freshwater animals are saltier than the surrounding water, so excess water floods in and must be removed to maintain the correct internal balance.

Regulating temperature

Thermoregulation—control of body temperature—is carried out by animals in two ways. Endotherms ("warm-blooded" animals) actively control their temperature by internal processes such as sweating or shivering. Ectotherms ("cold-blooded" animals) rely on external heat or cooling to regulate body temperature. As a result, endotherms can stay active in colder conditions and exploit more habitats.

SOME **DINOSAURS** HAD **FEATHERS,** SUGGESTING THAT THEY WERE **ENDOTHERMS**

	STAYING WARM	STAYING COOL
Endotherms ("warm blooded")	Body insulation—such as fur, feathers, or fat—reduces heat loss. Shivering muscles release heat into the rest of the body. Blood vessels in the skin constrict to reduce the amount of warm blood reaching the body's outer surface.	Sweating and panting make the liquid sweat or saliva evaporate into vapor, which cools the body. Blood vessels in the skin dilate, which results in more blood reaching the outer surface of the body, where it loses heat.
Ectotherms ("cold blooded")	Sunbathing on hot rocks.	Resting in the shade.

HOMEOSTASIS IN PLANTS

When water levels are low, plants close their stomata—pores that let gases and water vapor in and out of the leaves (see p.147). The pores are between two guard cells that change shape as water passes into or out of their vacuoles to open or close the pores.

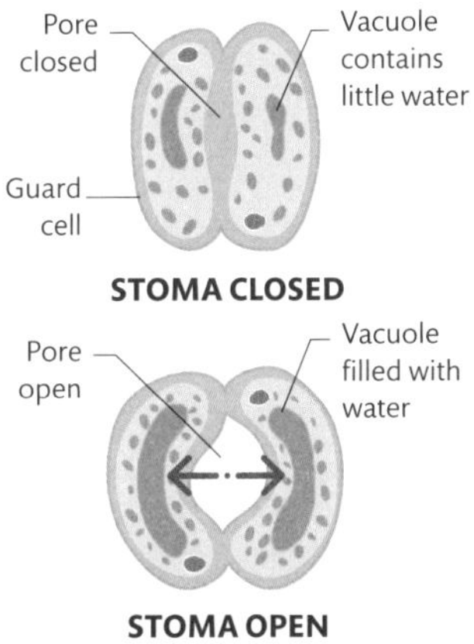
Pore closed
Vacuole contains little water
Guard cell
STOMA CLOSED
Pore open
Vacuole filled with water
STOMA OPEN

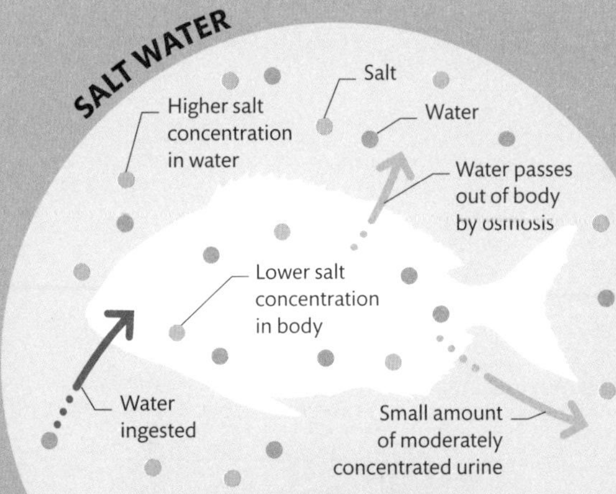
SALT WATER
Salt
Water
Higher salt concentration in water
Water passes out of body by osmosis
Lower salt concentration in body
Water ingested
Small amount of moderately concentrated urine

SEA BREAM

In marine habitats, the bodies of animals are less salty than the surrounding water so osmosis constantly draws water out of the animal, which must take in water continually to replace it.

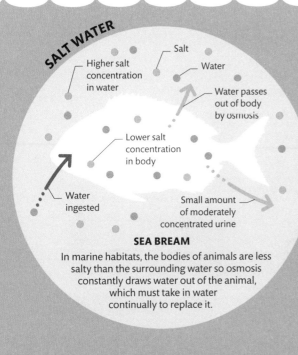

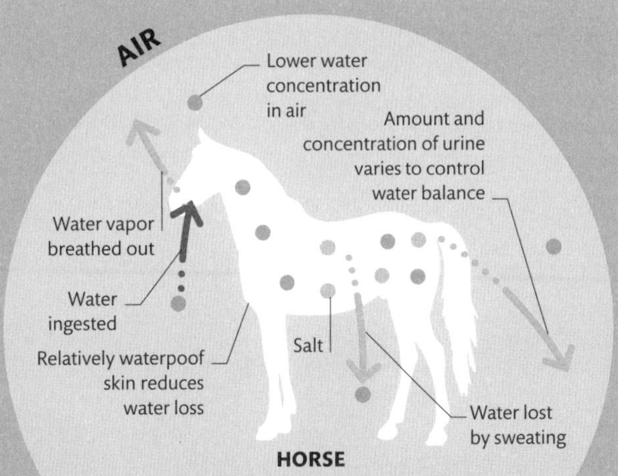
AIR
Lower water concentration in air
Amount and concentration of urine varies to control water balance
Water vapor breathed out
Water ingested
Relatively waterpoof skin reduces water loss
Salt
Water lost by sweating

HORSE

For land animals, water is continually lost to the environment. Animals must therefore conserve, regulate, and take in water to maintain the correct internal balance.

Kingdoms of life

To make sense of the diversity of life, scientists classify organisms into groups called taxons, whose members share specific features. Taxons are ranked by size, starting with domains and then kingdoms.

The five kingdoms

The originator of the most widely used classification system, Swedish taxonomist Carl Linnaeus envisaged that life was divided into just two kingdoms: animals and plants. However, the discovery of microorganisms and advances in cell biology have now revealed that life can be grouped into five kingdoms. While the kingdoms have certain distinct large-scale features, most of the defining characteristics are seen at the cellular level.

THREE DOMAINS

In 1977, American microbiologist Carl Woese found that life forms fall into three domains by their cell structure and the form of their ribosomal RNA (protein-making molecules inside cells). Eukarya are organisms whose cells have a nucleus. Archaea and bacteria have no cell nucleus, and each group has a distinct cell membrane structure.

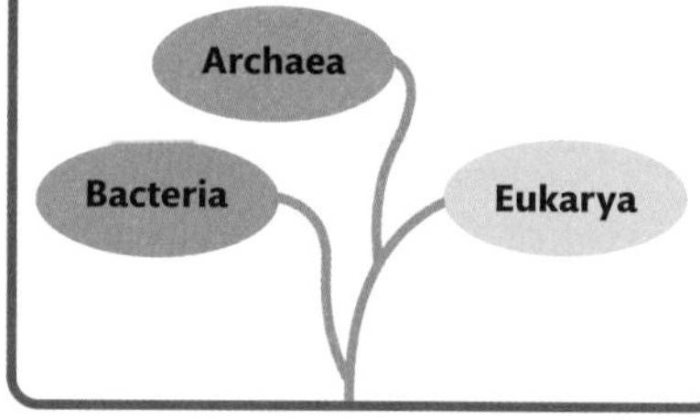

PROKARYOTES (PROKARYOTAE)

This kingdom includes the bacteria and the archaea (a distinct group of primeval organisms). Prokaryotes are single-celled microorganisms. Their cells are smaller than those of life forms in other kingdoms, and largely have simple shapes. The cells do not contain a nucleus or other membrane-bound structures, and the DNA is clustered in a single, simple chromosome.

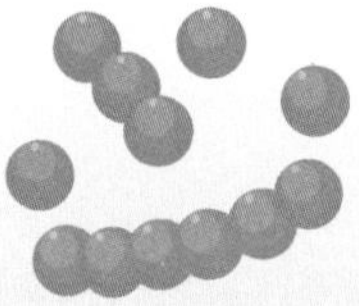
COCCUS

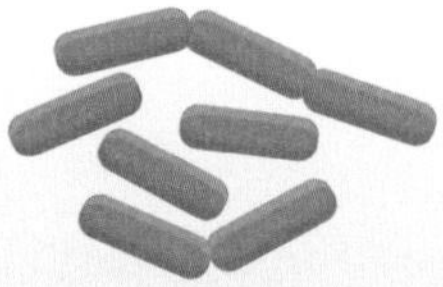
BACILLUS

SPIRILLUM

PROTOCTISTS (PROTOCTISTA)

These diverse organisms may have cell walls, or mobile cilia or flagella, or be enclosed in shell-like tests. All protoctist cells have a nucleus that contains DNA in chromosomes, and internal organelles. Some protoctists, such as amoebas, are like animals in that they ingest foods; others, such as diatoms, are plantlike algae that get their energy by photosynthesis.

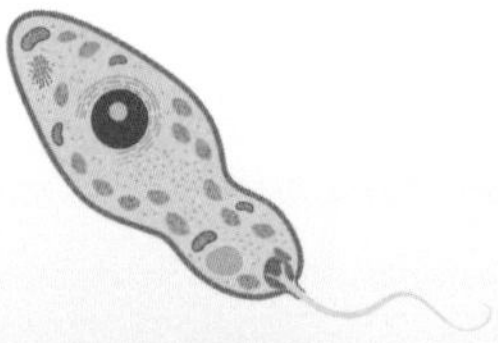
FLAGELLATE

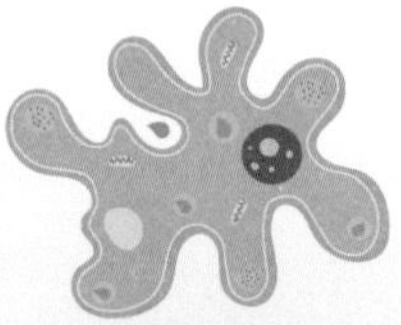
AMOEBA

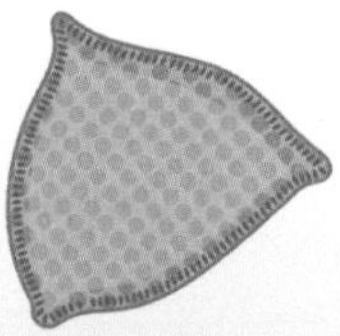
DIATOM

PLANTS (PLANTAE)

Plants are multicellular organisms that survive using photosynthesis to provide their source of energy. They include simple mosses and ferns and larger and more complex seed-bearing plants, such as conifers and angiosperms (flowering plants). They have cell walls (typically made of cellulose) and a cell nucleus, as well as internal organelles such as chloroplasts.

TREE

FERN

FLOWERING PLANT

Kingdom Chromista

Some scientists say that there is a sixth kingdom, named Chromista. It includes single-celled organisms drawn from Protoctista, and multicellular seaweeds, such as kelp, currently in the plant kingdom. The cells contain organelles that have a certain kind of chlorophyll, or structures that evolved from them; as a result, Chromista also includes the malaria parasite, *Plasmodium*.

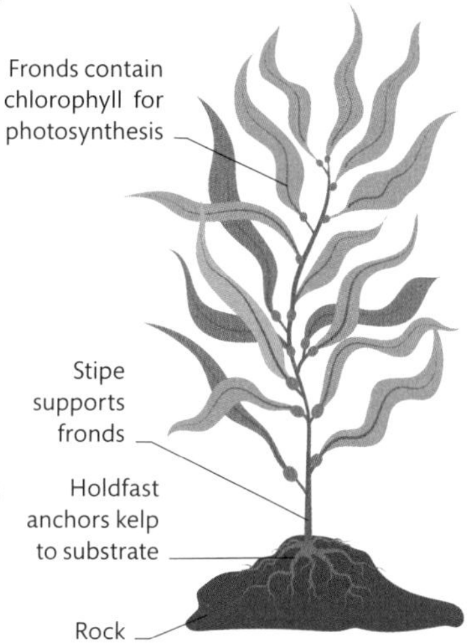
Fronds contain chlorophyll for photosynthesis
Stipe supports fronds
Holdfast anchors kelp to substrate
Rock substrate
KELP

Naming organisms

Species are classified using a binomial system: a pair of scientific names for the genus (overall group) and then the species (distinct organism). This does away with confusing local names. For example, Europe and North America both have birds commonly known as robins but these are distinct species and are classified differently.

EUROPEAN ROBIN
ERITHACUS RUBECULA

AMERICAN ROBIN
TURDUS MIGRATORIUS

FUNGI

All fungal cells use the biological polymer chitin to build their cell walls, a chemical also widely used by invertebrate animals. Fungi are saprophytes, which means they grow on their food source, often in barely noticeable networks of filaments, and secrete enzymes to digest it externally. Mushrooms and toadstools are temporary fruiting bodies that spread spores.

MUSHROOM

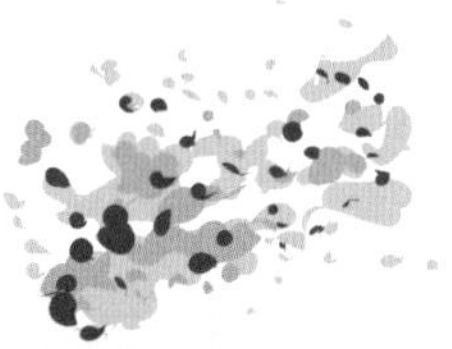
YEAST

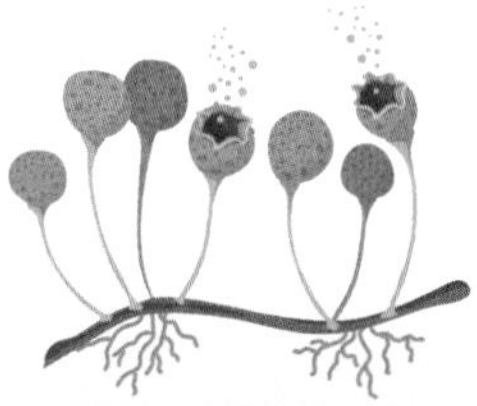
MOLD

ANIMALS (ANIMALIA)

Animals are multicellular, often highly complex organisms that rely on ingesting other organisms for food to provide energy. Their cells have a nucleus and internal organelles, such as mitochondria, but no rigid cell wall. Many animals are able to move around freely for at least part of their life cycle, and most need oxygen for metabolism (the cellular processes that sustain life).

ANNELID WORM

REPTILE

FISH

BIRD

MAMMAL

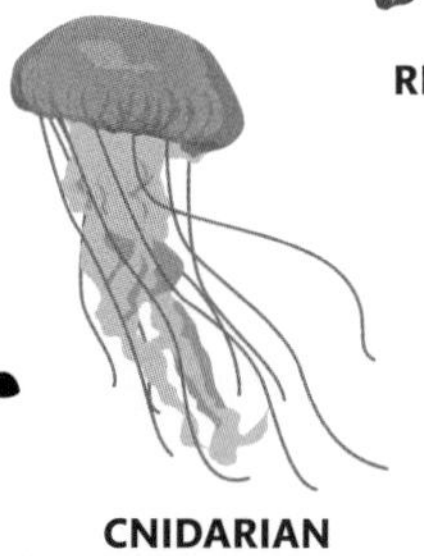
CNIDARIAN

INSECT

Viruses

A virus is a packet of parasitic DNA or RNA (genetic material). A parasite steals the resources of a host, and viruses do this by taking over the machinery of a cell to replicate the viral genes. In the process, this causes disease in the host.

ARE VIRUSES ALIVE?

Viruses are not generally considered to be alive, because they cannot carry out all of the essential life processes by themselves, such as reproduction and feeding.

Structure of viruses

Viruses are not cells, and they do not have the cellular structures needed to carry out life processes. Viruses essentially comprise a strand of genetic material, either DNA or RNA, enclosed within a protective coat of proteins. Most viruses are tiny compared to a human body cell—typically, about 50 times smaller.

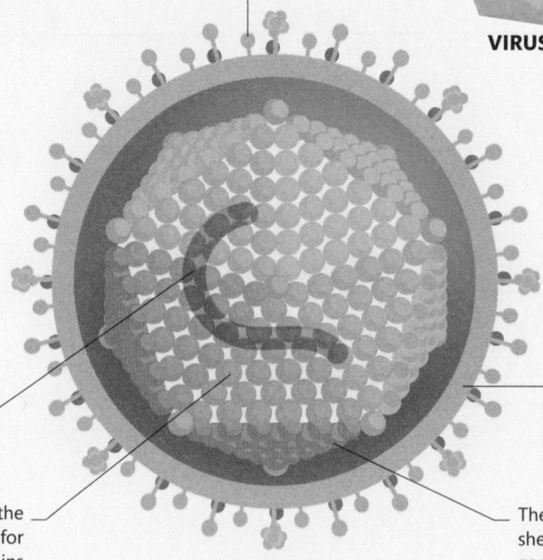
Surface or "spike" proteins lock onto proteins on the surface of a host cell
A single strand of RNA or a double strand of DNA contains the virus's genetic material
Capsomer protein units make up the capsid; the viral genes carry the code for the capsomer and other viral proteins
The outer envelope, present in most viruses, is made up of lipids and proteins
The capsid—a protein shell—surrounds the genetic material
VIRUS

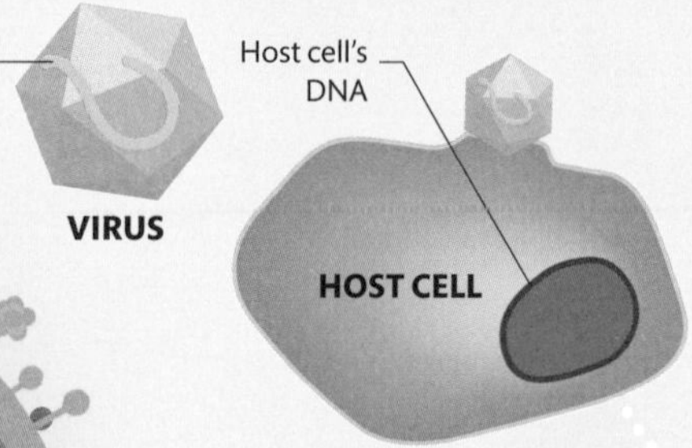
Virus's genetic material
VIRUS
Host cell's DNA
HOST CELL

1 Virus attacks host cell
A virus enters the body and locks onto a host cell. It targets the particular cells that have receptors matching its surface proteins.

Virus shapes

There is a great diversity in size and shape of viruses, and the number of viruses is unknown. Many thousands have been identified in mammals, birds, and plants, but there are likely to be millions more types yet to be investigated that attack other organisms.

THERE ARE **MORE VIRUSES** IN **2 PINTS (1 LITER)** OF **SEA WATER** THAN THERE ARE **PEOPLE** ON EARTH

Bacilliform
This lozenge- or bullet-shaped structure is limited to a group of RNA viruses, which includes the virus that causes rabies.

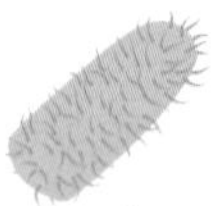

Helical rod
Viruses with this shape tend to attack plants. The DNA or RNA and protein coat are coiled in a spring shape.

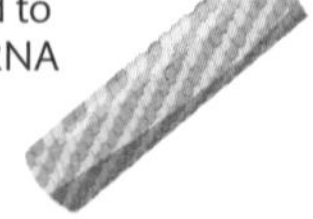

Filamentous
These are forms of helical viruses that form into long, thin strands, such as the human Ebola virus.

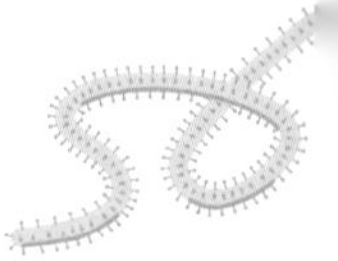

Complex
Known as bacteriophages, because they attack bacteria, these viruses stand on bacteria with their "legs" before injecting their DNA.

Spherical
Many common human viruses have this shape. The outer envelope is spherical but the inner capsid is often helical.

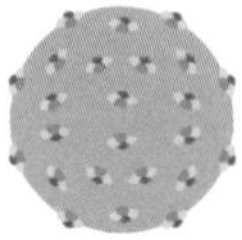

Polyhedral
The outer envelopes of these viruses have many sides, most commonly 20 (an icosahedron), as in adenoviruses.

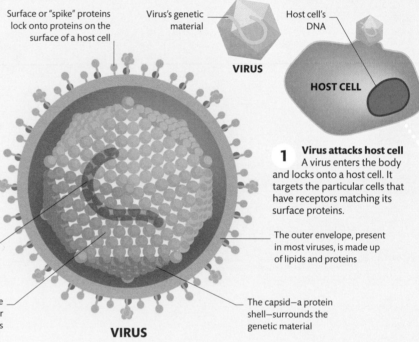

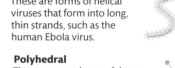

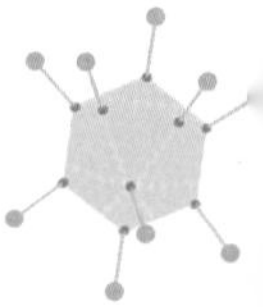

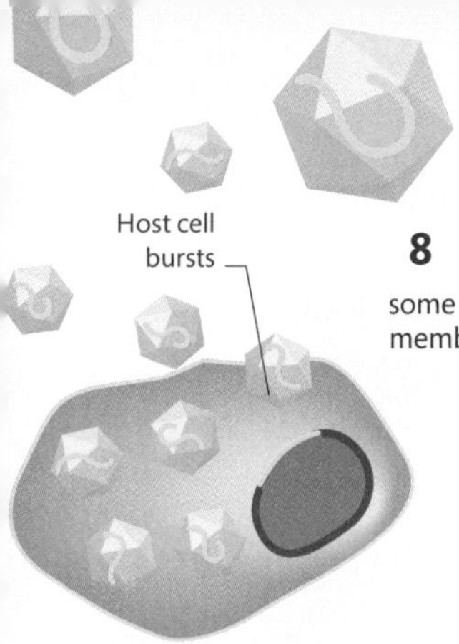
Host cell bursts

8 New virus particles released
The cell bursts and releases new viruses; some viruses may steal part of the cell's membrane as they leave.

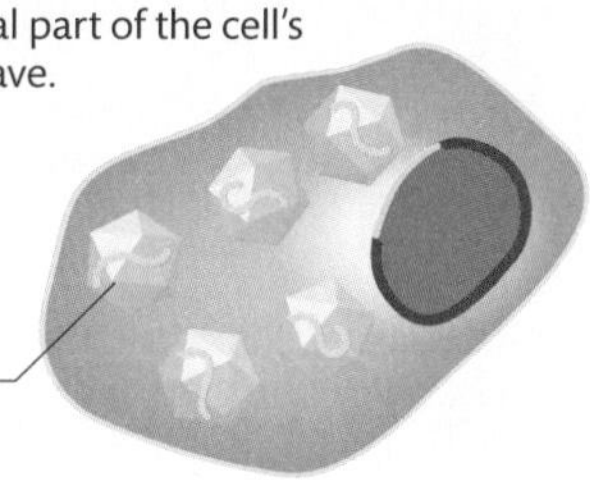
New virus particle

7 New virus particles formed
All the parts of the virus are assembled to make new virus particles that are capable of attacking new host cells.

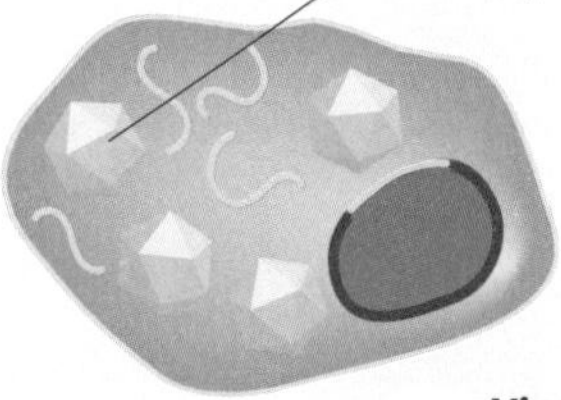
Viral coat proteins

Virus replication

A virus cannot reproduce by itself. To do so, it needs to take over a host cell and use that cell to replicate the virus's genetic material. In the process, the cell will make many copies of the virus but will eventually be killed. This destruction of cells is what causes diseases, such as flu and COVID-19.

6 Viral coat produced
The viral genes instruct the cell to produce proteins that new virus particles will need to make a capsid around their DNA.

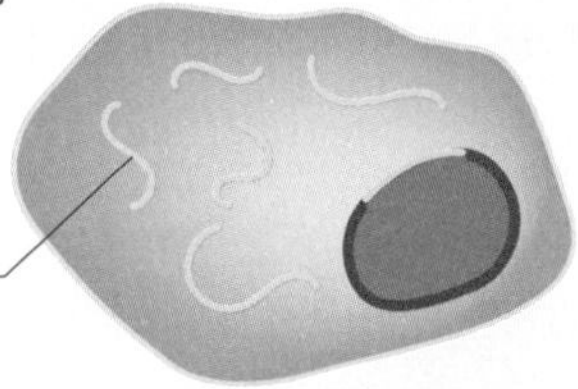
Copy of viral genes

5 Host cell replicates viral genetic material
Alternatively, the host cell produces many copies of the virus's DNA, which fill the cell's cytoplasm.

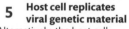

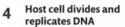

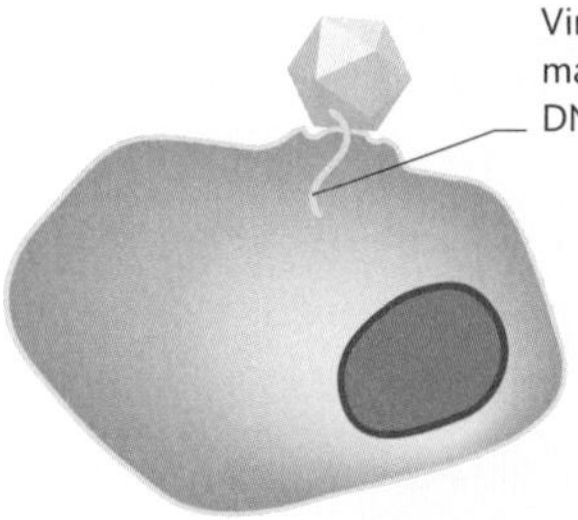
Virus's genetic material may be DNA or RNA

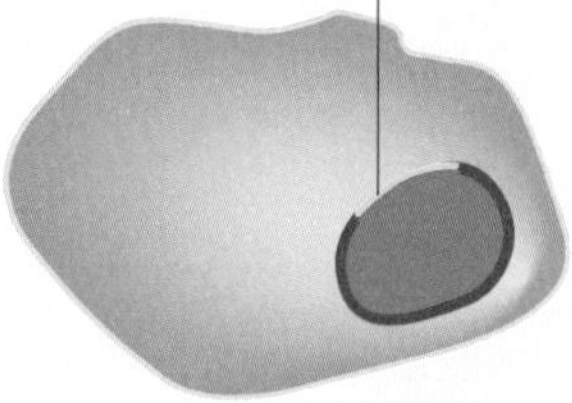
Viral genes in cell's DNA

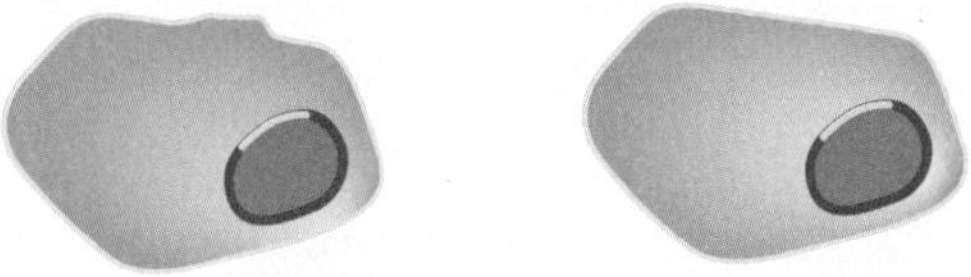

2 Viral genes enter host cell
The viral envelope stays outside the host cell but the virus's genetic material, usually with the capsid, is injected through the cell membrane.

3 Viral genes join host's DNA
In some cases, the viral DNA may copy itself using material from the cell's cytoplasm. In other cases, the viral DNA is added to the cell's genetic material.

4 Host cell divides and replicates DNA
The viral DNA is copied with the host cell's DNA when the cell divides. The daughter cells carry the virus with them.

DISEASE-CAUSING VIRUSES

Adenoviruses
A group of polyhedral viruses that can cause various illnesses in humans, including some types of common colds.

HIV
A spherical virus that damages the human immune system, potentially leading to other severe infections.

Coronaviruses
A group of crownlike spherical viruses that can affect the airways in humans. They include COVID-19 and some colds.

Rabies virus
A bacilliform virus that attacks the brain of humans and many other mammals, causing death if not treated promptly.

Varicella-zoster virus
A polyhedral virus that can cause chickenpox and shingles in humans. Similar viruses cause cold sores and herpes.

Ebola virus
A filamentous virus that can affect humans and other mammals, causing severe, often fatal, internal bleeding.

Human papillomavirus
A group of polyhedral viruses, some types of which cause warts or certain cancers, such as cervical cancer.

Tobacco mosaic virus
A helical rod virus that attacks tobacco plants; the first virus to be discovered, in the late 1800s.

The biosphere

The biosphere is the region on or close to our planet's surface in which it is possible for life to sustain itself. By definition, all life on Earth exists within the biosphere, and only humans have ever managed to travel beyond it.

Energy and resources

Life forms do not survive alone but instead associate in interdependent communities called ecosystems (see pp.182–83). The biosphere is the region of Earth that contains all these ecosystems. Every ecosystem has a set of ecological factors that govern how it functions, including nonliving, or abiotic, factors. These abiotic factors include energy sources, such as light, and supplies of useful chemicals, such as oxygen, liquid water, and rocky minerals. They are provided by three other spheres, or layers, around the planet: the lithosphere, hydrosphere, and atmosphere. The biosphere is the region where these other spheres combine to create conditions that are compatible with life.

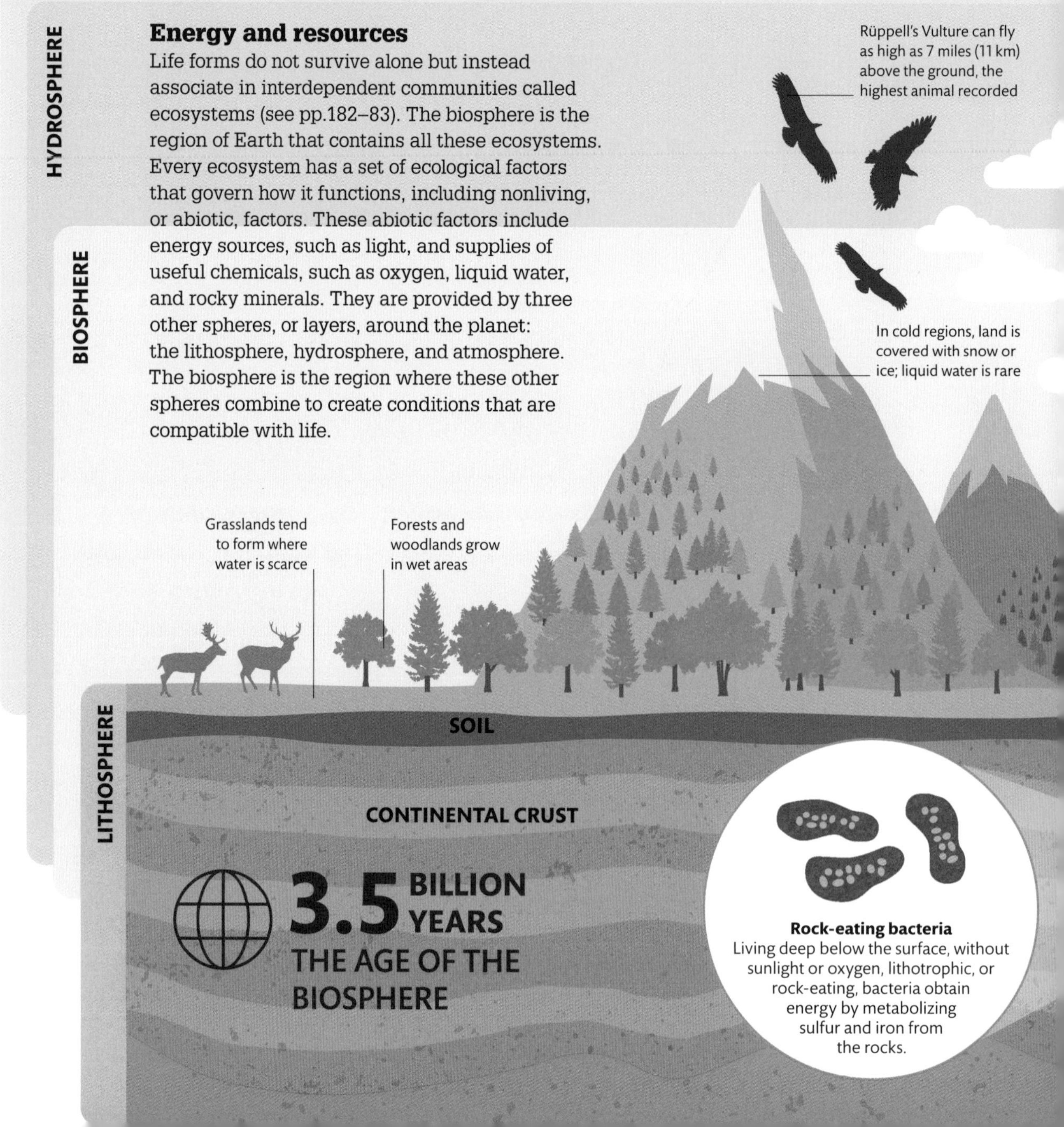

ATMOSPHERE
HYDROSPHERE
BIOSPHERE
LITHOSPHERE
Rüppell's Vulture can fly as high as 7 miles (11 km) above the ground, the highest animal recorded
In cold regions, land is covered with snow or ice; liquid water is rare
Grasslands tend to form where water is scarce
Forests and woodlands grow in wet areas
SOIL
CONTINENTAL CRUST

3.5 BILLION YEARS
THE AGE OF THE BIOSPHERE

Rock-eating bacteria
Living deep below the surface, without sunlight or oxygen, lithotrophic, or rock-eating, bacteria obtain energy by metabolizing sulfur and iron from the rocks.

Limits of the biosphere

The limits of the biosphere are primarily determined by the supply of energy and nutrients. Other critical factors include the oxygen level and temperature. Above an altitude of about 4 miles (6 km), the atmosphere is too thin for most animals to breathe. The lower limit is determined by temperature, which rises with increasing depth. Once deep rocks reach about 250°F (120°C), even extreme-living bacteria cannot survive.

Atmosphere
As well as containing oxygen and carbon dioxide, the atmosphere is the ultimate source of nitrogen, used in proteins and other essential life chemicals.

Hydrosphere
This layer contains water, mostly as liquid, although it may freeze into ice or evaporate into vapor. As well as filling the oceans, the hydrosphere also includes large amounts of groundwater in rocks.

Biosphere
The most crowded part of the biosphere is dry land. This forms a solid surface for dense ecosystems, such as forests, to grow. Oceans account for about 97 percent of the habitable space in the biosphere but contain only about 10 percent of the life seen on land, due to the relative lack of nutrients in the large volume of water.

Atmospheric winds spread fungal spores, plant seeds, microoganisms, and even tiny animals around the biosphere

Mineral nutrients from the land are washed into the oceans

Most ocean animals live within about 1,600 ft (500 m) of the surface

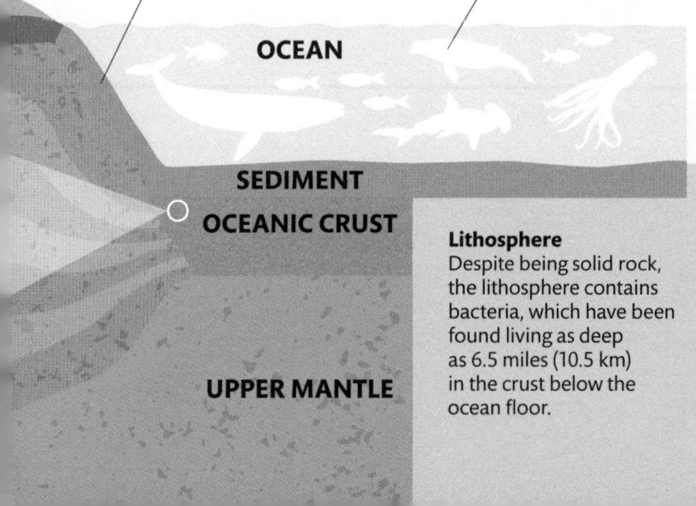
OCEAN
SEDIMENT
OCEANIC CRUST
UPPER MANTLE

Lithosphere
Despite being solid rock, the lithosphere contains bacteria, which have been found living as deep as 6.5 miles (10.5 km) in the crust below the ocean floor.

HOW LARGE IS THE BIOSPHERE?

The exact size of the biosphere is uncertain, but it is generally considered to occupy a volume about twice as large as all Earth's oceans combined.

ARTIFICIAL BIOSPHERES

If humans are to live for long periods away from Earth, we will need to develop an artificial biosphere. This environment will need to provide a breathable atmosphere as well as other essentials, such as food and water. Crucially, it will need to be self-sustaining. In the 1990s, an experimental artificial biosphere (called Biosphere 2) was built in the Arizona desert in the southwest US. Most of Biosphere 2 was, in effect, a vast greenhouse, with a range of habitat zones inside. The project also studied how humans coped with prolonged isolation. However, after more than a decade, the project ended without demonstrating that an artificial environment was viable.

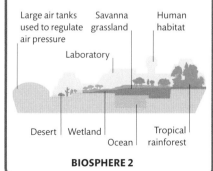

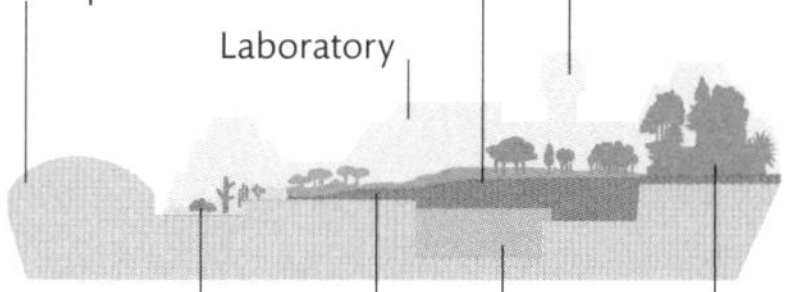
Large air tanks used to regulate air pressure
Savanna grassland
Human habitat
Laboratory
Desert
Wetland
Ocean
Tropical rainforest

BIOSPHERE 2

Nutrient cycles

About 95 percent of living bodies are made of four elements: hydrogen, oxygen, carbon, and nitrogen. These substances constantly cycle through the biosphere. Hydrogen and oxygen combine to make water, and the water cycle is a largely physical process. In contrast, the carbon and nitrogen cycles are driven by biological action.

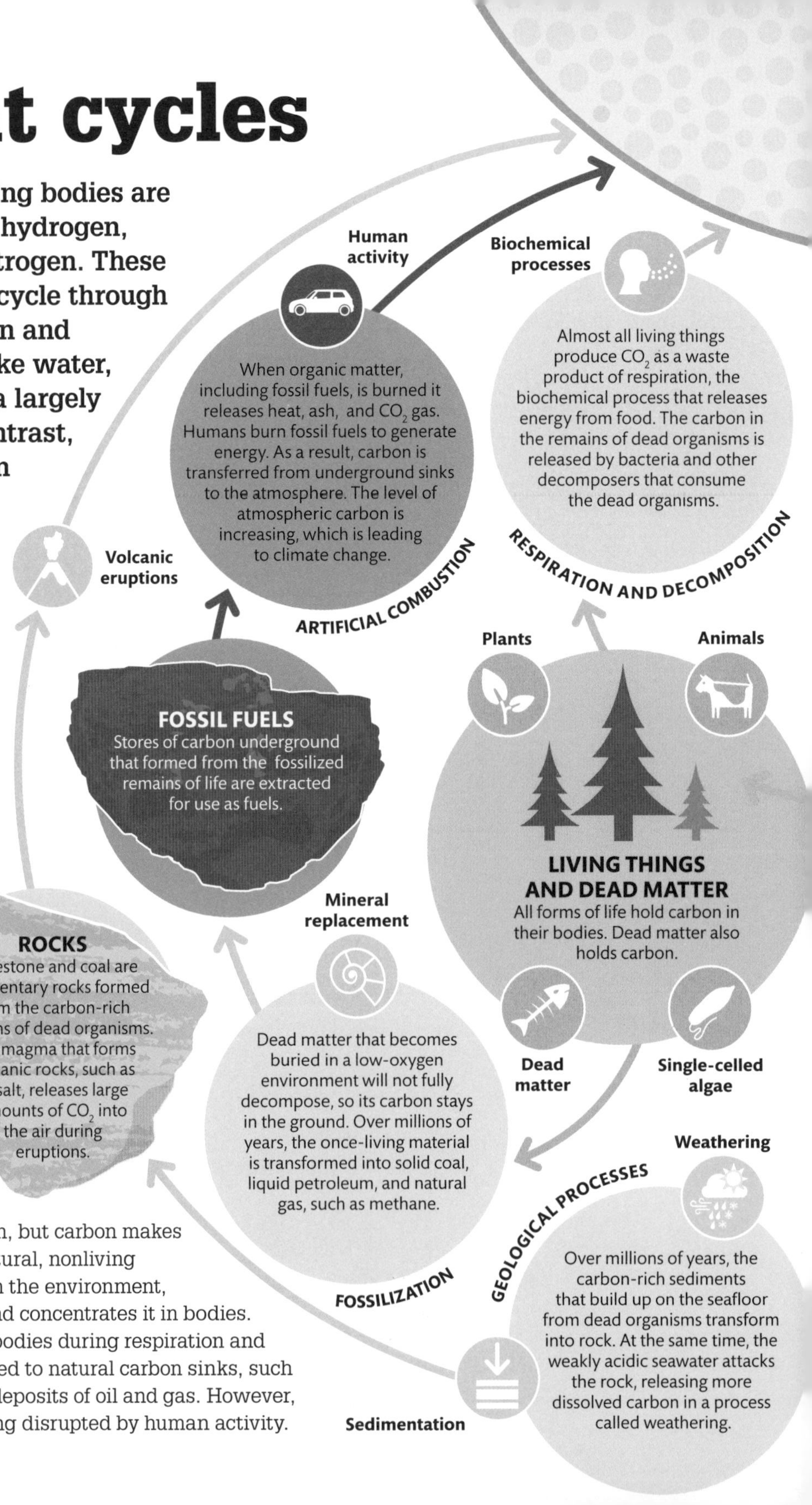

Human activity
When organic matter, including fossil fuels, is burned it releases heat, ash, and CO_2 gas. Humans burn fossil fuels to generate energy. As a result, carbon is transferred from underground sinks to the atmosphere. The level of atmospheric carbon is increasing, which is leading to climate change.
ARTIFICIAL COMBUSTION
Biochemical processes
Almost all living things produce CO_2 as a waste product of respiration, the biochemical process that releases energy from food. The carbon in the remains of dead organisms is released by bacteria and other decomposers that consume the dead organisms.
RESPIRATION AND DECOMPOSITION
Volcanic eruptions
FOSSIL FUELS
Stores of carbon underground that formed from the fossilized remains of life are extracted for use as fuels.
Plants
Animals
LIVING THINGS AND DEAD MATTER
All forms of life hold carbon in their bodies. Dead matter also holds carbon.
Dead matter
Single-celled algae
ROCKS
Limestone and coal are sedimentary rocks formed from the carbon-rich remains of dead organisms. The magma that forms volcanic rocks, such as basalt, releases large amounts of CO_2 into the air during eruptions.
Mineral replacement
Dead matter that becomes buried in a low-oxygen environment will not fully decompose, so its carbon stays in the ground. Over millions of years, the once-living material is transformed into solid coal, liquid petroleum, and natural gas, such as methane.
FOSSILIZATION
Weathering
GEOLOGICAL PROCESSES
Over millions of years, the carbon-rich sediments that build up on the seafloor from dead organisms transform into rock. At the same time, the weakly acidic seawater attacks the rock, releasing more dissolved carbon in a process called weathering.
Sedimentation

KEY
Parts of the carbon cycle occur in our lifetime. Other parts take millions of years.

- Slow (millions of years)
- Fast, natural (in our lifetime)
- Fast, artificial (in our lifetime)

The carbon cycle

About 20 percent of a living organism is composed of carbon, but carbon makes up only about 0.2 percent of natural, nonliving material. Life takes carbon from the environment, primarily by photosynthesis, and concentrates it in bodies. Carbon is released from living bodies during respiration and when they die; it is then returned to natural carbon sinks, such as the atmosphere, rocks, and deposits of oil and gas. However, this natural carbon cycle is being disrupted by human activity.

Atmosphere
Carbon dioxide (CO_2) makes up only about 0.04 percent of the atmosphere but plays a vital role in many life processes and also has a major effect on the global climate.

The nitrogen cycle

Nitrogen is an essential component of proteins, which are vital for life. Nitrogen is the most abundant gas in the atmosphere, but is unreactive. Most living things rely on bacteria to assimilate, or "fix," nitrogen from the air, converting it into nitrates for other organisms to use.

3.5 BILLION PEOPLE DEPEND ON **SYNTHETIC NITRATE FERTILIZERS** TO GROW **FOOD**

Taking in CO_2

Plants on land use energy from sunlight to lock CO_2 into bigger, more complex molecules, such as sugars. Single-celled algae perform an equivalent feat in the surface waters of the ocean. The organic carbon then passes through food chains.

PHOTOSYNTHESIS

CO_2 exchange

AIR-SEA TRANSFER

Atmospheric CO_2 dissolves easily in the oceans. The process is reversible so there is a slow, equal exchange between the air and water. Marine life uses the dissolved carbon to make carbonate shells, which eventually sink and create rock-forming sediments on the seabed.

OCEANS
Carbon is stored in ocean water as CO_2, carbonic acid, hydrogen carbonate, and carbonate.

Death and excretion

Plants eaten by animals

Some plants have nitrogen-fixing bacteria in their roots

ORGANIC WASTE

NITROGEN GAS IN THE AIR

Decomposers, such as bacteria and fungi, release simple nitrogen compounds, such as ammonia

Breakdown by denitrifying bacteria

Nitrifying bacteria fix nitrogen gas from the air. Lightning combines nitrogen and oxygen in the air

AMMONIA

Collection of nitrates by plant roots; the nitrates are used for growth and repair

NITRATES IN THE SOIL

Bacteria in soil convert ammonia into organic nitrate compounds

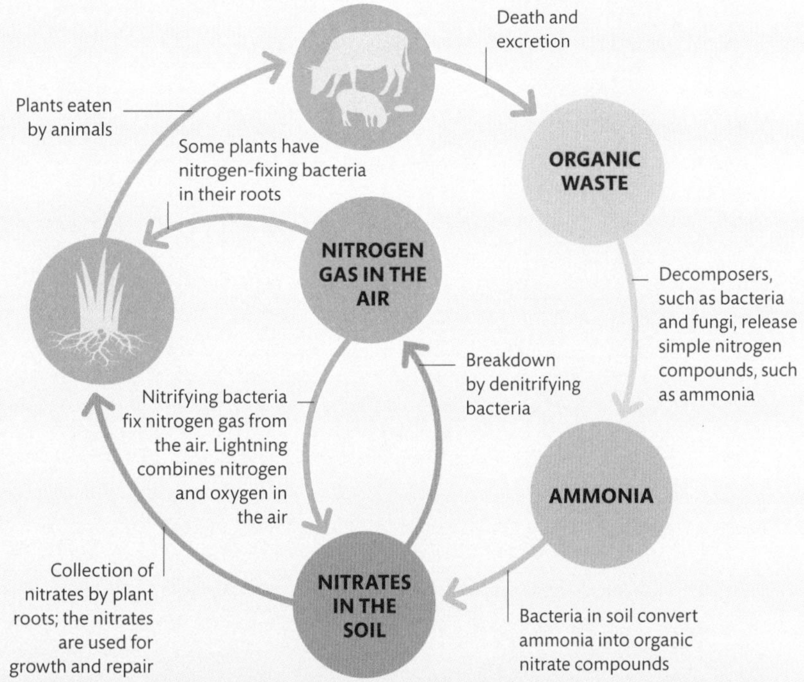

HUMAN INFLUENCES ON THE CARBON CYCLE

Over the past two centuries, human burning of fossil fuels has transferred several billion tons of carbon from underground carbon sinks to the atmosphere. The added carbon in the air traps heat, which is changing the global climate and creating extreme weather.

Total global CO_2 emissions

CO_2 emissions (billion tons)

40
35
30
25
20
15
10
5
0

1860 1880 1900 1920 1940 1960 1980 2000 2020

Year

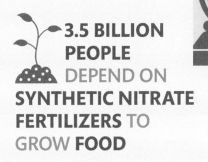

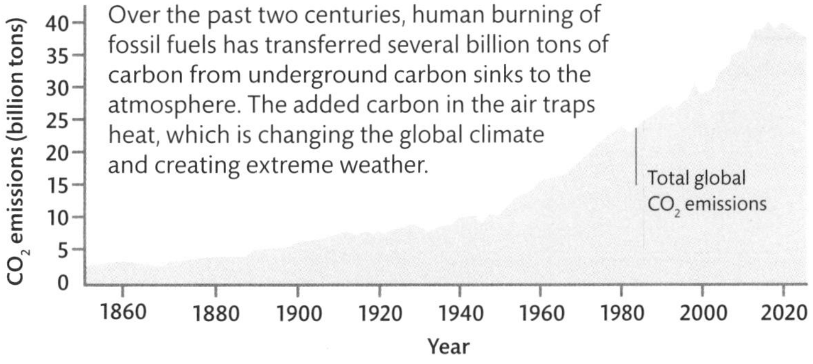

Life in extreme environments

The average air temperature on Earth is currently 57°F (14°C), and the average ocean surface temperature is 68°F (20°C). Most organisms are comfortable in these conditions. However, a few life forms can survive severe cold or heat or tolerate damaging chemicals; these are the extremophiles.

SURVIVING RADIOACTIVITY

Radioactivity damages the chemical makeup of a cell, so when the cell tries to divide it is destroyed in the process. Dangerous levels of radioactivity are largely a human-made phenomenon; levels in nature are relatively low. Nevertheless, some life forms are more resistant than others. For example, radioactivity has less impact on insects as their cells divide less frequently than cells in mammals; the latter produce billions of new cells each day.

Extremophiles

The name "extremophile" means "one that loves extreme conditions." Most are bacteria and archaea, which evolved early in Earth's history when the biosphere was a much harsher place to live, although a few types of frog, fish, insect, and crustacean are also found in extreme habitats. Extremophiles are adapted to a specific extreme. Some survive heat that would destroy the enzymes of ordinary organisms or withstand freezing conditions in which regular cell metabolism would stop. Others can cope with high levels of chemicals that would ordinarily disrupt an organism's ability to regulate its normal functions.

THE BACTERIUM ***DEINOCOCCUS RADIODURANS*** IS ONE OF THE **TOUGHEST MICROORGANISMS** KNOWN; IT CAN EVEN **SURVIVE** THE **VACUUM OF SPACE**

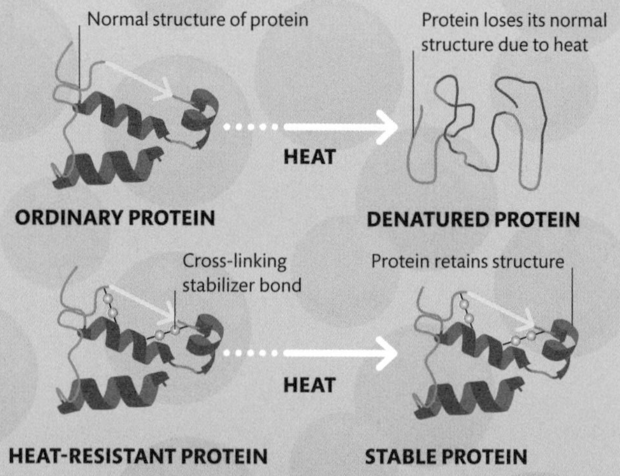
Normal structure of protein
Protein loses its normal structure due to heat
HEAT
ORDINARY PROTEIN
DENATURED PROTEIN
Cross-linking stabilizer bond
Protein retains structure
HEAT
HEAT-RESISTANT PROTEIN
STABLE PROTEIN

Thermophiles

These "heat loving" microorganisms can survive in water from above 100°F up to about 176°F (40°–100°C). Normally, proteins such as enzymes, when exposed to high heat, become denatured: they lose their normal structure and function. Thermophiles have heat-resistant proteins reinforced with cross-linking bonds so they hold their shape and keep working at high temperatures.

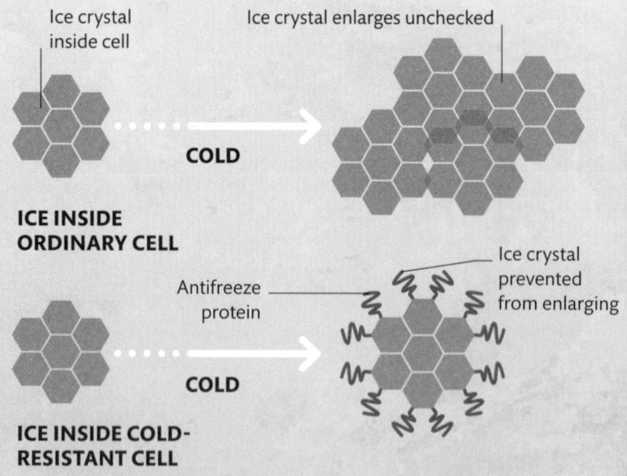
Ice crystal inside cell
Ice crystal enlarges unchecked
COLD
ICE INSIDE ORDINARY CELL
Antifreeze protein
Ice crystal prevented from enlarging
COLD
ICE INSIDE COLD-RESISTANT CELL

Psychrophiles

These "cold lovers" are resistant to freezing. They include frogs that spend the winter in frozen river water. When a normal cell is exposed to freezing conditions, large ice crystals form in the cytoplasm (see p.54), destroying the cell. By contrast, cytoplasm in psychrophiles contains cryoprotectants, or antifreeze proteins, which ensure that the ice only forms tiny crystals.

Tardigrades

These microscopic animals can survive boiling water, being frozen to –328°F (–200°C), and even outer space. Tardigrades normally live in various moist environments, including in soil and in marine or freshwater sediments. However, in dry, oxygen-poor, or other extreme conditions, some species transform their body into a variety of inactive states that can survive for years or even centuries.

Active state
In favorable conditions, a tardigrade is fully functional, able to eat, move, grow, and reproduce.

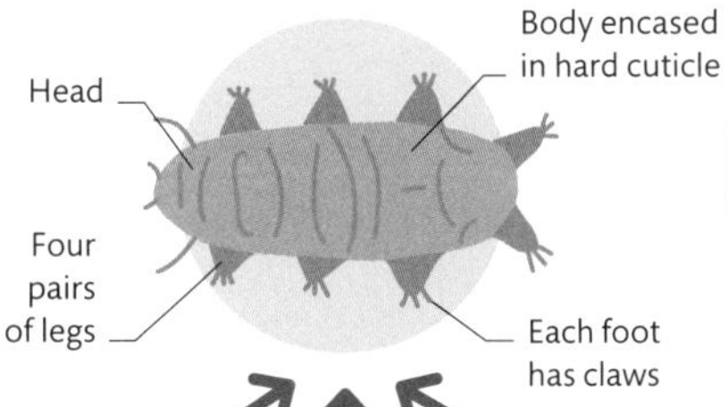
Head
Body encased in hard cuticle
Four pairs of legs
Each foot has claws

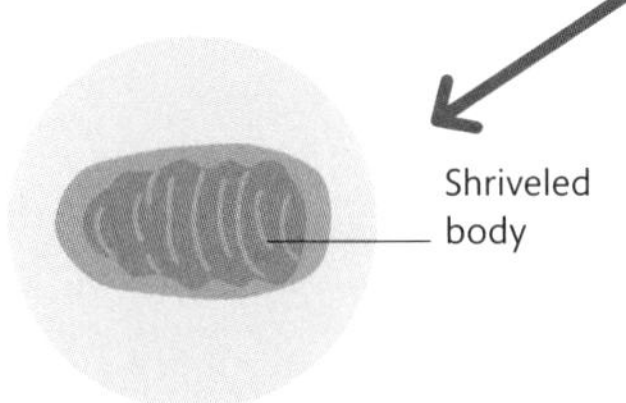
Shriveled body

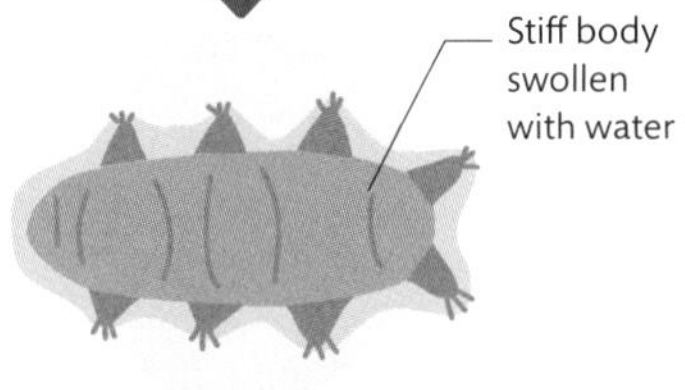
Stiff body swollen with water

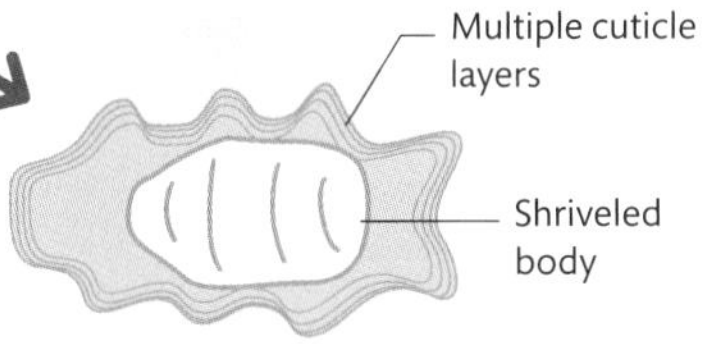
Multiple cuticle layers
Shriveled body

Tun
In extremely cold, salty, or dry conditions, the processes of living are suspended. The tardigrade's body dries out, shrinks, and curls up; this state is called the tun state.

Anoxybiosis
Low-oxygen conditions cause the tardigrade's body to absorb increased levels of water, becoming swollen, stiff, and immobile. The tardigrade becomes dormant.

Cyst
To withstand slow changes such as seasonal changes, the tardigarde builds up extra layers of cuticle. Its body shrivels inside the hardened, thick cuticle, producing a cyst.

CAN EXTREMOPHILES BE USEFUL?

Heat-resistant enzymes from thermophilic bacteria are used in polymerase chain reactions (PCR, see p.204), in which DNA is heated for analysis. PCR is used to detect infections such as COVID-19.

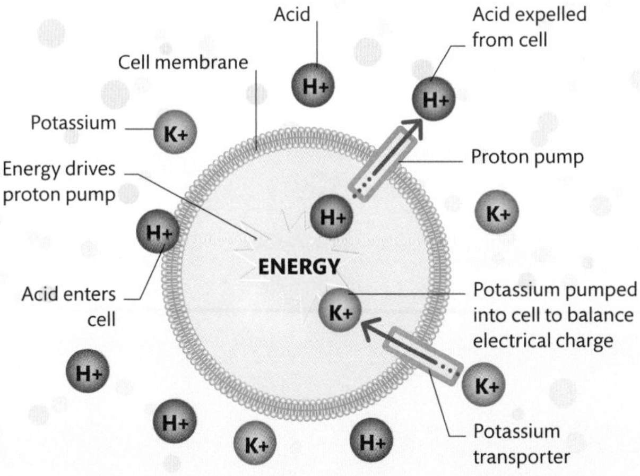
Acid
Acid expelled from cell
Cell membrane
H+
Potassium
K+
Proton pump
Energy drives proton pump
ENERGY
Acid enters cell
Potassium pumped into cell to balance electrical charge
Potassium transporter

Acidophiles
An acid is a chemical that, when mixed with water, increases the concentration of positively charged hydrogen ions (H+). These cause internal damage to most cells, but acidophiles have a "proton pump" that ejects them, keeping the cell's internal pH neutral. To stop cells from becoming negatively charged and attracting more H+ ions, potassium ions (K+) are pumped in.

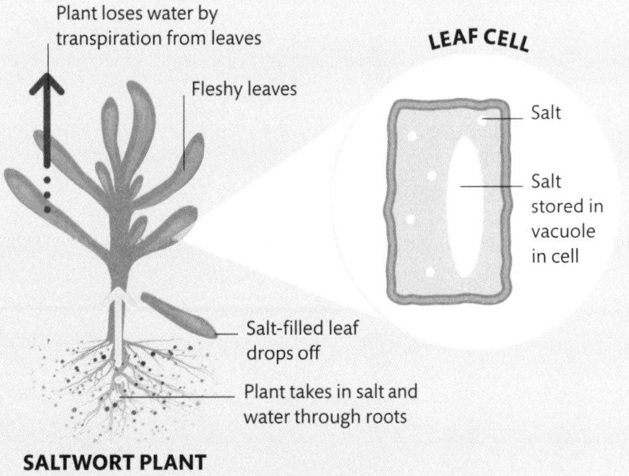
Plant loses water by transpiration from leaves
Fleshy leaves
LEAF CELL
Salt
Salt stored in vacuole in cell
Salt-filled leaf drops off
Plant takes in salt and water through roots
SALTWORT PLANT

Halophiles
These salt-loving extremophiles mostly live where seawater has evaporated, leaving high concentrations of salt. For example, saltworts are plants that grow near coasts where salt levels are too high for other plants. The plant draws in salty water and sequesters the salt in a vacuole (see p.63). Older leaves contain more salt and are shed to get rid of the chemical.

Origins of life

Life on Earth is thought to have begun around 3.7 billion years ago, only a few hundred million years after the planet formed. The process by which the first life forms were generated from nonliving material, known as abiogenesis, is not known, but several theories have been proposed.

INORGANIC INGREDIENTS

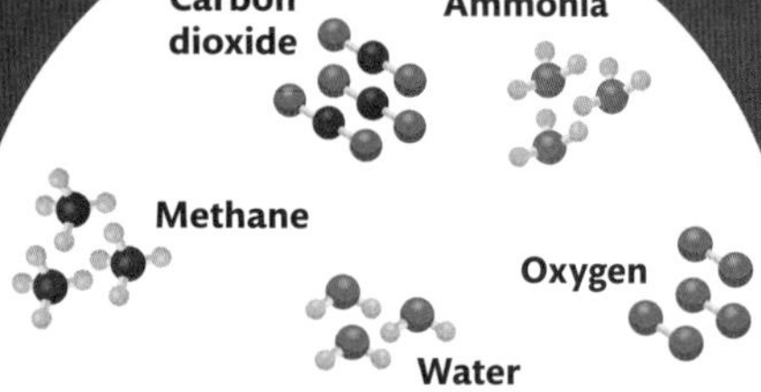
Carbon dioxide
Ammonia
Methane
Oxygen
Water

1 Earth's early atmosphere was full of simple, carbon- and nitrogen-rich compounds released by volcanic activity. These dissolved into water in the oceans.

ENERGY FROM GEOTHERMAL HEAT AND LIGHTNING

SIMPLE ORGANIC MOLECULES

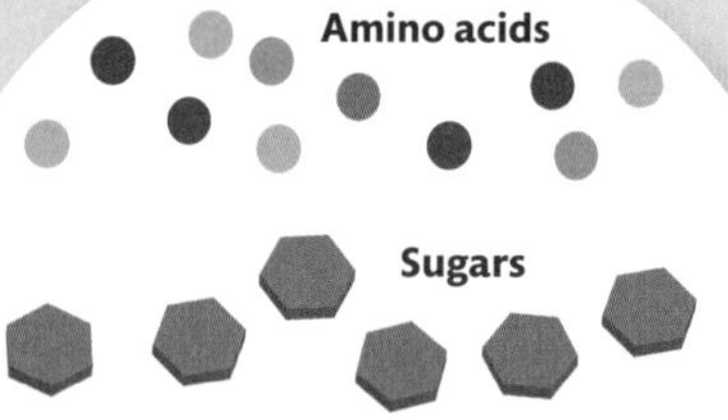
Amino acids
Sugars

2 Triggers such as lightning, volcanic eruptions, sunlight, and ultraviolet rays caused the compounds to join together to form simple organic substances such as sugars and amino acids.

OXYGEN WAS **POISONOUS** TO THE **EARLIEST** FORMS OF **LIFE**

COMPLEX ORGANIC MOLECULES

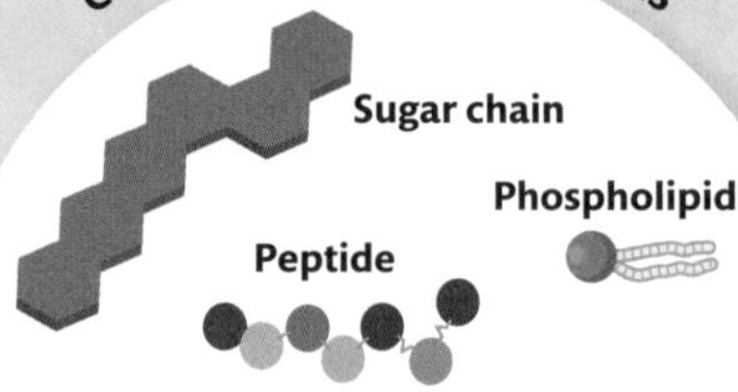
Sugar chain
Phospholipid
Peptide

3 The process of creating more complex molecules continued, and the simple organic molecules formed polymers (long molecules built up from chains of small units), creating primitive proteins (peptides), carbohydrates (sugars), and lipids (fats).

The primordial soup

The first major scientific theory on the origin of life, the "primordial soup" theory, was set out in the early 20th century. This theory proposes that Earth's early oceans contained the raw materials for life, and that complex chemical structures, such as DNA and proteins, were "cooked" from them over a long period of random chemical interactions.

ARE NEW FORMS OF LIFE EVOLVING NOW?

Yes, new species are constantly developing, and existing species, even humans, are still evolving, although the rate of such changes may be slow.

CHIMNEY
Chimney formed by deposition of minerals from hot, chemical-rich water rising through crust

REPLICATORS

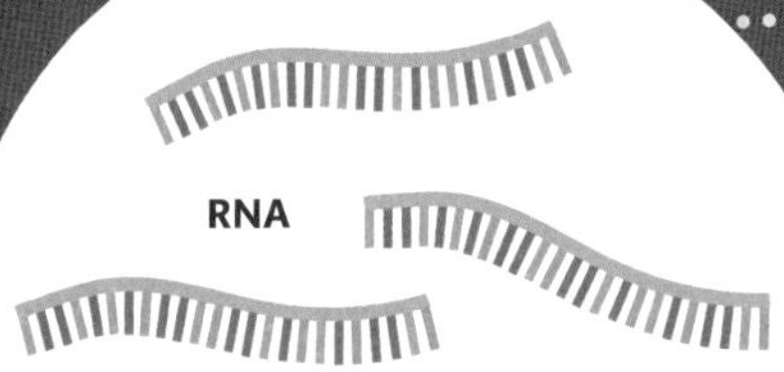
RNA

4 It is thought that some polymers could autocatalyze, which means that one molecule acted as a template to build a second. These replicating chemicals competed for raw materials and evolved by natural selection until RNA evolved.

CELLS

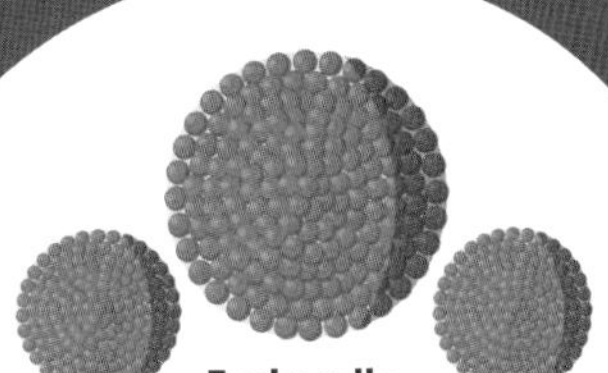
Early cells

5 The first protocells may have been simply a replicating polymer enclosed in a membranous vesicle, able to store raw materials for making copies of itself and chemicals involved in chains of reactions—the earliest form of metabolism.

MEMBRANES

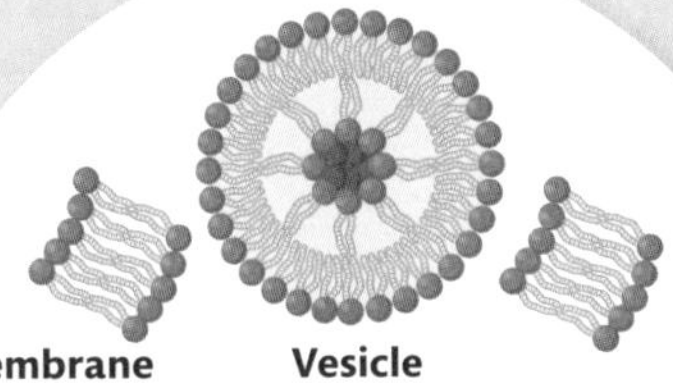
Membrane
Vesicle

4 In some organic molecules, one end repels water while the other end is attracted to it. In water, such molecules may have clustered into membranes with the water-loving ends as the outer layer. Some membranes formed sacs called vesicles.

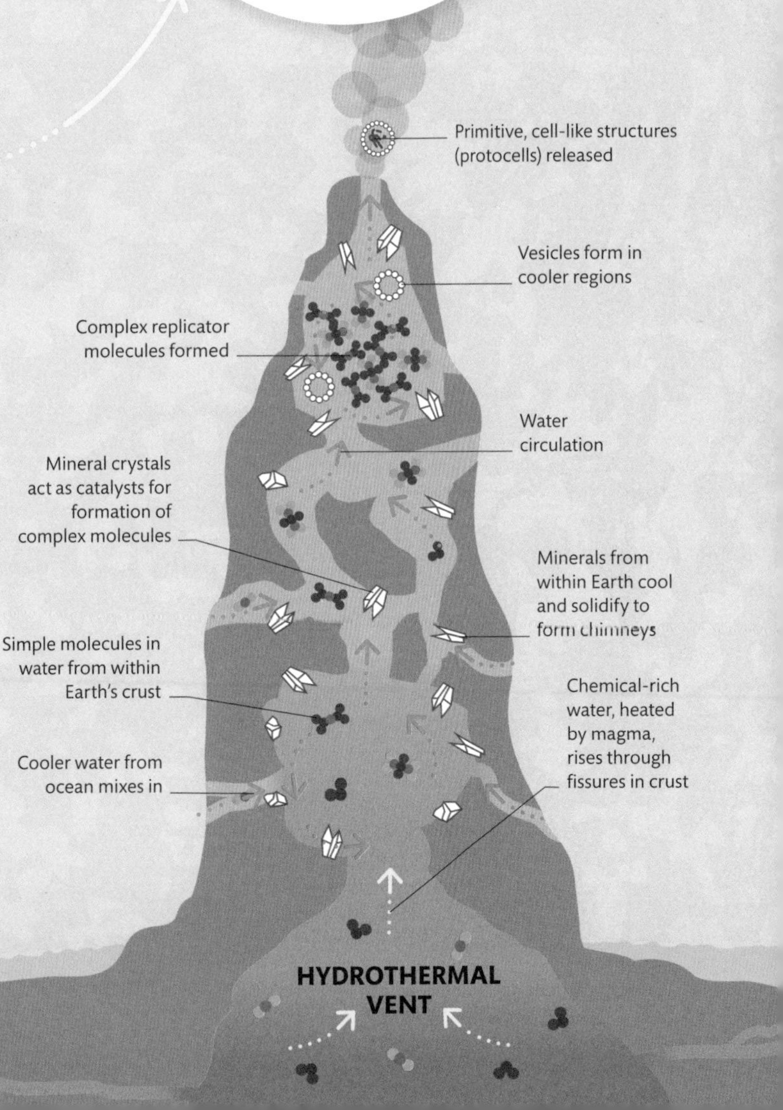
Primitive, cell-like structures (protocells) released
Vesicles form in cooler regions
Complex replicator molecules formed
Water circulation
Mineral crystals act as catalysts for formation of complex molecules
Minerals from within Earth cool and solidify to form chimneys
Simple molecules in water from within Earth's crust
Chemical-rich water, heated by magma, rises through fissures in crust
Cooler water from ocean mixes in
HYDROTHERMAL VENT

Hydrothermal vents

In the hydrothermal vent theory, it is proposed that life first arose at hydrothermal vents—places on the deep ocean bed where heat and minerals well up from inside Earth's crust. The heat and pressure provided the right conditions for the creation of complex substances, which could then combine to form protocells. The rich supply of chemicals and mineral catalysts may have powered chemical reactions in the protocells.

Life on other planets

Astrobiology is the search for organisms living beyond Earth. The science is focused on finding planets, moons, and zones around stars where the conditions for harboring life might exist.

The habitable zone

A region around a star within which the temperature of any orbiting body allows water to remain constantly liquid on that body's surface is called a habitable zone. It is assumed that alien life will need water, as life does on Earth, so a planet or moon with liquid water is more likely to be hospitable to life. Earth occupies the habitable zone in the Solar System. Habitable zones around other stars depend on the stars' size and temperature. Most stars in the Universe are red dwarfs, which are cooler and smaller than the Sun, so habitable zones around them will be much closer to those stars than Earth is to the Sun.

COULD WE EVER VISIT ALIENS?

The nearest possibly habitable planet is more than 4 light-years away. Even the fastest spacecraft that we have today would take 73,000 years to get there.

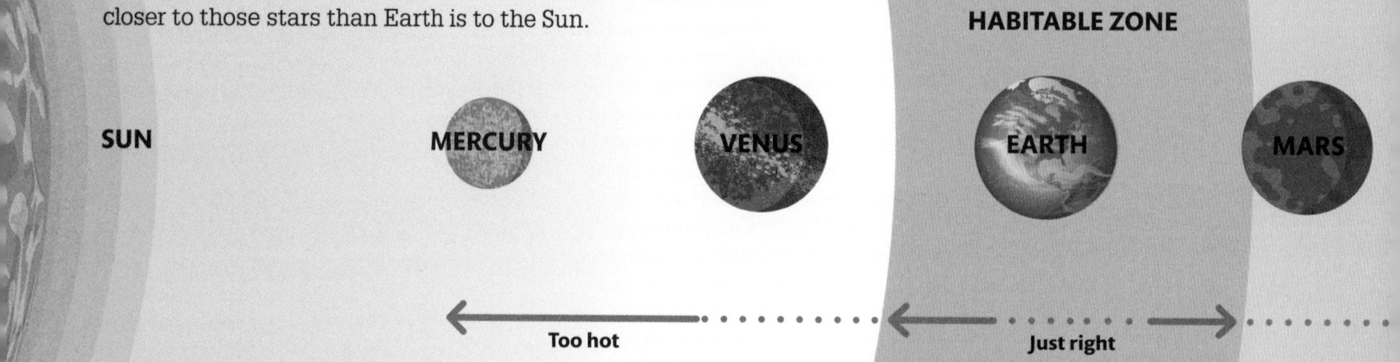
HABITABLE ZONE
SUN
MERCURY
VENUS
EARTH
MARS
Too hot
Just right

Factors that could sustain complex life

Our galaxy contains an estimated 400 billion planets. It is likely that many of them are in habitable zones, hosting the conditions for life to arise. However, additional factors (see below) are needed for a planet to enjoy long periods of stability, so that complex, diverse species can develop as on Earth.

Surface temperature
The average should be above 32°F (0°C), the freezing point of water, but less than 104°F (40°C), above which delicate molecules such as proteins start to break apart.

Surface water
The planet needs a permanent body of water. It is assumed that alien life will use water as the medium in which metabolic processes take place.

Stable star
Unstable stars undergo significant changes in brightness over their lifetimes, and emit powerful solar storms that irradiate planets, creating mass extinction events.

Location in galaxy
A planet should be near enough to the galaxy's center to gain elements for a solid planet but far enough away to avoid being blasted by lethal radiation.

Massive neighbor
Earth benefits from Jupiter's gravity, which pulls in comets and asteroids before they enter the inner Solar System, thus reducing the risk of Earth being hit by meteors.

Large moon
Earth's moon is large relative to the size of Earth. One result is a large tidal range; this created coastal habitats in which life could emerge from water onto land.

THERE ARE 6 BILLION POTENTIAL EARTHLIKE PLANETS IN OUR GALAXY

Europa
Jupiter's moon Europa has a solid ice surface covering an ocean that contains at least twice as much liquid water as Earth's oceans. It is thought that life could exist around hydrothermal vents on Europa's ocean floor, as it does on Earth.

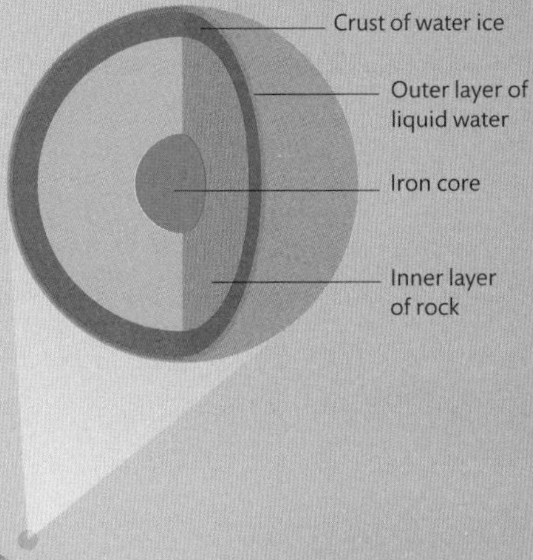
Crust of water ice
Outer layer of liquid water
Iron core
Inner layer of rock

SEARCHING FOR LIFE

Astrobiologists are hoping new telescopes will reveal chemical activity that shows life processes in action in the atmospheres of distant planets. Radio emissions from space are also being scanned for signs of active communication from an alien civilization.

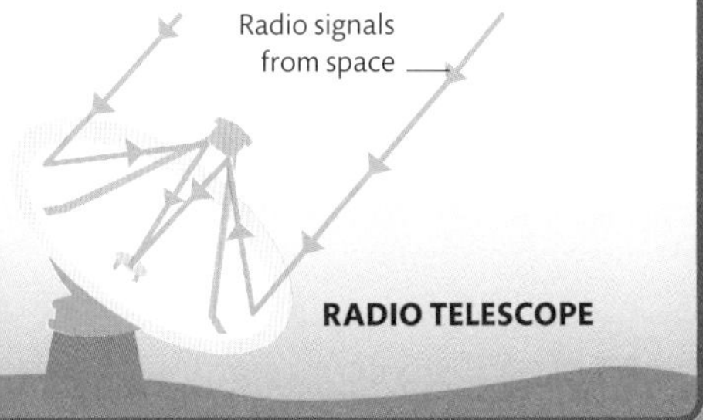
Radio signals from space
RADIO TELESCOPE

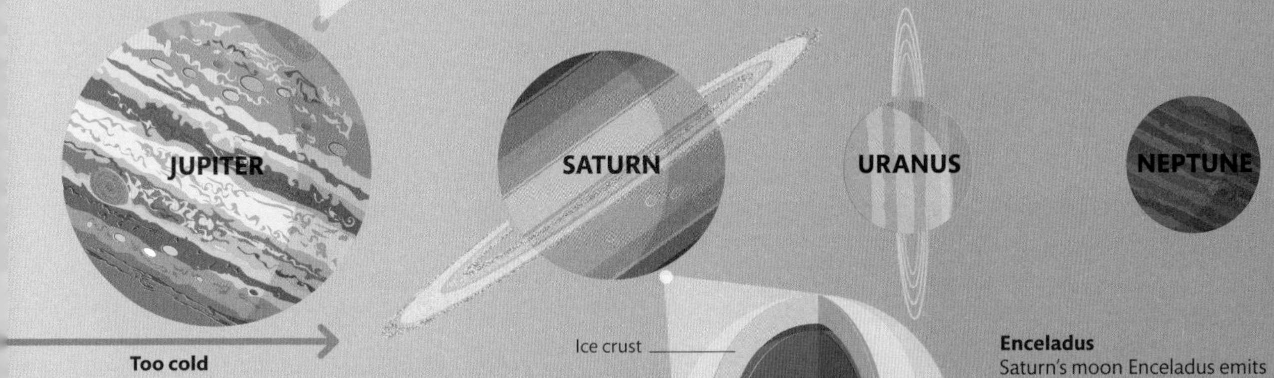
JUPITER
SATURN
URANUS
NEPTUNE
Too cold

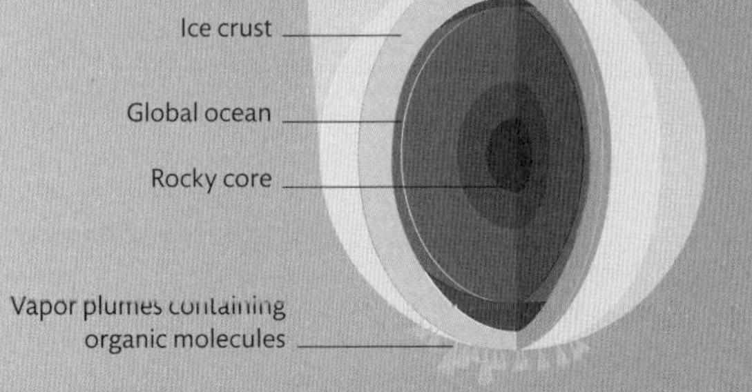
Ice crust
Global ocean
Rocky core
Vapor plumes containing organic molecules

Enceladus
Saturn's moon Enceladus emits plumes of salty water including simple carbon and nitrogen compounds. The plumes may result from heating of water under the icy crust, perhaps by volcanic activity, which could create the conditions for life.

Large, hot, metallic core
Earth's core generates a powerful magnetic field; this produces the magnetosphere, which protects Earth from electrified particles streaming out from the Sun.

Sufficient mass
The planet must be large and dense enough to have a gravitational force able to hold on to a thick atmosphere, which traps heat and cycles nutrients.

Atmosphere
The atmosphere must contain large amounts of carbon and nitrogen as raw materials for life, as well as water, which can fall as rain to create oceans.

Spin and tilt
The Moon's gravity locks Earth into a stable rotation and minimizes wobbles around Earth's axis, so day length and seasonal changes do not fluctuate excessively.

Plate tectonics
An active surface, with continents and oceans in constant motion, may boost biodiversity by isolating groups of wildlife so they evolve in different ways.

Carbon compounds
The planet must orbit inside the "soot line," within which complex carbon compounds are broken up by the star's heat, creating simple carbon molecules for use by life.

THE CHEMISTRY OF LIFE

Metabolism

Originating from the Greek word for "change," metabolism is the collective term given to all the chemical reactions and processes that occur within an organism to keep it alive.

Vital processes

All organisms are constantly carrying out thousands of metabolic processes to stay alive. The reactions allow an organism to obtain energy from its food and perform essential body functions (such as breathing and moving), maintain and repair its cells and tissues, manage its hormones, and regulate its temperature.

How metabolism works

A cow needs a range of inputs, such as food, to perform the essential metabolic processes for life. As well as useful products that can be used in the body, these reactions also produce waste.

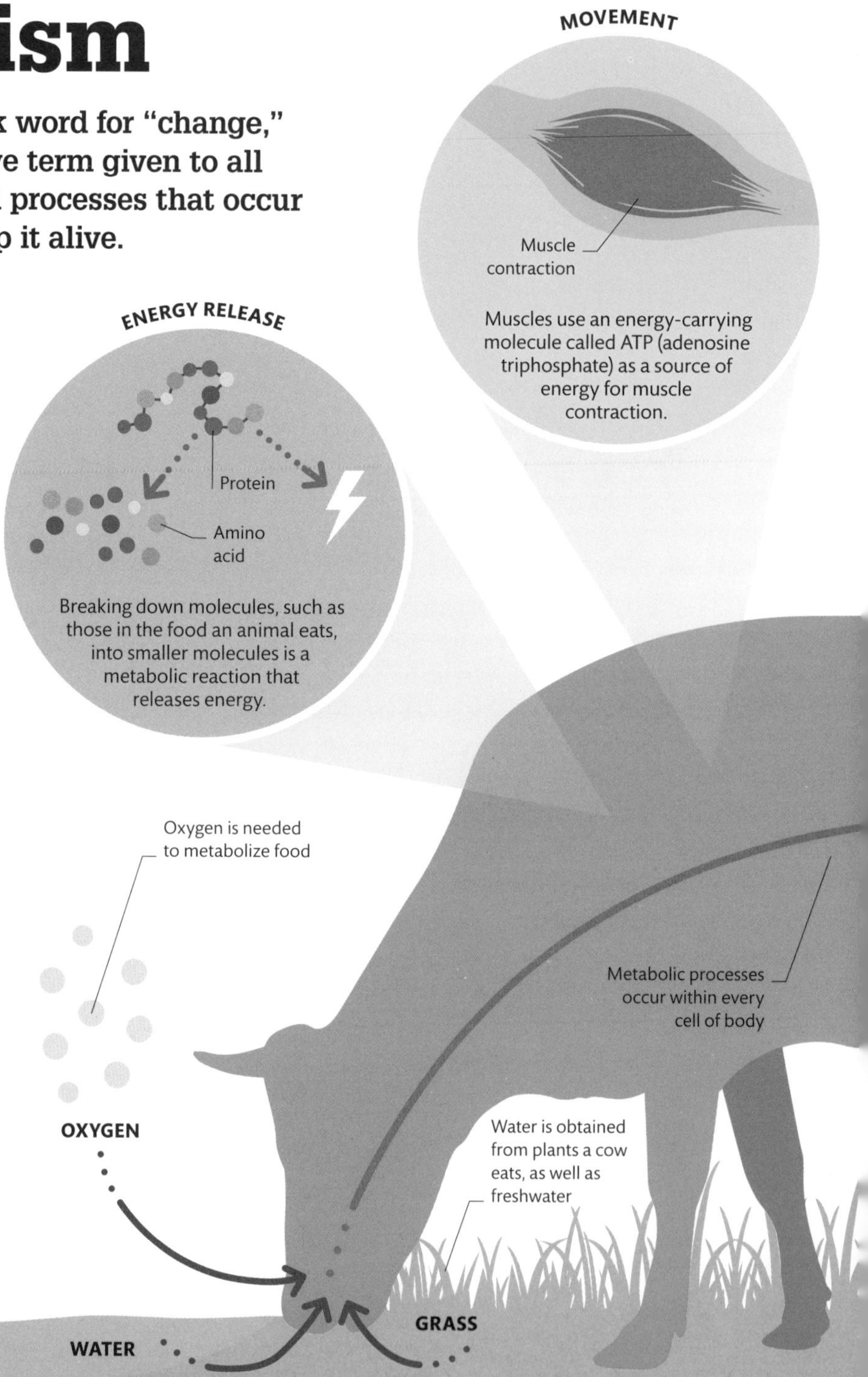
MOVEMENT
Muscle contraction
Muscles use an energy-carrying molecule called ATP (adenosine triphosphate) as a source of energy for muscle contraction.
ENERGY RELEASE
Protein
Amino acid
Breaking down molecules, such as those in the food an animal eats, into smaller molecules is a metabolic reaction that releases energy.
Oxygen is needed to metabolize food
Metabolic processes occur within every cell of body
OXYGEN
Water is obtained from plants a cow eats, as well as freshwater
GRASS
WATER

1 Inputs
The oxygen, water, and energy necessary to power the cow's metabolic reactions come from the air it breathes, the water it drinks, and the sugars and fibers contained in the plants it eats.

2 Processes
Food is broken down into its smallest elements: amino acids, fatty acids, and sugars. Amino acids are used to build proteins for cell maintenance, while sugars and fatty acids provide the body with energy.

METABOLIC RATE

The amount of energy an organism needs to perform metabolic processes is called its metabolic rate. It can be measured in calories per day (kcal/day).

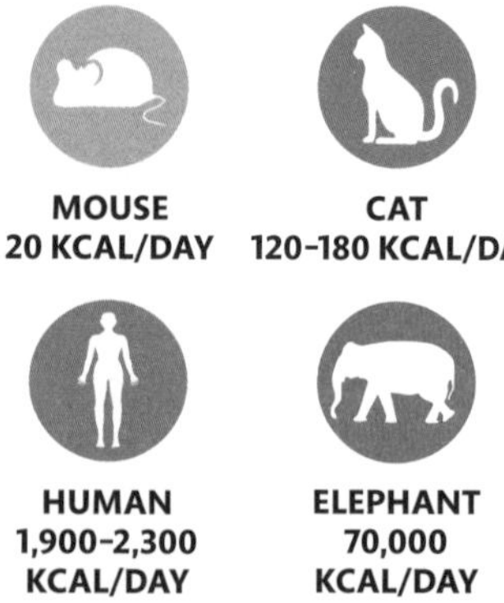
MOUSE
20 KCAL/DAY
CAT
120–180 KCAL/DAY
HUMAN
1,900–2,300
KCAL/DAY
ELEPHANT
70,000
KCAL/DAY

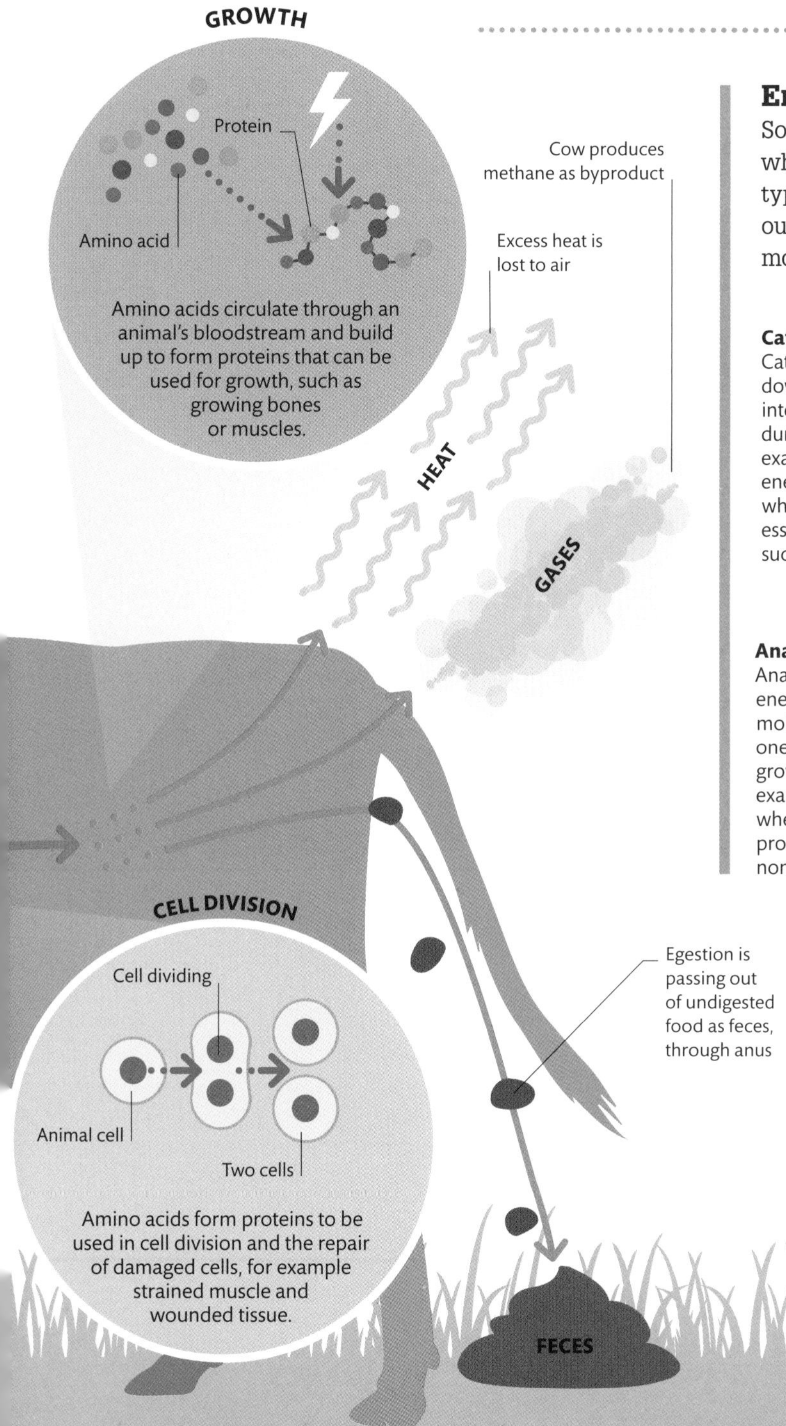
GROWTH
Protein
Amino acid
Amino acids circulate through an animal's bloodstream and build up to form proteins that can be used for growth, such as growing bones or muscles.
Cow produces methane as byproduct
Excess heat is lost to air
HEAT
GASES
CELL DIVISION
Cell dividing
Animal cell
Two cells
Amino acids form proteins to be used in cell division and the repair of damaged cells, for example strained muscle and wounded tissue.
Egestion is passing out of undigested food as feces, through anus
FECES

Energy exchange

Some metabolic processes require energy, while others release it. There are two main types of reaction: building complex molecules out of simpler units (anabolism) and breaking molecules into smaller units (catabolism).

Catabolism
Catabolic reactions break down larger molecules into smaller, simple ones, during digestion for example. This releases energy in the process, which can be used in essential bodily processes, such as respiration.

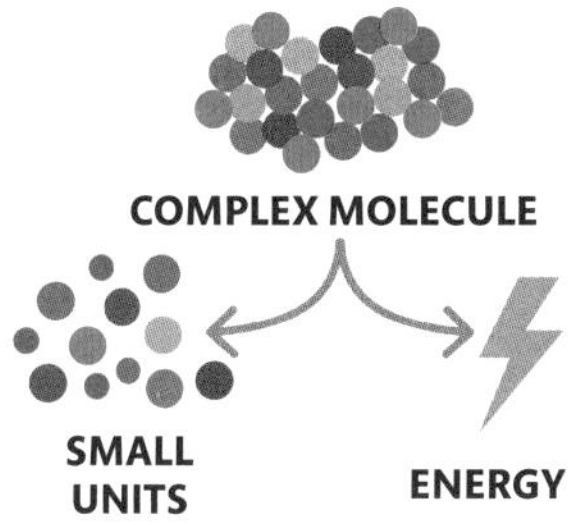
COMPLEX MOLECULE
SMALL UNITS
ENERGY

Anabolism
Anabolic reactions consume energy while linking simple molecules into complex ones in order to support growth and building. One example is gluconeogenesis, where the liver and kidneys produce glucose from non-carbohydrate sources.

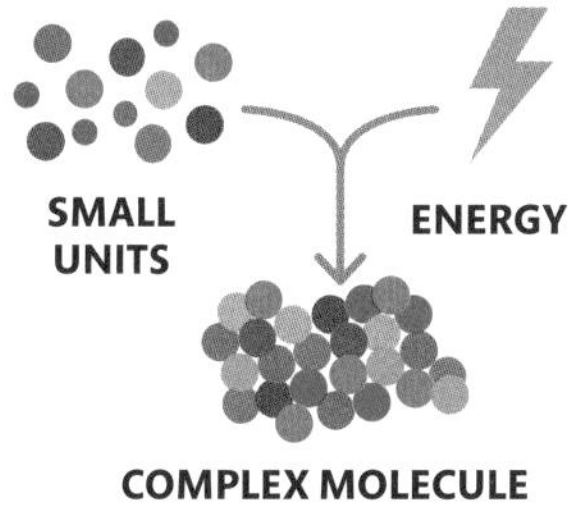
SMALL UNITS
ENERGY
COMPLEX MOLECULE

DOES A SLOW METABOLISM LEAD TO OBESITY?

While metabolic rate varies from person to person, it does not predict body mass. Obese people can have the same daily expenditure as slim people.

3 Byproducts
The products of metabolic reactions that cannot be used or stored in the cow's body are called byproducts. They are removed as waste, including heat, feces, urine, and gases.

WHALES **CONSUME 20–50 MILLION** CALORIES A DAY – EQUIVALENT TO **60,000 SALMON FILLETS**

Carbohydrates

Along with lipids and proteins, carbohydrates are one of the three main nutrients essential to the diet of all living things. The most common carbohydrate is glucose, a type of simple sugar and the main source of energy for cells, tissues, and organs.

What are carbohydrates?

Present in both plants and animals, carbohydrates are central to all life. They are a large group of compounds that contain carbon, hydrogen, and oxygen atoms. Plants create carbohydrates during photosynthesis (see pp.46–47) and store them in long polysaccharide chains called starch, while animals must obtain most of their carbohydrates from what they eat and store them as glycogen.

Functions of carbohydrates

Carbohydrates have wide-ranging functions in the bodies of animals like this brown bear. When it eats food containing carbohydrates, such as pine cones, the bear's digestive system breaks the carbohydrates down into the sugar glucose, the body's primary energy source, to be used throughout the body.

BRAIN

4 Brain
Carbohydrates are vital for brain function because the brain requires a lot of energy.

Pine cones are a source of carbohydrate, protein, and lipids

2 Heart and blood
Sugars from digestion are absorbed into the blood and pumped by the heart to all the body's tissues.

HEART

LIVER

Sugar is converted into fatty acids, which act as a store of energy around the body

STOMACH

3 Liver and storage
Residual sugars released by the digestion of carbohydrates are stored as glycogen in the liver, ready to be converted back into glucose for energy when needed.

SMALL INTESTINE

Sugars are absorbed through walls of small intestine

HOW DO CARNIVORES GET CARBOHYDRATES?

Carnivorous animals have a high-fat, low-carbohydrate diet. They obtain the carbohydrates they need by eating nutrient-rich, herbivorous prey.

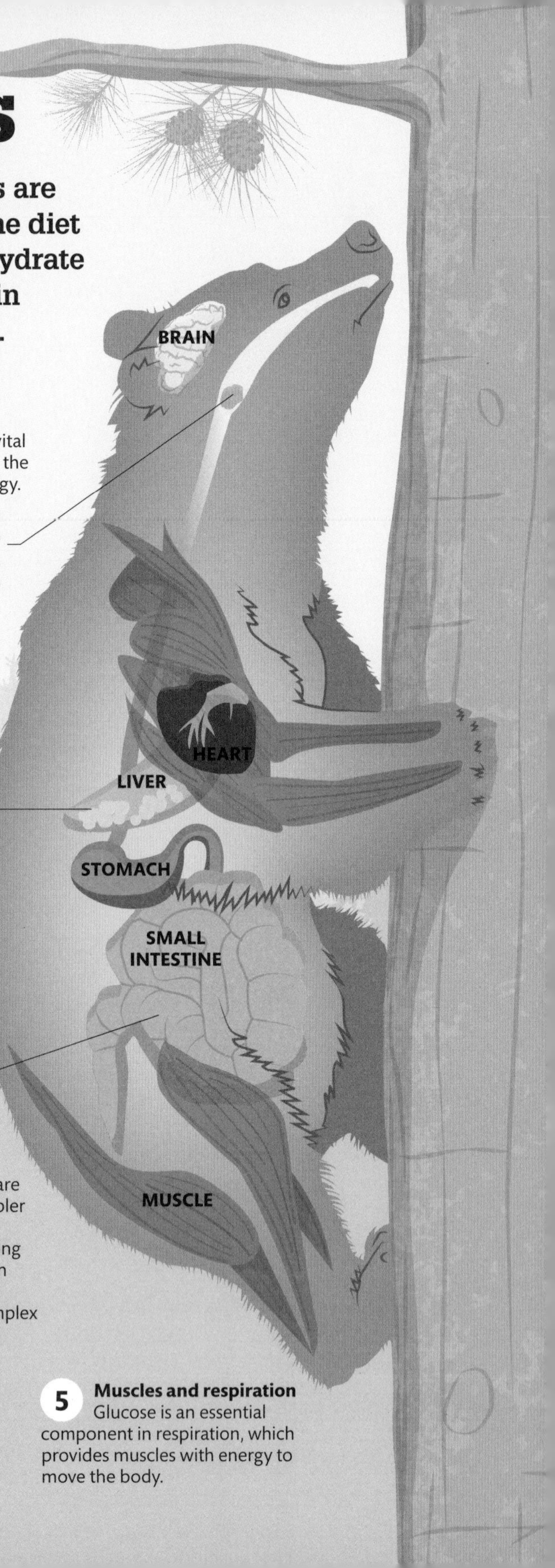

1 Digestion
Carbohydrates are broken down into simpler units during digestion (see pp.162–63), releasing carbon atoms for use in biochemical synthesis (the production of complex molecules) and sugars.

MUSCLE

5 Muscles and respiration
Glucose is an essential component in respiration, which provides muscles with energy to move the body.

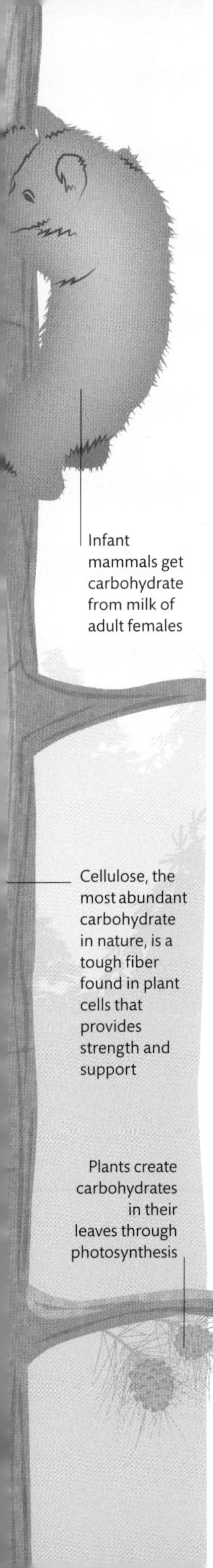
Infant mammals get carbohydrate from milk of adult females
Cellulose, the most abundant carbohydrate in nature, is a tough fiber found in plant cells that provides strength and support
Plants create carbohydrates in their leaves through photosynthesis
Branches contain digestible starches and sugar, as well as fiber, which cannot be digested but gives structure to feces

Types of carbohydrate

Chemically, a carbohydrate can be classified as a monosaccharide, disaccharide, oligosaccharide, or polysaccharide, depending on the number of sugar molecules it is composed of. Monosaccharides function as energy storage molecules and as the building blocks of more complex sugars, being used as structural elements. Disaccharides are mainly used in cell transport. Polysaccharides serve as a stored source of energy and release energy slower than monosaccharides and disaccharides.

Monosaccharide
Monosaccharides, such as glucose and fructose, consist of one unit of sugar. They are the building blocks of more complex sugars and are used to store and release energy.

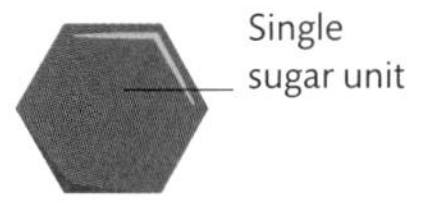
Single sugar unit
MONOSACCHARIDE

Disaccharide
Sucrose (produced by plants), lactose (found in milk), and maltose are disaccharides. They are made up of two monosaccharides that are linked together.

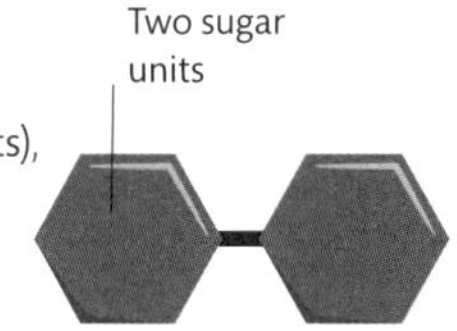
Two sugar units
DISACCHARIDE

Oligosaccharide
Carbohydrates made from three to six sugar units are called oligosaccharides. Human breast milk contains oligosaccharides as well as lactose (a disaccharide).

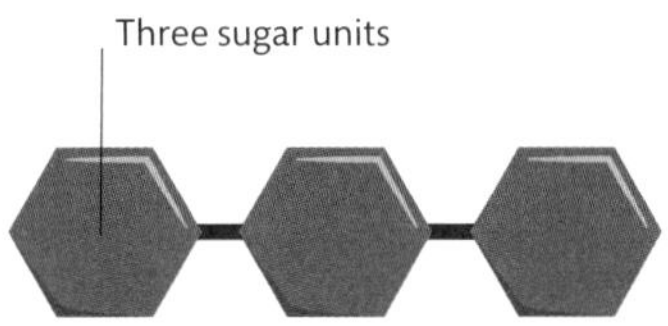
Three sugar units
OLIGOSACCHARIDE

Polysaccharide
Polysaccharides, such as cellulose, starch, and glycogen, are large polymers made up of many sugar molecules. Their structure can be branched or linear.

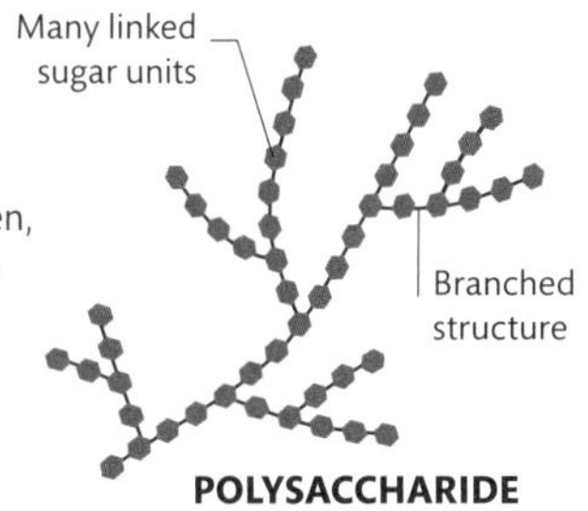
Many linked sugar units
Branched structure
POLYSACCHARIDE

Sources of carbohydrates

Carbohydrates come in three forms: simple sugars, starches, and fiber. Simple sugars (monosaccharides and disaccharides) are the body's main source of energy and the building blocks of more complex sugars. Starches and fibers (polysaccharides) are complex carbohydrates that serve as a stored source of energy. They are found in milk and plants, including fruits, grains, and vegetables.

TYPE	SOURCE
Simple sugars	Simple sugars are naturally found in fruit. They are also produced for candy, desserts, and processed foods—either to sweeten or to improve texture.
Starches	Starches are found in leaves, fruits, roots, and stems. Cereals, grains, and rice are sources of dietary starch, as well as some vegetables like potatoes, peas, and corn.
Fiber	Sources of dietary fiber include beans, nuts, avocados, and broccoli. Fiber is also found in cotton, hemp, bamboo, wool, silk, and mohair.

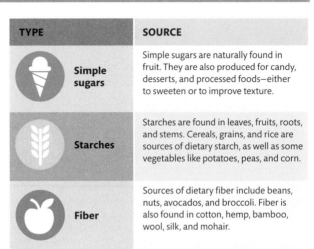

LACTOSE FOUND IN MILK IS THE ONLY **CARBOHYDRATE** OF **ANIMAL ORIGIN**

Lipids

Lipids, such as fats and oils, are essential for the functioning of all organisms. As well as providing texture to many of the foods animals eat, lipids perform the more vital roles of storing energy, forming cell membranes, and acting as chemical messengers.

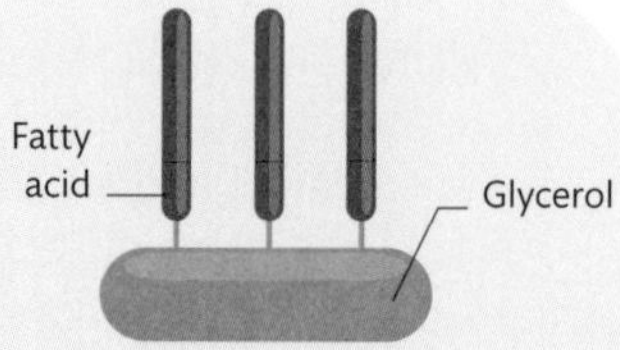
Fatty acid
Glycerol

Triglycerides
A triglyceride is made up of a glycerol unit and three chains of fatty acids. Fatty acids can be saturated or unsaturated. Most saturated fats are solid at room temperature, while unsaturated fats tend to be liquid.

What are lipids?

Lipids are more than just the fats and oils familiar to us in our diets. They are a diverse group of molecules that include waxes, some vitamins, and hormones, and they form most of the membranes in an organism's cells. Lipids are composed of long strings of carbon atoms with hydrogen and oxygen molecules attached. Some lipids also contain nitrogen and phosphorus. Despite having different chemistries and unlike almost all the other molecules in living cells, all lipids are insoluble in water. Animals can make some lipids themselves, but others they must obtain from the food they eat, such as plants.

Sesame seeds, like all seeds, contain high levels of healthy fats
Avocado is high in healthy fats
Lipids make up one-third of a slice of cheese
As well as protein, beef contains high levels of lipids

BEEF BURGER

Types of lipid
There are four main groups of lipid: triglycerides, phospholipids, steroids, and waxes. They have different molecular structures, which make them suited to a range of vital roles in the bodies of plants, animals, and other living things.

Triglycerides
Also known as "fats and oils," triglyceride molecules can be broken down by digestion and stored as fat in the body. Fats are solid at room temperature and are used by animals for insulation, protection, and long-term energy storage. Oils are liquid at room temperature and are used by plants as long-term energy storage.

Saturated and unsaturated fats

A saturated fat is one that is "saturated" with hydrogen atoms, which means it holds the maximum possible number of hydrogen atoms and contains no double carbon–carbon bonds in its chemical structure. An unsaturated fat has at least one double carbon–carbon bond. Saturated fats are mainly found in animal foods, such as cheese, meats, and butter, but a few plant foods are also high in saturated fats, such as coconut oil and palm oil. Unsaturated fats are found in nuts, seeds, and plant oils.

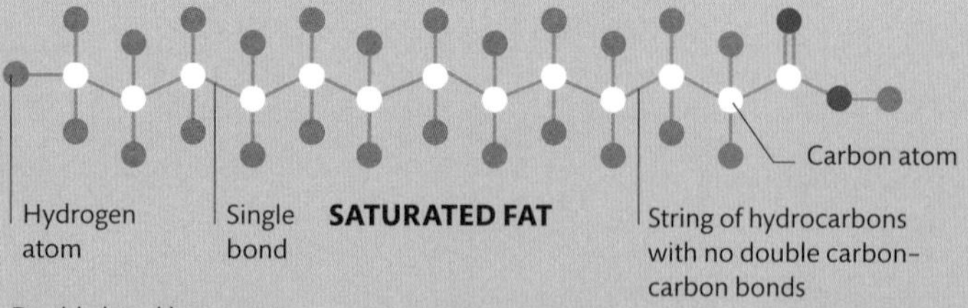
Carbon atom
Hydrogen atom
Single bond
String of hydrocarbons with no double carbon-carbon bonds

SATURATED FAT

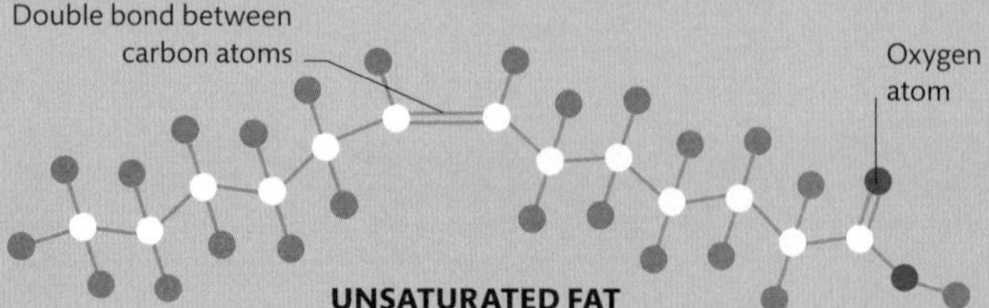
Double bond between carbon atoms
Oxygen atom

UNSATURATED FAT

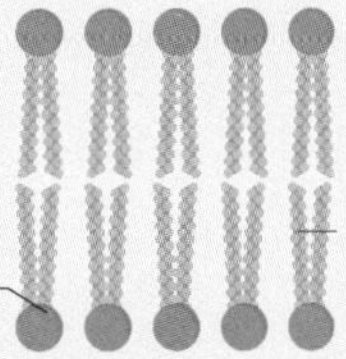
Hydrophilic head
Hydrophobic tail

Phospholipid bilayer
In cell membranes, phospholipids are arranged in a double layer. The hydrophilic heads face outward, contacting the watery liquid inside and outside the cell, while the hydrophobic tails point inward.

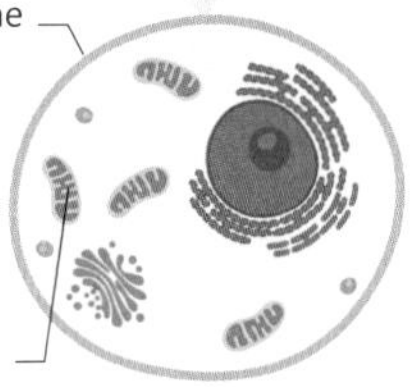
Cell membrane
Organelle

ANIMAL CELL

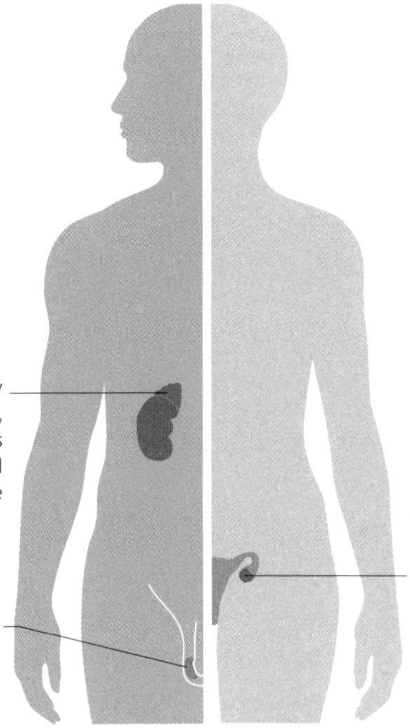
Produced by adrenal glands, cortisol regulates metabolism and immune response
Testes secrete sex hormones
Ovaries secrete sex hormones

MALE FEMALE

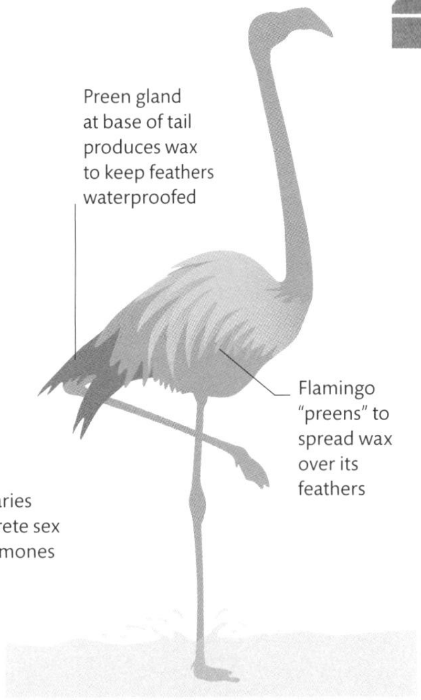
Preen gland at base of tail produces wax to keep feathers waterproofed
Flamingo "preens" to spread wax over its feathers

FLAMINGO

Phospholipids
These lipids form the membranes of almost all cells (see pp.56–57) and the organelles within them. Each phospholipid molecule is composed of a hydrophobic ("water-hating") lipid tail, with a hydrophilic ("water-loving") head. The molecules form a barrier between a cell's contents and its surroundings.

Steroids
Steroid hormones, including cortisol and the sex hormones estrogen and testosterone, are derived from a waxy lipid called cholesterol. The hormones act as chemical messengers, communicating between the cells, tissues, and systems of the human body and regulating various bodily processes.

Waxes
Waxes are lipids that coat the feathers of some aquatic birds, the leaves of some plants, and the cuticles of some insects. They are also found in the ears of some animals to protect the eardrums. The hydrophobic properties of waxes help these organisms to repel water, keeping them dry.

CHOLESTEROL

Cholesterol is a waxy lipid found in the blood. It is needed to build healthy cells, but high levels of "bad" cholesterol (called LDL) can build up as fatty deposits in blood-vessel walls. This can restrict blood and oxygen from reaching the heart, increasing the risk of heart disease.

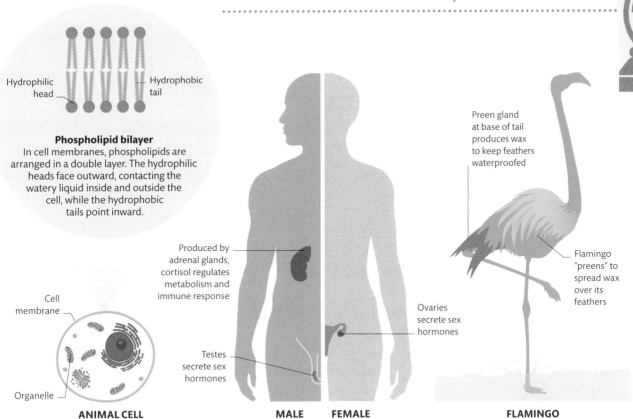

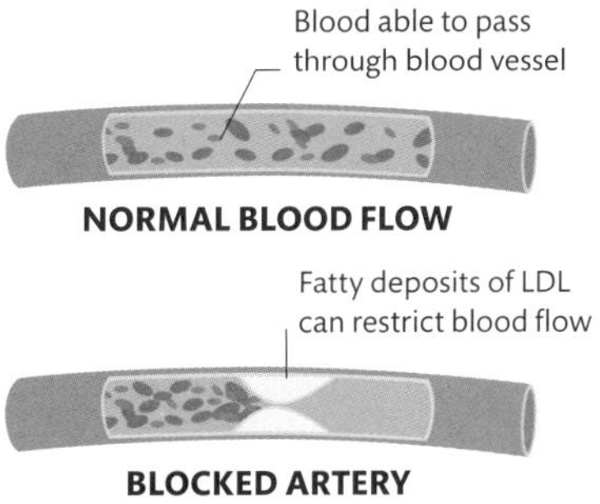
Blood able to pass through blood vessel
NORMAL BLOOD FLOW
Fatty deposits of LDL can restrict blood flow
BLOCKED ARTERY

HOW DO ANIMALS STORE FATS?

Animals store fats in several different ways: insects use a special organ called a "fat body," sharks store fats in their liver, and fish store fats around and within muscle fibers.

RNA

The main role of RNA (ribonucleic acid) is to translate the DNA code to manufacture proteins. There are three different types of RNA: messenger RNA (mRNA), transfer RNA (tRNA), and ribosomal RNA (rRNA). RNA is found in the nucleus and cytoplasm of plant and animal cells. A chain of RNA is single-stranded and shorter than DNA. It can be shaped like a single helix or a straight molecule, or it may be twisted upon itself. Like in DNA, there are four possible bases attached to each sugar in an RNA molecule: adenine, cytosine, guanine, and uracil (which takes the place of thymine). Adenine always binds with uracil, while cytosine always binds with guanine.

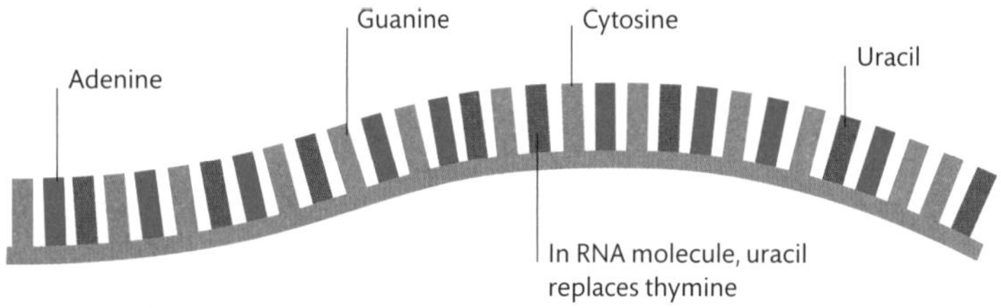
Guanine
Cytosine
Uracil
Adenine
In RNA molecule, uracil replaces thymine

Messenger RNA
DNA cannot leave the nucleus, so mRNA is a genetic template that acts as a messenger between DNA and protein assembly units called ribosomes (see p.59).

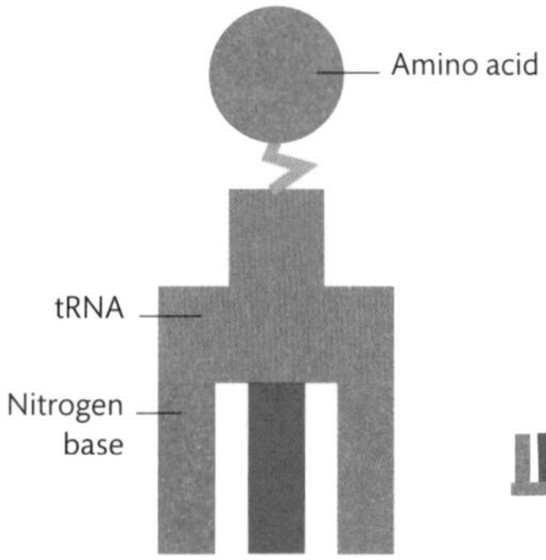
Amino acid
tRNA
Nitrogen base

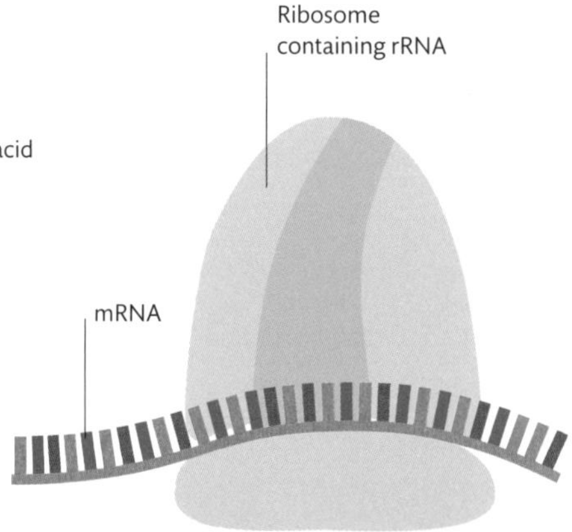
Ribosome containing rRNA
mRNA

Transfer RNA
tRNA is responsible for bringing protein building blocks called amino acids together during translation (see p.91) to form the peptide chain that becomes a protein.

Ribosomal RNA
rRNA is the main component of ribosomes. A ribosome reads an mRNA sequence and translates it into amino acids to build protein molecules.

RNA MAKES UP **5 PERCENT OF HUMAN BODY WEIGHT**, WHILE **DNA** FORMS **JUST 1 PERCENT**

MITOCHONDRIAL DNA

Mitochondrial DNA (mtDNA) contains the 37 genes needed for mitochondria to function. While nuclear DNA is inherited from both parents, mtDNA is inherited only from the mother. By analyzing mtDNA samples from thousands of people, scientists have traced a single common female ancestor for everyone alive today. In forensic science, mtDNA is often used to analyze old teeth, bones, or hair that have low DNA content.

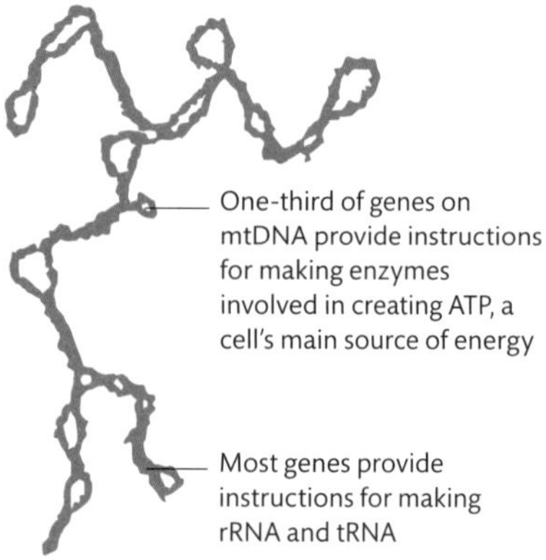
One-third of genes on mtDNA provide instructions for making enzymes involved in creating ATP, a cell's main source of energy
Most genes provide instructions for making rRNA and tRNA

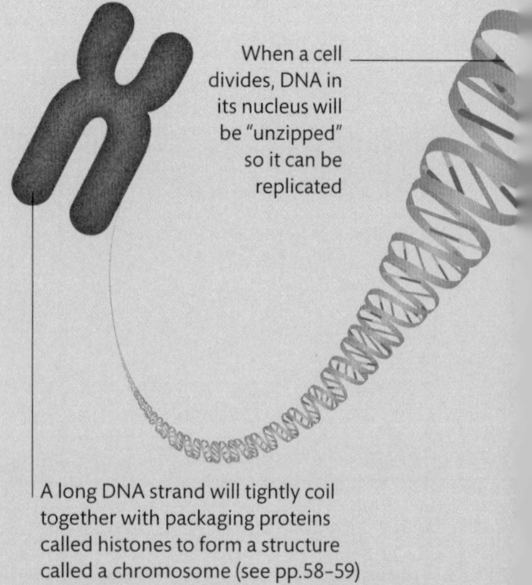
When a cell divides, DNA in its nucleus will be "unzipped" so it can be replicated
A long DNA strand will tightly coil together with packaging proteins called histones to form a structure called a chromosome (see pp.58–59)

Nucleic acids

All living things have nucleic acids in their cells. DNA is essential for life. Within its twisting shape it contains, in coded form, all the instructions an organism needs to grow and develop. Also required for life, RNA translates instructions from DNA to make proteins for organisms.

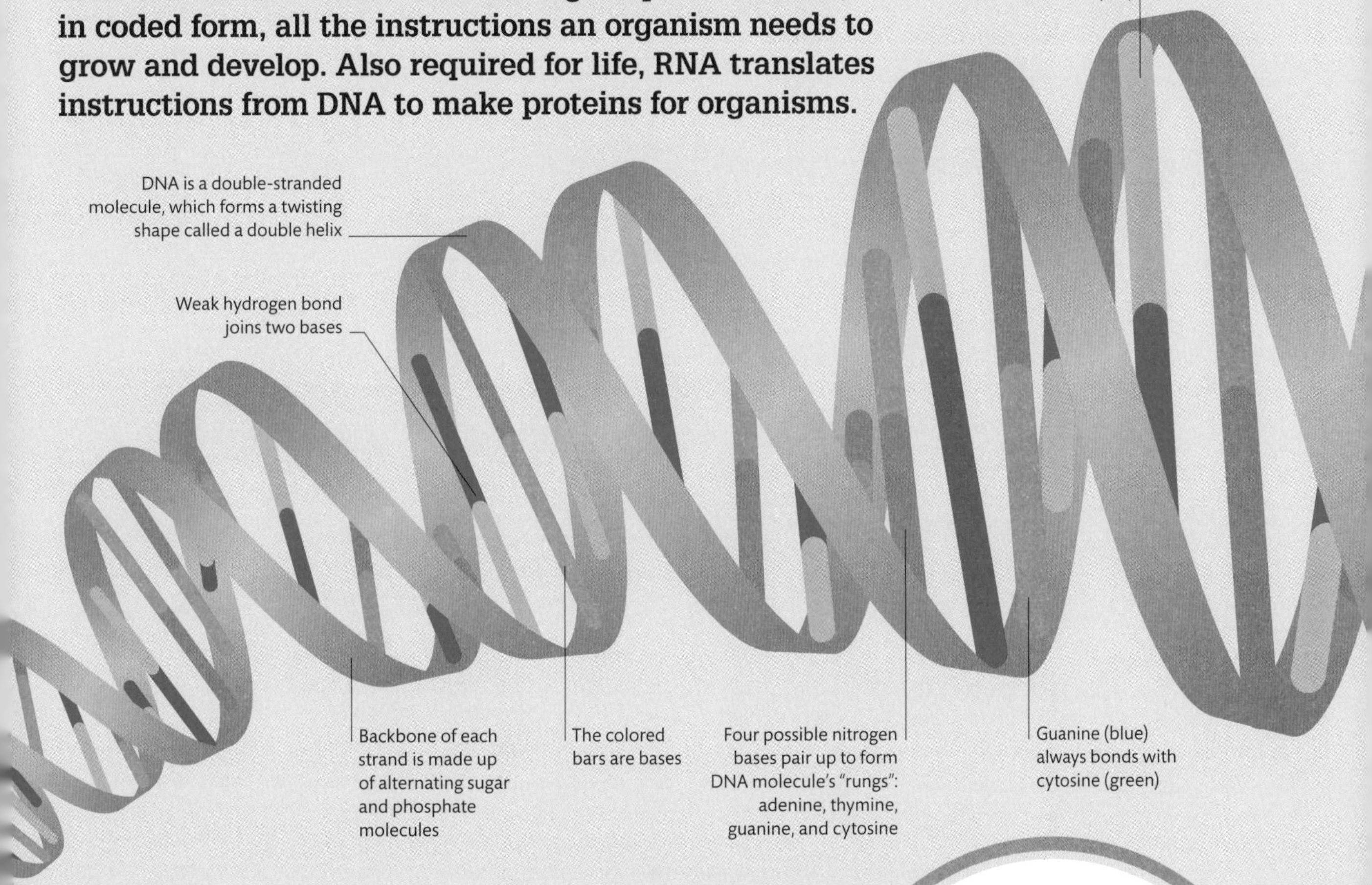
Thymine (yellow) always bonds with adenine (red)
DNA is a double-stranded molecule, which forms a twisting shape called a double helix
Weak hydrogen bond joins two bases
Backbone of each strand is made up of alternating sugar and phosphate molecules
The colored bars are bases
Four possible nitrogen bases pair up to form DNA molecule's "rungs": adenine, thymine, guanine, and cytosine
Guanine (blue) always bonds with cytosine (green)

DNA

Deoxyribonucleic acid (DNA) is the molecule that carries the genetic information that allows all organisms to develop and function. It determines whether a living thing will develop into a plant, an animal, or something else, and encodes instructions for making large molecules called proteins (see pp.38–39).
In animals and plants, most of the DNA is found coiled up as chromosomes inside cell nuclei. When cells divide, the DNA molecules are copied so all cells contain a copy of the vital code.
A DNA molecule is structured like a coiled ladder, with sugar and phosphate forming the backbone and pairs of nitrogen bases making up the rungs.

HOW MUCH DNA DO WE HAVE?

If every DNA molecule in the human body was laid out end-to-end, it would stretch farther than the distance between the Sun and Pluto—3.9 billion miles (6.2 billion km).

Proteins

Found in all living things, proteins are large biological molecules that perform thousands of metabolic roles. Each protein's function is defined by its shape, which is based on an exact chemical makeup. The information needed to build a body's many proteins is stored in its DNA.

CAN WE DESIGN NEW PROTEINS?

Recent advances in computing mean we can predict the shape of a protein from its primary structure. This means we can now figure out what a protein does and adjust its design.

The structure of proteins

A protein is a polymer—a long molecule built up from a chain of smaller units. The small units that make up proteins are called amino acids, which are simple organic compounds that include a nitrogen atom. Human proteins are built using 20 possible amino acids, but there are a few more used in other life forms. Every protein has a precise order of amino acids chained together. A typical protein includes about 500 amino acid molecules.

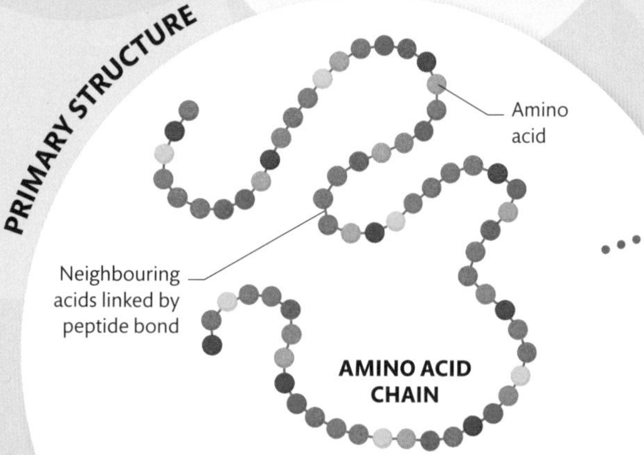

PRIMARY STRUCTURE
Amino acid
Neighbouring acids linked by peptide bond
AMINO ACID CHAIN

1 A protein's primary structure is the order of the amino acids within it. Adjacent amino acids are linked by a peptide bond–a bond between a carbon atom of one acid and a nitrogen atom of another. A chain of several dozen amino acids is called a polypeptide.

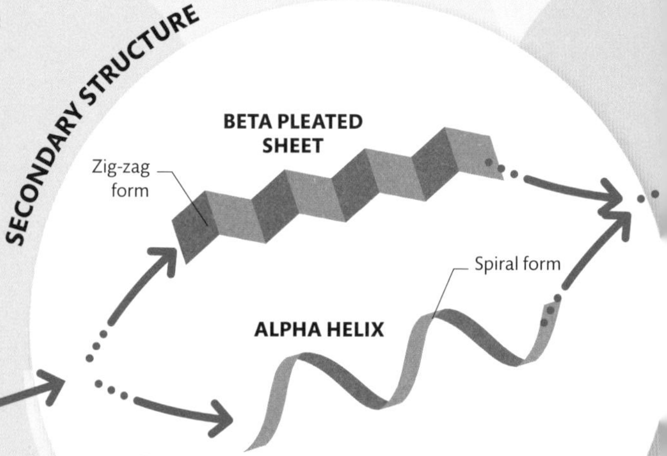

SECONDARY STRUCTURE
BETA PLEATED SHEET
Zig-zag form
Spiral form
ALPHA HELIX

2 The amino acids in a polypeptide twist and fold to create a secondary structure, held in place by weak hydrogen bonds that form between the atoms in its backbone. This structure is generally either a twisting helix or a pleated sheet.

Protein folding

There are four levels to a protein's ultimate form. This is because the many components in a long protein molecule attract and repel other parts of the chain, so the protein folds into a complex three-dimensional shape.

PRIONS

A prion is a misfolded protein that can trigger disease by making other variants of the same protein, which also change their shape. Prion diseases are very rare.

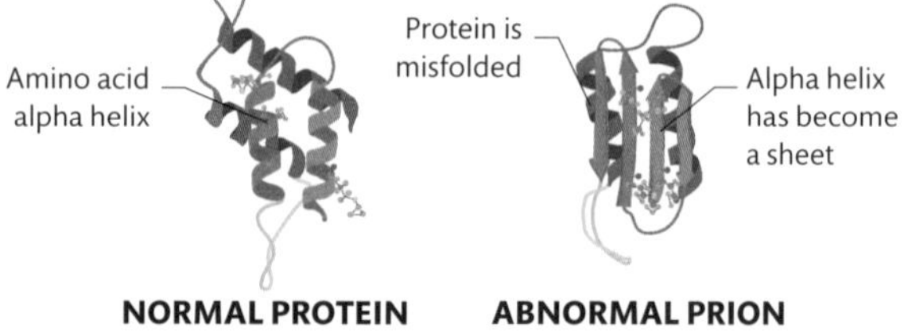

Amino acid alpha helix
Protein is misfolded
Alpha helix has become a sheet
NORMAL PROTEIN
ABNORMAL PRION

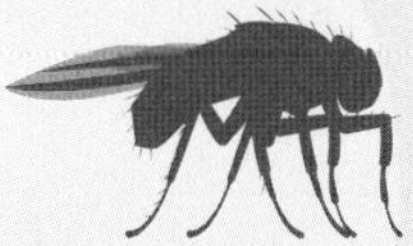

THE TINY BODY OF A FRUIT FLY INCLUDES ABOUT 10,000 TYPES OF PROTEINS

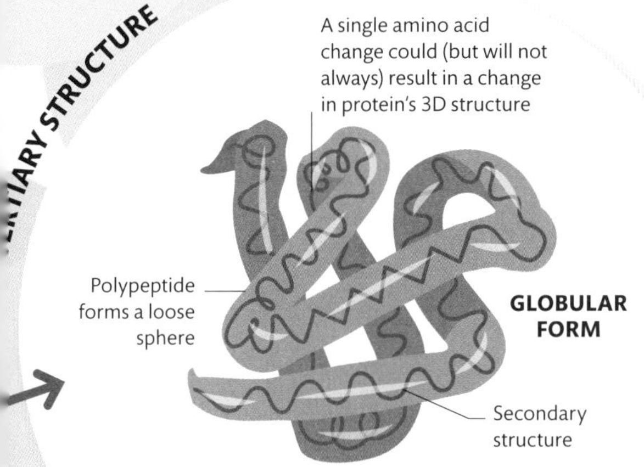
TERTIARY STRUCTURE
A single amino acid change could (but will not always) result in a change in protein's 3D structure
Polypeptide forms a loose sphere
GLOBULAR FORM
Secondary structure

3 The secondary protein structure folds in three dimensions to create a globular form, called the tertiary structure. The shape is highly dependent on the strength of the bonds between different parts of the secondary structure.

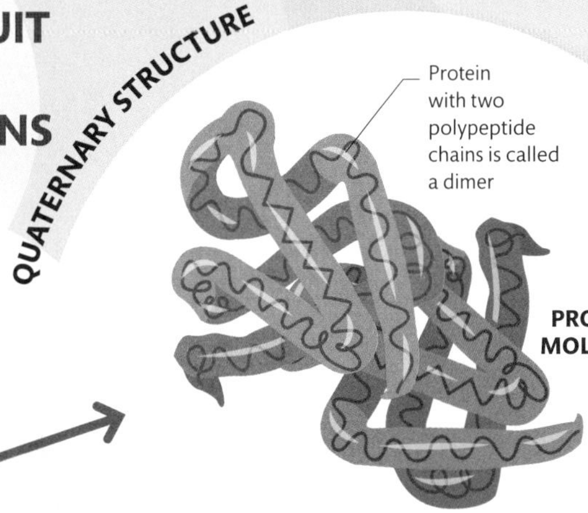
QUATERNARY STRUCTURE
Protein with two polypeptide chains is called a dimer
PROTEIN MOLECULE

4 Some proteins are made up of more than one polypeptide chain, which join to form a single structure known as a quaternary structure. Many proteins are made up of just one polypeptide chain, so do not have this fourth level to their form.

How proteins are used

Proteins are used in all parts of a living body, from structural tissues like skin and cartilage, to major organs like muscle, as well as working at a molecular level as enzymes (see pp.40–41). The precise shapes of proteins mean they can be put to use in highly specific ways. If the primary structure is correct, then any quaternary shape will also be correct, making it possible to manufacture proteins in large quantities.

KEY FUNCTIONS OF PROTEINS

Cell structure	The cytoskeleton is made of strands of protein. Proteins also control movement across cell membranes.	**Blood**	The oxygen carrier in blood, called hemoglobin, is a globular protein made from four polypeptides.
DNA	Chromosomes organize DNA by coiling it around proteins called histones. Enzymes manage gene replication and translation.	**Digestive enzymes**	Nutrients in food are broken down, or digested, into simple constituents by protein enzymes.
Hormones	Many hormones that circulate through a body are globular proteins. For example, insulin is a small protein.	**Neurons**	The wirelike extensions of neurons have protein pores that allow charged ions in and out to create electrical pulses.
Storage	Storage proteins are found in egg whites, plant seeds, and milk. They keep amino acids in reserve until needed for growth.	**Muscle proteins**	Proteins work together to make muscles contract: one protein will pull itself along another to reduce the total muscle length.

Enzymes

An enzyme is a biological catalyst—a substance that facilitates or speeds up (catalyzes) a chemical reaction without being changed itself. Living organisms use thousands of different enzymes to drive consecutive series of chemical reactions, known as metabolic pathways, which maintain life. All enzymes are based on protein molecules, but the function of each type of enzyme depends on the shape of the molecule.

The lock-and-key model

All enzymes are believed to work through a mechanism called the lock-and-key model. Each enzyme is able to handle a specific set of raw materials, known as a substrate. In this analogy, the enzyme protein is the lock because its molecule contains a zone, called the active site, which is shaped in a way that allows the molecules of the substrate—the key—to bond to it. When the substrate molecules link to the active site (like a key in a lock), a chemical reaction is triggered, altering the chemical properties of the substrate.

1 Enzyme and substrate
The shape of an enzyme complements that of the substrate it acts upon. This diagram shows an anabolic enzyme, which converts two substrate chemicals into a single product. In contrast, catabolic enzymes (such as digestive enzymes) break a single substrate chemical into multiple products.

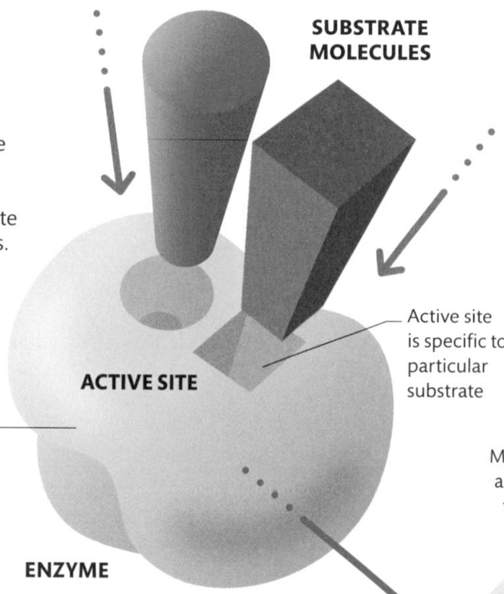

SUBSTRATE MOLECULES
Active site is specific to particular substrate
ACTIVE SITE
Enzyme is uniquely shaped to allow active site to link efficiently to both substrate molecules
ENZYME

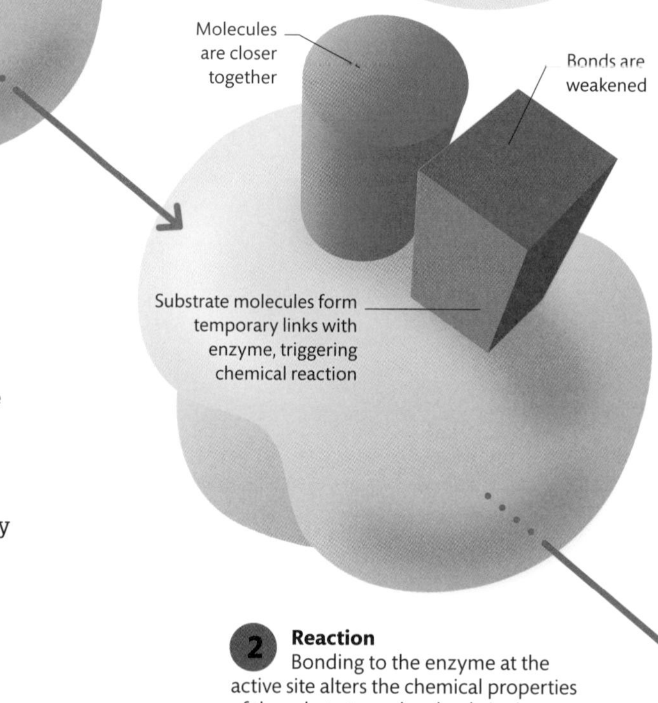

Molecules are closer together
Bonds are weakened
Substrate molecules form temporary links with enzyme, triggering chemical reaction

2 Reaction
Bonding to the enzyme at the active site alters the chemical properties of the substrate molecules, bringing them closer and weakening some of the bonds within them. The molecules now break these bonds and recombine to form a new, larger product.

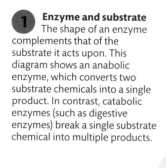

3 Product
The newly formed product has different chemical properties from the original substrate molecules. This means that it is no longer able to bond to the enzyme's active site, and so it is released. The enzyme has not been altered by the reaction.

DO PLANTS HAVE ENZYMES?

All organisms, plants included, use enzymes to control metabolism. For example, the enzymes that work with DNA are similar in every life form.

ALTERING ENZYMES

Enzymes work best at an optimum temperature. For most human enzymes, this is around 99°F (37°C). If it is too cold or too hot, the enzyme becomes less efficient. At temperatures higher than 131°F (55°C), the enzyme's structure changes, or denatures. This alters the shape of the active site, and the enzyme stops working altogether.

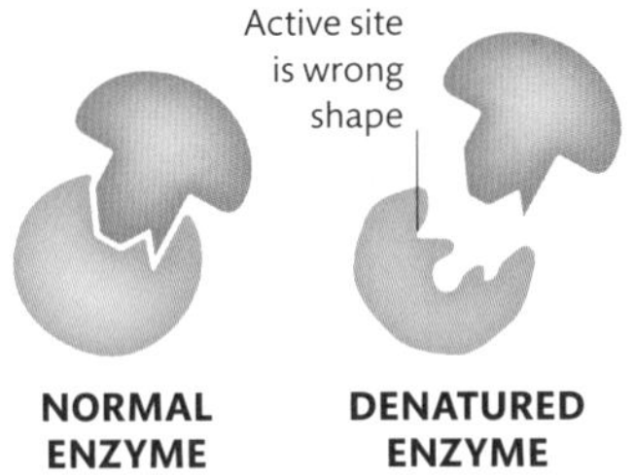
Active site is wrong shape
NORMAL ENZYME
DENATURED ENZYME

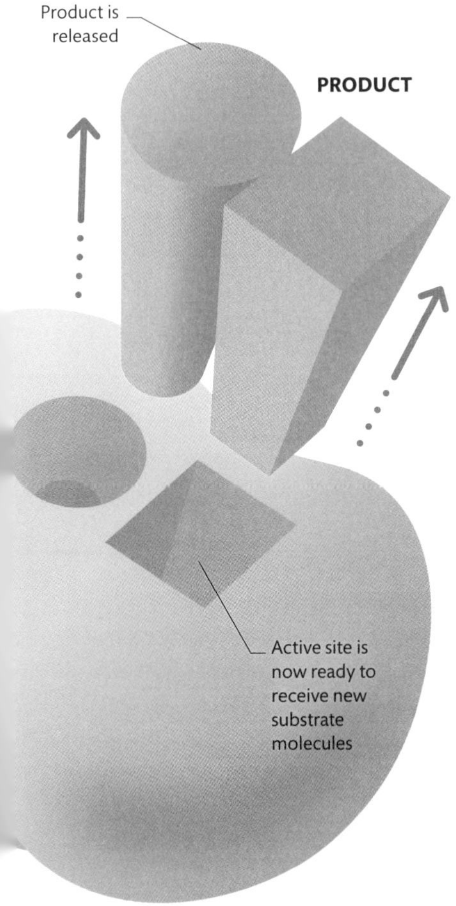
Product is released
PRODUCT
Active site is now ready to receive new substrate molecules

Digestive enzymes

Enzymes are a key part of an animal's digestive system, where large, complex molecules in food are broken down into simpler units that are easier for the body to absorb into the bloodstream. Each type of nutrient is handled by its own set of digestive enzymes. These enzymes are secreted by different parts of the digestive tract, creating digestive juices that mix with the food.

Carbohydrases
These enzymes break down complex carbohydrates such as starches into smaller sugar molecules. The intestines then absorb the digested sugars.

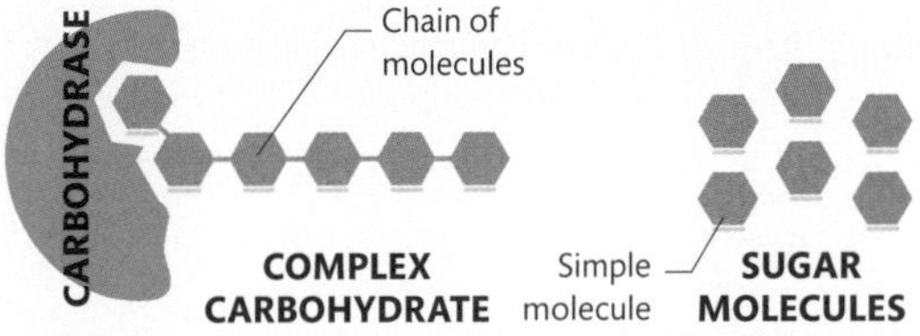
CARBOHYDRASE
Chain of molecules
COMPLEX CARBOHYDRATE
Simple molecule
SUGAR MOLECULES

Lipases
Fats and oils, known as lipids, are digested by lipase enzymes in the small intestine. Lipids are constructed from three fatty acids linked to a single glycerol.

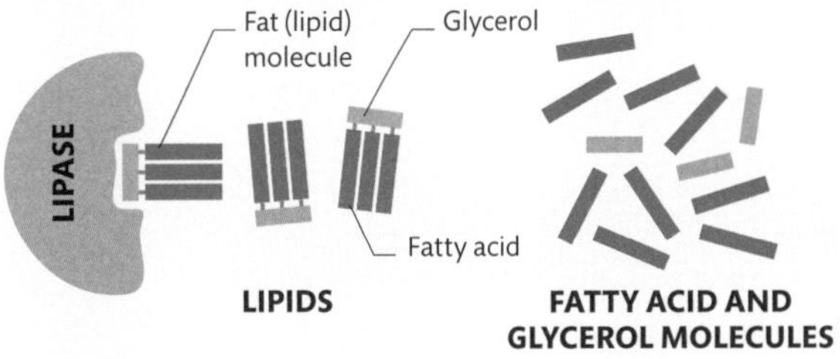
LIPASE
Fat (lipid) molecule
Glycerol
Fatty acid
LIPIDS
FATTY ACID AND GLYCEROL MOLECULES

Protease
Proteins are made up of long chains of smaller units called amino acids. Protease enzymes break up these chains so the individual units can be absorbed separately.

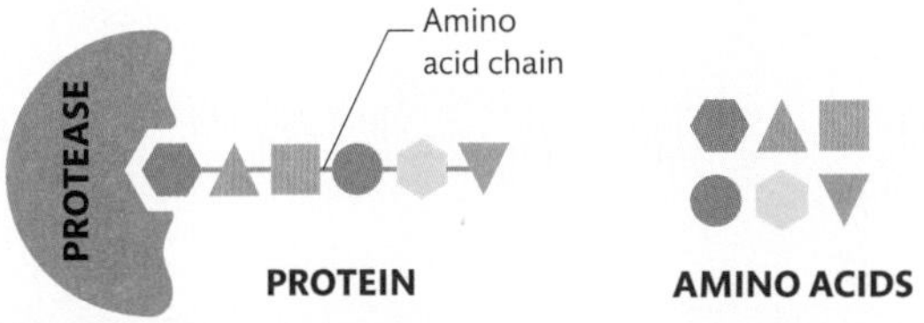
PROTEASE
Amino acid chain
PROTEIN
AMINO ACIDS

ENZYMES IN INDUSTRY

The catalytic power of enzymes means they can be used in manufacturing and as components of useful products, instead of less efficient and more polluting inorganic chemicals. In the future, artificial enzymes may be designed to enable chemical reactions that are not seen in living organisms.

Plastic
Enzymes found in some bacteria are able to digest certain plastics into harmless products.

Laundry soaop
Biological laundry products use enzymes to remove organic material from dirty fabrics.

Cheese
Stomach enzymes that convert milk into a digestible solid are used to make some cheeses.

Structural materials

Living bodies may be composed of thousands of distinct substances, but just a small handful of versatile materials are responsible for ensuring their structural integrity.

Building materials

Each of the different kingdoms of life uses its own unique combination of structural materials. The most important of these are cellulose, chitin, and collagen, despite each being found in different types of organisms. These three materials are all polymers—molecules that are built up from chains of smaller units. Their molecules can be easily combined to form strong fibers and sheets, making them fundamental in providing shape and support for body parts at all scales.

CELLULOSE IS THE MOST ABUNDANT NATURALLY OCCURRING ORGANIC COMPOUND

CELLULOSE

Cellulose is a polysaccharide (see pp.32–33) made from long chains of glucose molecules. It is the main material in the walls of all plant cells. Together the walls of the cells give the plant's leaves, stems, and roots a solid structure. Wood and bark is made from cellulose fibers cross-linked with other molecules.

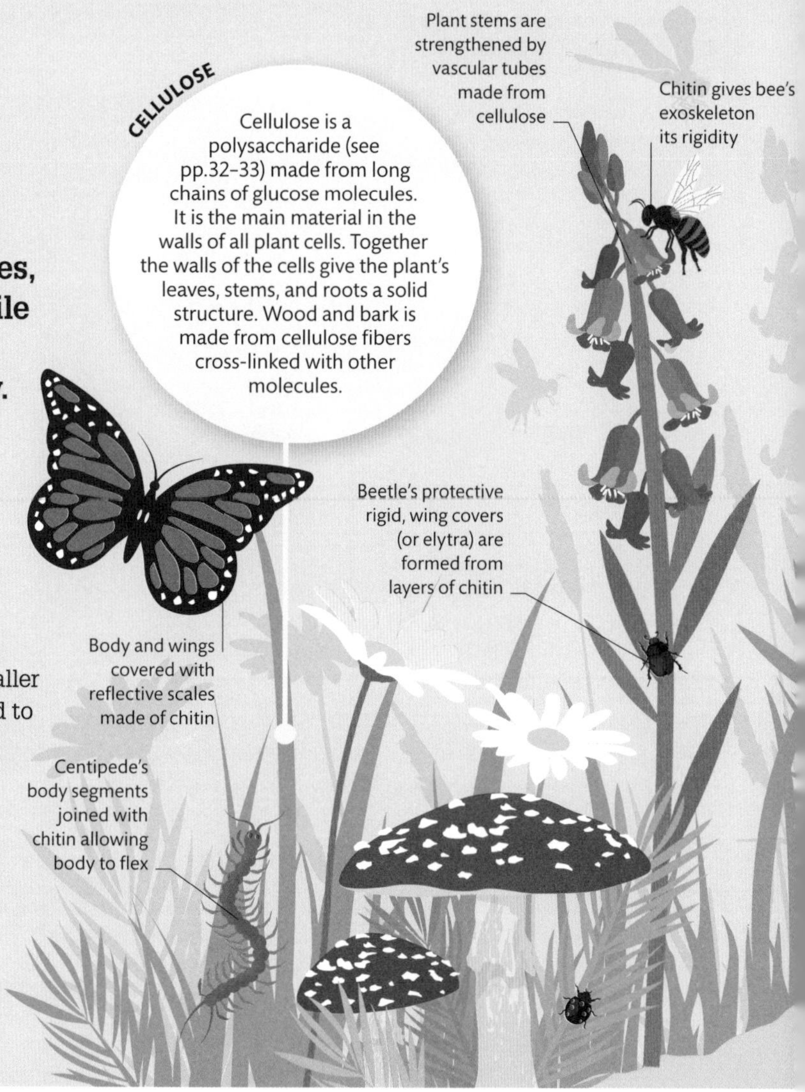
Plant stems are strengthened by vascular tubes made from cellulose
Chitin gives bee's exoskeleton its rigidity
Beetle's protective rigid, wing covers (or elytra) are formed from layers of chitin
Body and wings covered with reflective scales made of chitin
Centipede's body segments joined with chitin allowing body to flex

KERATIN

A fibrous protein, keratin is found in all kinds of animal. Its properties make it perfect for creating a flexible and waterproof coating to the body, so it is found in skin, scales, and hair. Keratin also helps fortify other external body parts, such as fingernails, claws, and horns.

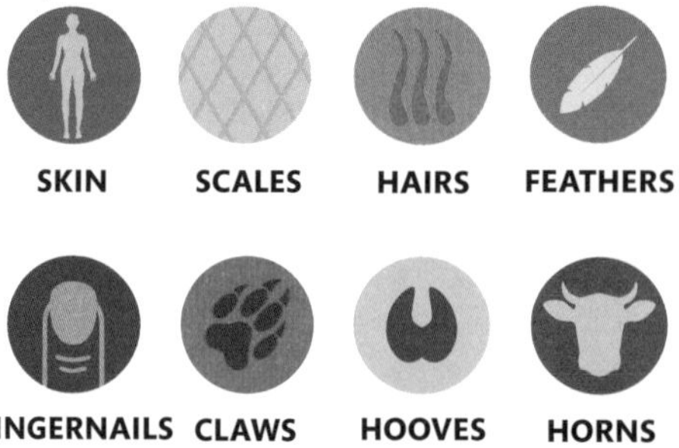
SKIN
SCALES
HAIRS
FEATHERS
FINGERNAILS
CLAWS
HOOVES
HORNS

ARE BONES ALIVE?

Bone is a living tissue. The calcium-rich minerals that create a light but stiff structure are being constantly renewed. Inside, the bone also has blood and nerve connections.

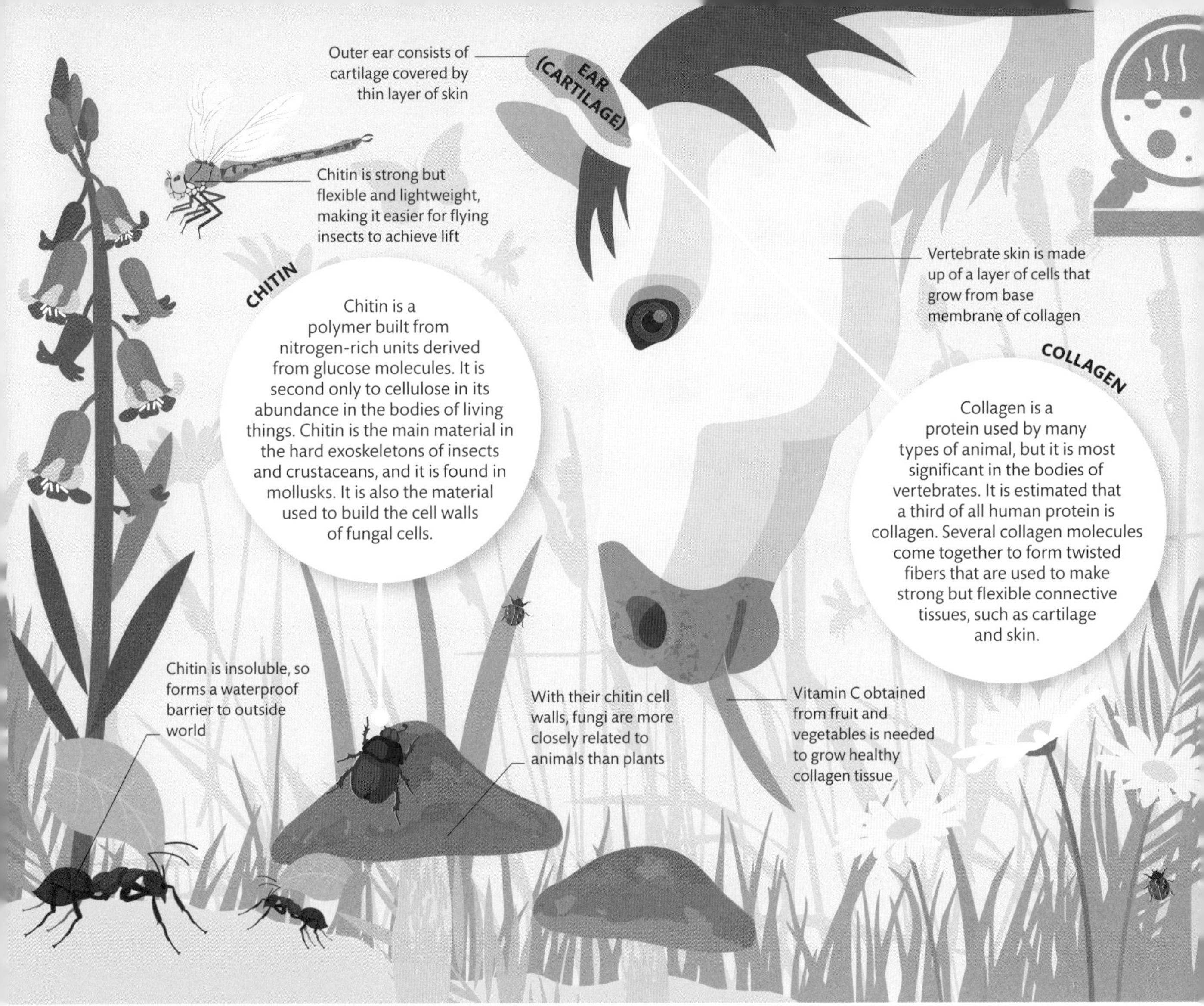
Outer ear consists of cartilage covered by thin layer of skin
EAR (CARTILAGE)
Chitin is strong but flexible and lightweight, making it easier for flying insects to achieve lift
Vertebrate skin is made up of a layer of cells that grow from base membrane of collagen
CHITIN
Chitin is a polymer built from nitrogen-rich units derived from glucose molecules. It is second only to cellulose in its abundance in the bodies of living things. Chitin is the main material in the hard exoskeletons of insects and crustaceans, and it is found in mollusks. It is also the material used to build the cell walls of fungal cells.
COLLAGEN
Collagen is a protein used by many types of animal, but it is most significant in the bodies of vertebrates. It is estimated that a third of all human protein is collagen. Several collagen molecules come together to form twisted fibers that are used to make strong but flexible connective tissues, such as cartilage and skin.
Chitin is insoluble, so forms a waterproof barrier to outside world
With their chitin cell walls, fungi are more closely related to animals than plants
Vitamin C obtained from fruit and vegetables is needed to grow healthy collagen tissue

Making hard skeletons

Animals require hard, stiff tissues to act as a protective outer armor, such as a shell, or to provide internal structures and anchor points for muscles, which is the purpose of bone. The formation of solid crystals in the spaces between cells gives strength to these vital tissues. The mineral constituents can be collected as soluble forms in the food or water an animal consumes, then converted into insoluble solids.

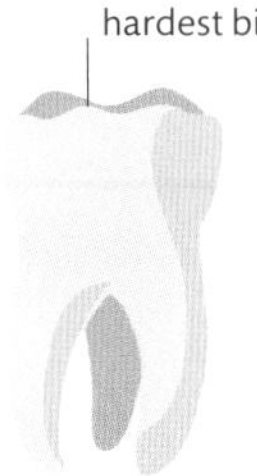
Tooth enamel, which contains hydroxyapatite, is hardest biological material

Hydroxyapatite
A natural form of calcium phosphate, hydroxyapatite is used in vertebrate bones and teeth. The mineral makes up about 70 percent of the human skeleton's weight.

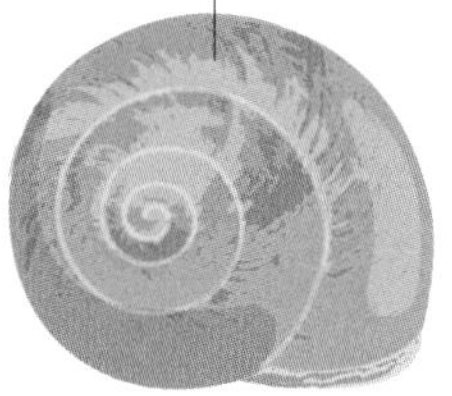
Shell is made from layers of tiny crystals

Aragonite
Mollusk, crustacean, and other shellfish shells are hardened by aragonite, a form of calcium carbonate. The raw ingredients are extracted from sea water.

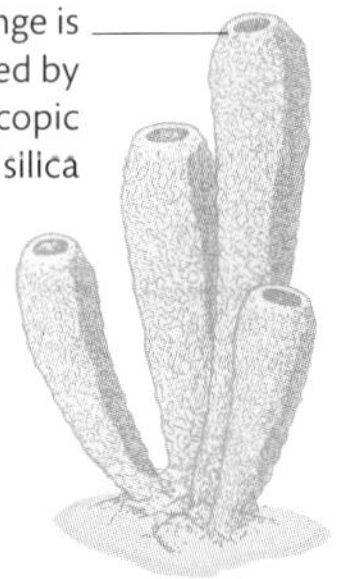
Sponge is protected by microscopic spikes of silica

Silica
The bodies of some sponges and corals are constructed from particles made of silica. This glasslike substance is the naturally occurring form of silicon dioxide.

Vitamins

The human body requires small amounts of 13 different essential chemicals, called vitamins. They are wide ranging in their forms and functions but all play an important role in the body's metabolic processes. The human body cannot build vitamins from raw ingredients, but they can be extracted from the foods we eat as part of a balanced diet.

Fat-soluble vitamins

These vitamins are absorbed from the intestines while encapsulated in globules of fat. They can be stored in the body's fat deposits. Fat-soluble vitamins will not reach toxic levels through a normal diet but consuming large quantities as supplements can cause health problems.

Water-soluble vitamins

These vitamins are taken directly into the bloodstream from the digestive juices. The body will only take what it needs, and any excess in the diet (or in supplements) is left to leave the body. Water-soluble vitamins cannot be stored, so they need to be consumed daily.

VITAMIN A

Vitamin A has roles in many processes, including vision, immune function, and bone formation. Deficiency can cause eye problems such as night blindness.

VITAMIN E

Vitamin E is an important component of the immune system and protects cell membranes in the skin and eyes. Deficiencies are rare.

B VITAMINS

B vitamins are a mixture of eight molecules—B1 to B3, B5 to B7, B9, and B12. They are involved in all areas of good health and are easily available in a balanced diet.

VITAMIN D

Vitamin D contributes to uptake of calcium and phosphorus. A deficiency leads to rickets, where the bones become soft and may grow malformed.

VITAMIN K

Vitamin K is important in the way blood clots. Deficiencies are rare but are linked to skin that bruises easily and frequent nosebleeds.

VITAMIN C

Vitamin C is closely associated with fending off infections. A lack of vitamin C causes scurvy, a nasty condition that impacts many parts of the body.

KEY

Meat	Milk	Oily fish	Tomatoes	Nuts	Ready meals
Poultry	Leafy greens	Peanuts	Bananas	Bacon	Lettuce
Liver	Broccoli	Eggs	Oranges	Whole grains	Olives
Fish	Avocado	Olive oil	Strawberries	Potato chips	Cheese

Micronutrients

Vitamins and minerals are essential substances called micronutrients, which organisms need in small amounts to grow and stay healthy. Micronutrients often cannot be easily stored, so a small but constant supply is required.

ELEMENTS IN THE BODY

Just four elements make up more than 96 percent of human body weight. Sodium, potassium, chlorine, sulfur, and magnesium are present in only tiny quantities, but the body cannot survive without them.

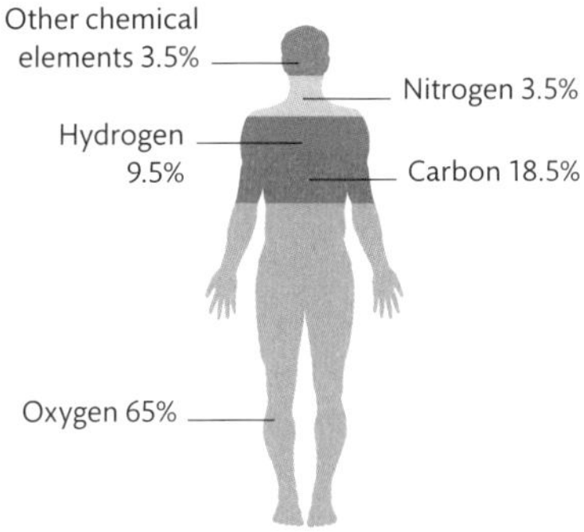
Other chemical elements 3.5%
Nitrogen 3.5%
Hydrogen 9.5%
Carbon 18.5%
Oxygen 65%

Minerals

The many millions of biochemicals used by all forms of life are primarily composed of four elements: hydrogen, oxygen, nitrogen, and carbon. However, metabolic processes require several other elements, which can be collected from naturally occurring compounds, called minerals, found in soils and water.

Magnesium
This element is the central component of chlorophyll, the green pigment used by plants and other photosynthetic organisms to collect energy from sunlight.

Sodium
Positively charged forms of sodium, called ions, are widely used in metabolic reactions and to make the electrical pulses used in nerves and muscles.

Chloride
The main negatively charged ions in a living body, chloride ions often work against sodium and potassium to neutralize electrical differences.

Potassium
Positively charged potassium ions are used in concert with sodium to create tiny electrical differences, or voltages, across membranes.

Sulfur
Sulfur is a component of several amino acids, used to make the proteins in skin, hair, feathers, and horns.

Phosphorus
This highly reactive element is a component of energy carriers, such as adenosine triphosphate (ATP), used by all cells.

Calcium
One of the most abundant minerals in living bodies, calcium ions are used in metabolism but are also the basis of structural materials like bone and shell.

Trace elements
There are many elements, such as zinc, selenium, and iodine, that are detected in tiny amounts in living tissue. Many appear to have a role in good health, although their exact functions are not always understood. It may be that some trace elements are simply impurities.

UNLIKE ANIMALS, **PLANTS DO NOT NEED VITAMINS,** THEY JUST NEED **MINERALS**

Photosynthesis

Tiny structures in plant cells capture light energy from the Sun and convert it to chemical energy, which is stored as sugar. This process is called photosynthesis. Directly or indirectly, it feeds the entire living world.

WHY DO PLANT LEAVES LOOK GREEN?

The chlorophyll molecules in chloroplasts absorb red and violet-blue light (the colors most effective at driving photosynthesis) and reflect green light.

From sunlight to sugar

Photosynthesis takes place mostly in plants' leaves. It requires light, water, and carbon dioxide (CO_2). The roots draw water from the soil, and CO_2 enters leaves through tiny pores called stomata. Organelles called chloroplasts (see p.61) inside leaf cells absorb light energy, and chemical reactions convert this into the energy that the plant needs to live and grow.

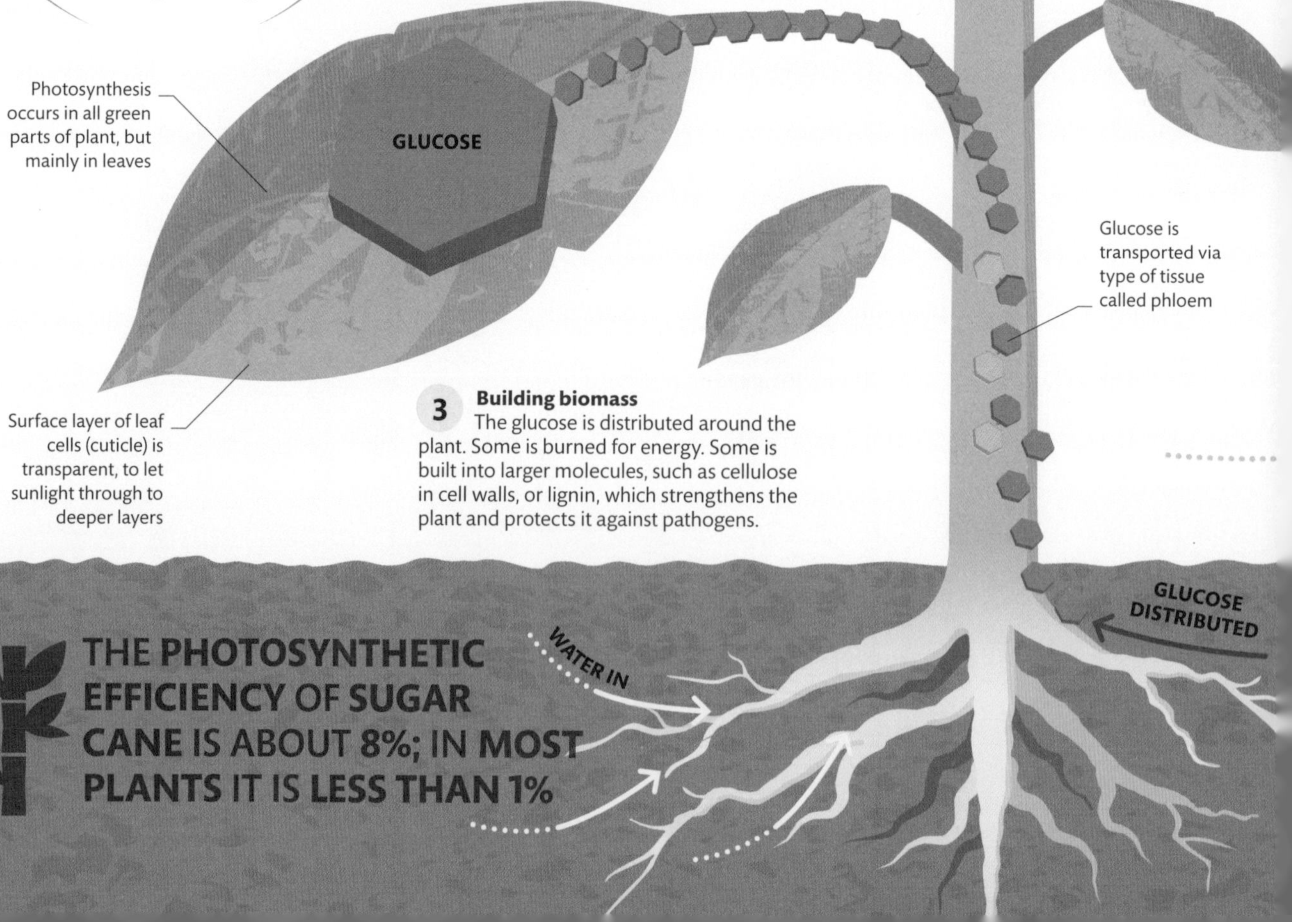

Glucose helps build lignin in woody parts of plant
Photosynthesis occurs in all green parts of plant, but mainly in leaves
GLUCOSE
Glucose is transported via type of tissue called phloem
Surface layer of leaf cells (cuticle) is transparent, to let sunlight through to deeper layers
GLUCOSE DISTRIBUTED
WATER IN

3 Building biomass
The glucose is distributed around the plant. Some is burned for energy. Some is built into larger molecules, such as cellulose in cell walls, or lignin, which strengthens the plant and protects it against pathogens.

THE PHOTOSYNTHETIC EFFICIENCY OF SUGAR CANE IS ABOUT 8%; IN MOST PLANTS IT IS LESS THAN 1%

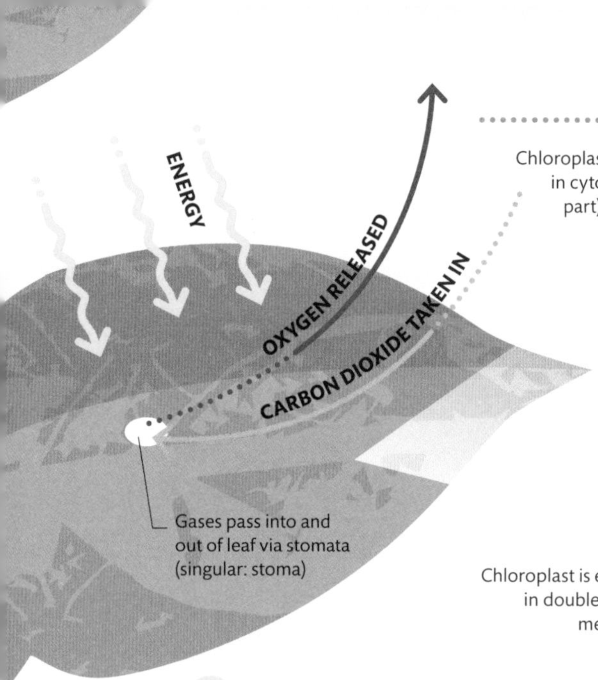
ENERGY
OXYGEN RELEASED
CARBON DIOXIDE TAKEN IN
Gases pass into and out of leaf via stomata (singular: stoma)

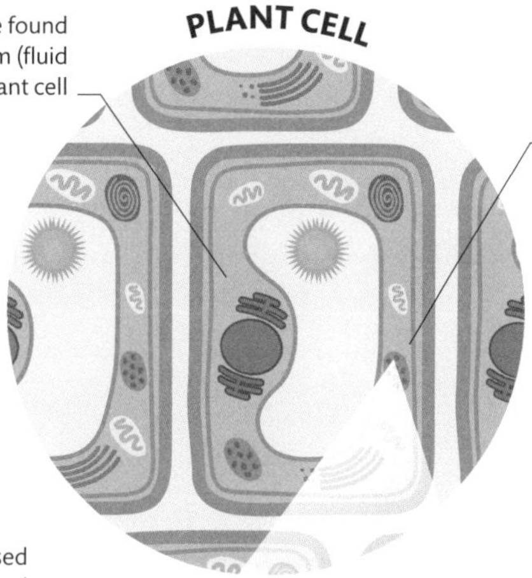
PLANT CELL
Chloroplasts are found in cytoplasm (fluid part) of plant cell
Chloroplasts are 2–10 micrometers in diameter and 1 micrometer thick

Plant cells
Each leaf cell (or cell in other green parts) typically contains 10–100 chloroplasts, where photosynthesis occurs. Chloroplasts lie near the surface of the plant, where they can easily absorb sunlight.

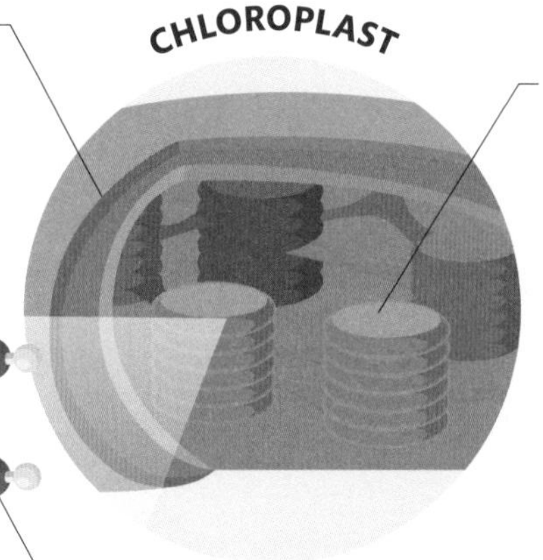
CHLOROPLAST
Chloroplast is enclosed in double-layered membrane
Thylakoids are grouped in stacks called grana, surrounded by fluid called stroma

Chloroplasts
Each chloroplast holds stacks of membranes called thylakoids, which absorb energy from sunlight. The thylakoids contain the pigment chlorophyll, which absorbs blue-violet light and red light.

1 Converting solar energy
Chloroplasts in cells absorb photons (particles of light) and release electrons. This causes a chain of reactions that produce the "energy molecules" ATP and NADPH, and split water molecules to release oxygen.

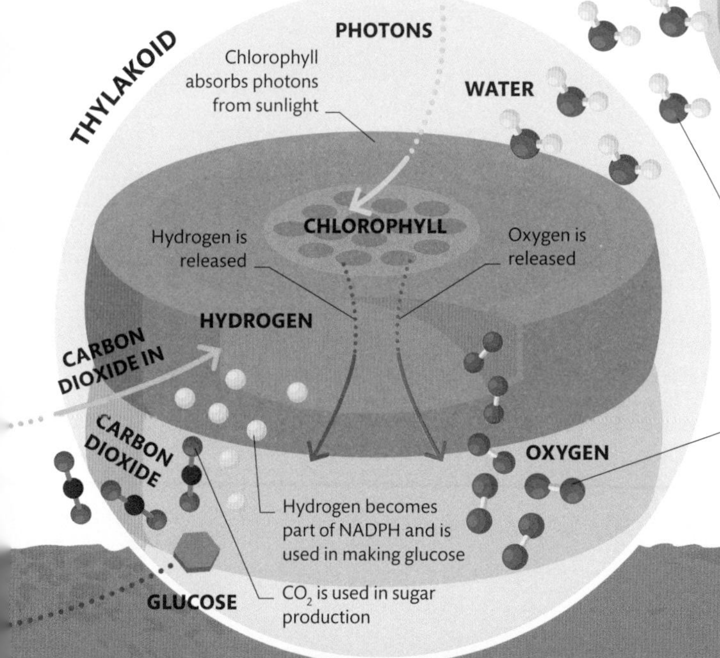
THYLAKOID
PHOTONS
Chlorophyll absorbs photons from sunlight
WATER
CHLOROPHYLL
Hydrogen is released
Oxygen is released
HYDROGEN
CARBON DIOXIDE IN
CARBON DIOXIDE
OXYGEN
Hydrogen becomes part of NADPH and is used in making glucose
GLUCOSE
CO_2 is used in sugar production
Water consists of two hydrogen and one oxygen molecule
Oxygen leaves cells and is released into air

2 Making sugar
In the fluid that surrounds the thylakoid stacks an enzyme incorporates CO_2 into an intermediate molecule and then into carbohydrate molecules such as glucose. In the process, energy is released from the ATP and NADPH molecules. This process is called the Calvin cycle.

CAM PLANTS

Most plants open their stomata in daytime, but cacti open theirs at night to conserve water. Cacti fix CO_2 into organic acids, which are converted into sugars. This process is called CAM (crassulacean acid metabolism).

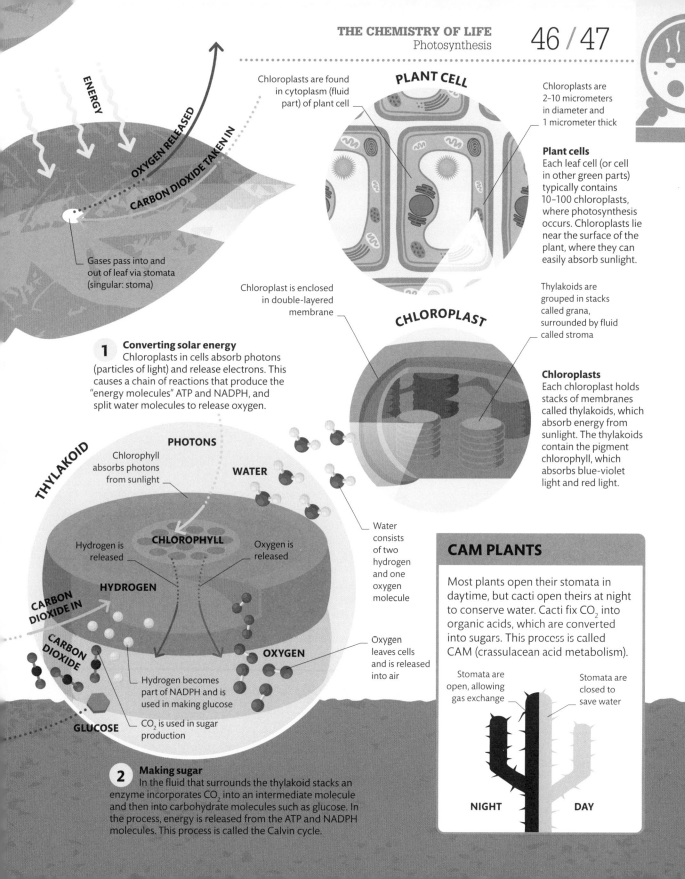

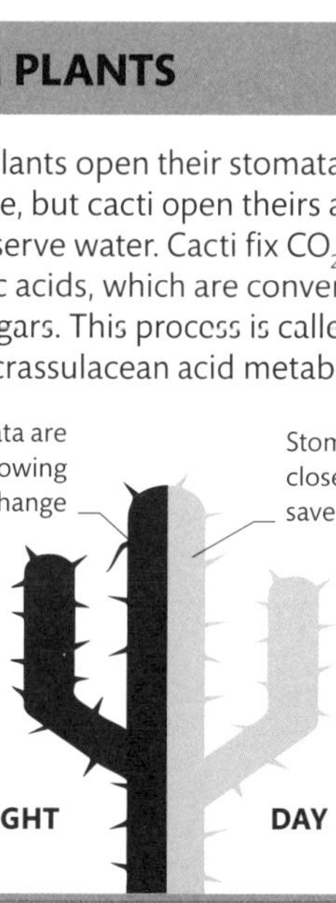
Stomata are open, allowing gas exchange
Stomata are closed to save water
NIGHT
DAY

Respiration

The process of respiration takes place inside cells. It involves chemically breaking down nutrients from food to produce energy. Respiration may be aerobic (using oxygen) or anaerobic (using no oxygen).

Aerobic respiration

Used by organisms that have a good oxygen supply, the aerobic form of respiration is the most efficient. Oxygen breathed in from the surrounding air reacts with a fuel, such as glucose (sugar), to release energy; this also generates the waste products carbon dioxide and water. This reaction is similar to combustion, where a fuel burns quickly in air. However, aerobic respiration slows that reaction into several small steps, each releasing a usable packet of energy.

GLUCOSE + OXYGEN → CARBON DIOXIDE + WATER + ENERGY

The chemical process of aerobic respiration
Aerobic respiration can be shown as an equation. Glucose and oxygen react to form carbon dioxide and water, releasing energy in the process.

1 Delivering the fuel
Respiration is shown here in muscle tissue, which uses a lot of energy. Blood brings a near-constant supply of oxygen and glucose to the tissue. Body cells can also store fuel as a complex carbohydrate called glycogen, made from chains of sugars.

2 Splitting glucose
The first stage is glycolysis, in which a glucose molecule is split into two molecules of pyruvate, releasing a small amount of energy.

3 Releasing energy
The pyruvate molecules enter the cell's mitochondria (see p.60), where they are combined with oxygen and broken down in several further steps, releasing energy at each step.

4 Waste products
The carbon dioxide and water generated from respiration diffuse out of the cell, into the surrounding tissue fluid, and from there into blood plasma. Carbon dioxide is carrried back to the lungs, where it is exhaled.

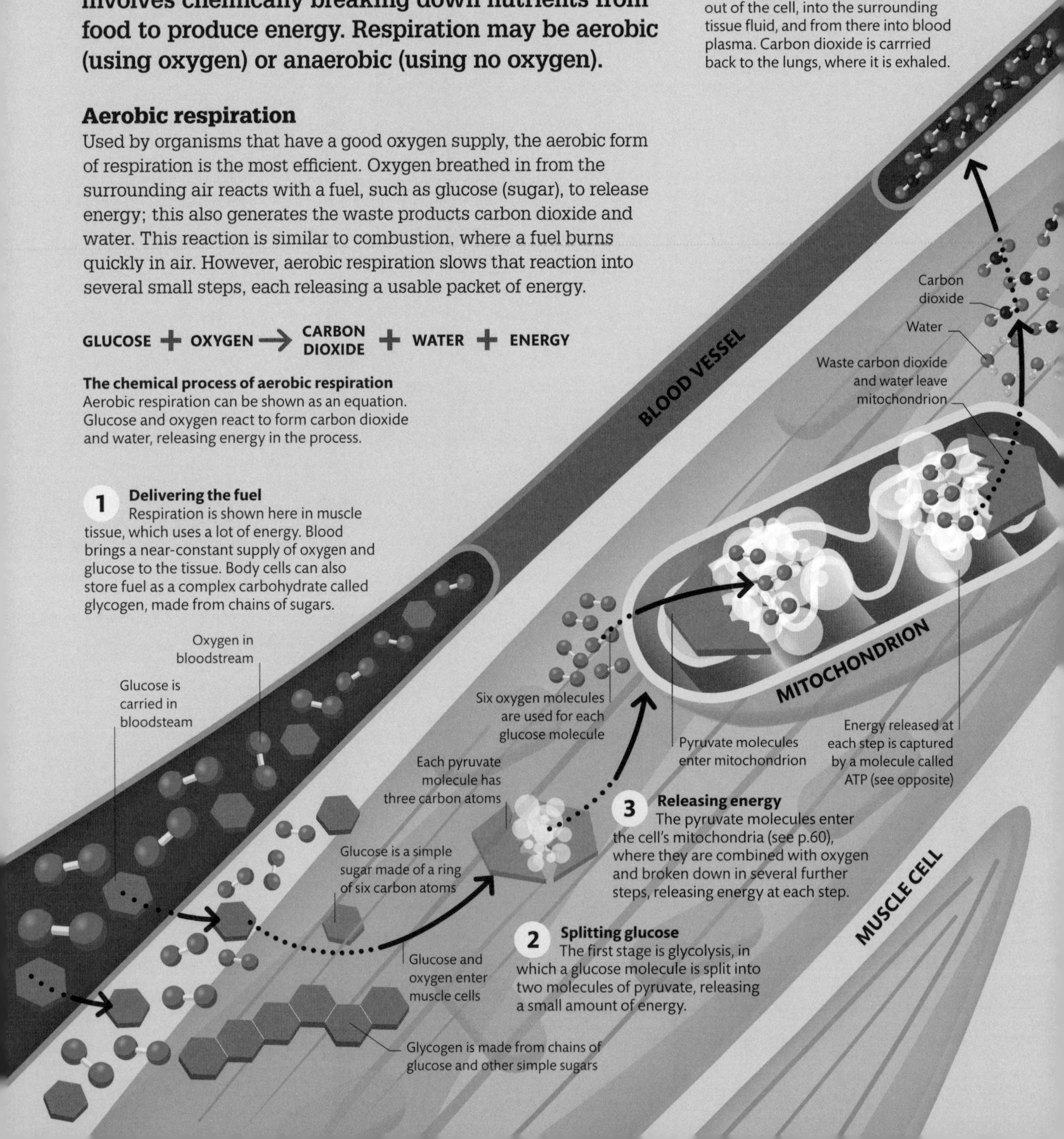
BLOOD VESSEL
MITOCHONDRION
MUSCLE CELL
Oxygen in bloodstream
Glucose is carried in bloodsteam
Glucose is a simple sugar made of a ring of six carbon atoms
Glucose and oxygen enter muscle cells
Glycogen is made from chains of glucose and other simple sugars
Each pyruvate molecule has three carbon atoms
Six oxygen molecules are used for each glucose molecule
Pyruvate molecules enter mitochondrion
Energy released at each step is captured by a molecule called ATP (see opposite)
Waste carbon dioxide and water leave mitochondrion
Water
Carbon dioxide

Anaerobic respiration

In situations where little oxygen is available, organisms rely on anaerobic respiration. This breaks up glucose and releases energy without using oxygen. Some organisms, such as soil bacteria and yeasts, use only this kind of respiration. Organisms that usually respire aerobically, such as humans, do it anaerobically during periods of hard physical activity, such as sprinting.

WINE, BREAD, AND BEER ARE MADE BY YEAST USING ANAEROBIC RESPIRATION

GLUCOSE → LACTIC ACID + ENERGY

The chemical process of anaerobic respiration
This simplified equation shows anaerobic respiration in animal cells. Enzymes break glucose into smaller lactic acid molecules, releasing just a small amount of energy.

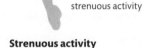

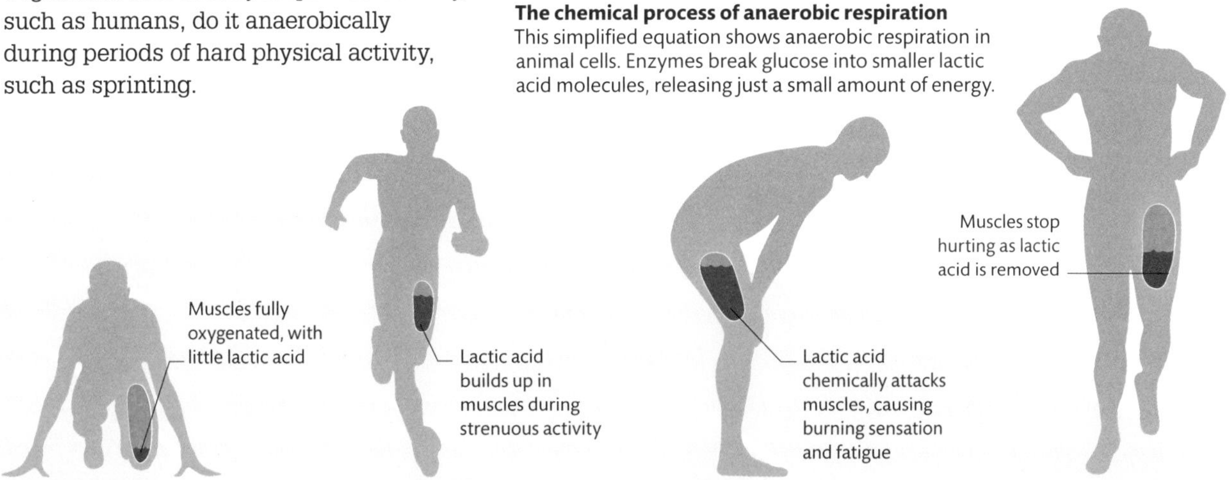
Muscles fully oxygenated, with little lactic acid
Lactic acid builds up in muscles during strenuous activity
Lactic acid chemically attacks muscles, causing burning sensation and fatigue
Muscles stop hurting as lactic acid is removed

Low activity
During rest or moderate activity, the body is able to take in all the oxygen it requires to respire aerobically and meet its energy needs.

Strenuous activity
The blood and lungs cannot supply the muscles with oxygen fast enough, so much of the respiration is anaerobic.

Oxygen debt
The body takes in extra oxygen by breathing deeply. The amount of oxygen needed to break down excess lactic acid is called the oxygen debt.

Recovery
As the oxygen level rises, the lactic acid falls. It is broken down by being oxidized during aerobic respiration.

THE ENERGY CARRIERS ADP AND ATP

Energy released from respiration is captured by energy-carrier molecules, which supply it to fuel other chemical reactions in the cell. The most important carriers are adenosine triphosphate (ATP) and adenosine diphosphate (ADP). Whenever a packet of energy is released in each step of respiration, this is used to add a phosphate group to ADP, converting it to ATP. To power other reactions within the cell, the ATP gives up its third phosphate, releasing its stored energy, and converts back to ADP.

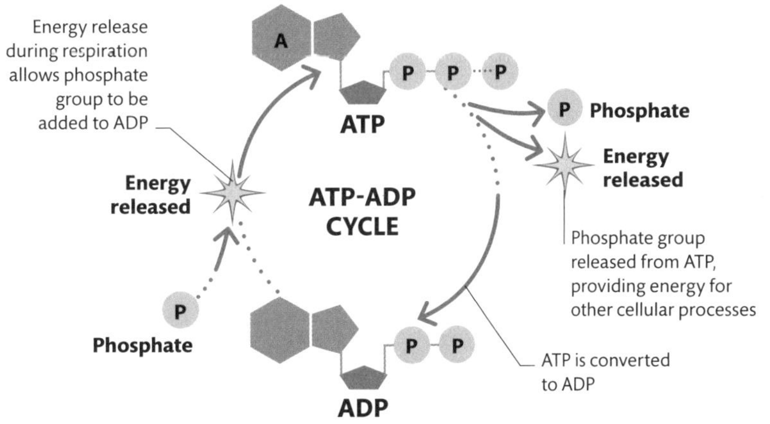
Energy release during respiration allows phosphate group to be added to ADP
A
P P P
ATP
P Phosphate
Energy released
Energy released
ATP-ADP CYCLE
Phosphate group released from ATP, providing energy for other cellular processes
P
Phosphate
P P
ADP
ATP is converted to ADP

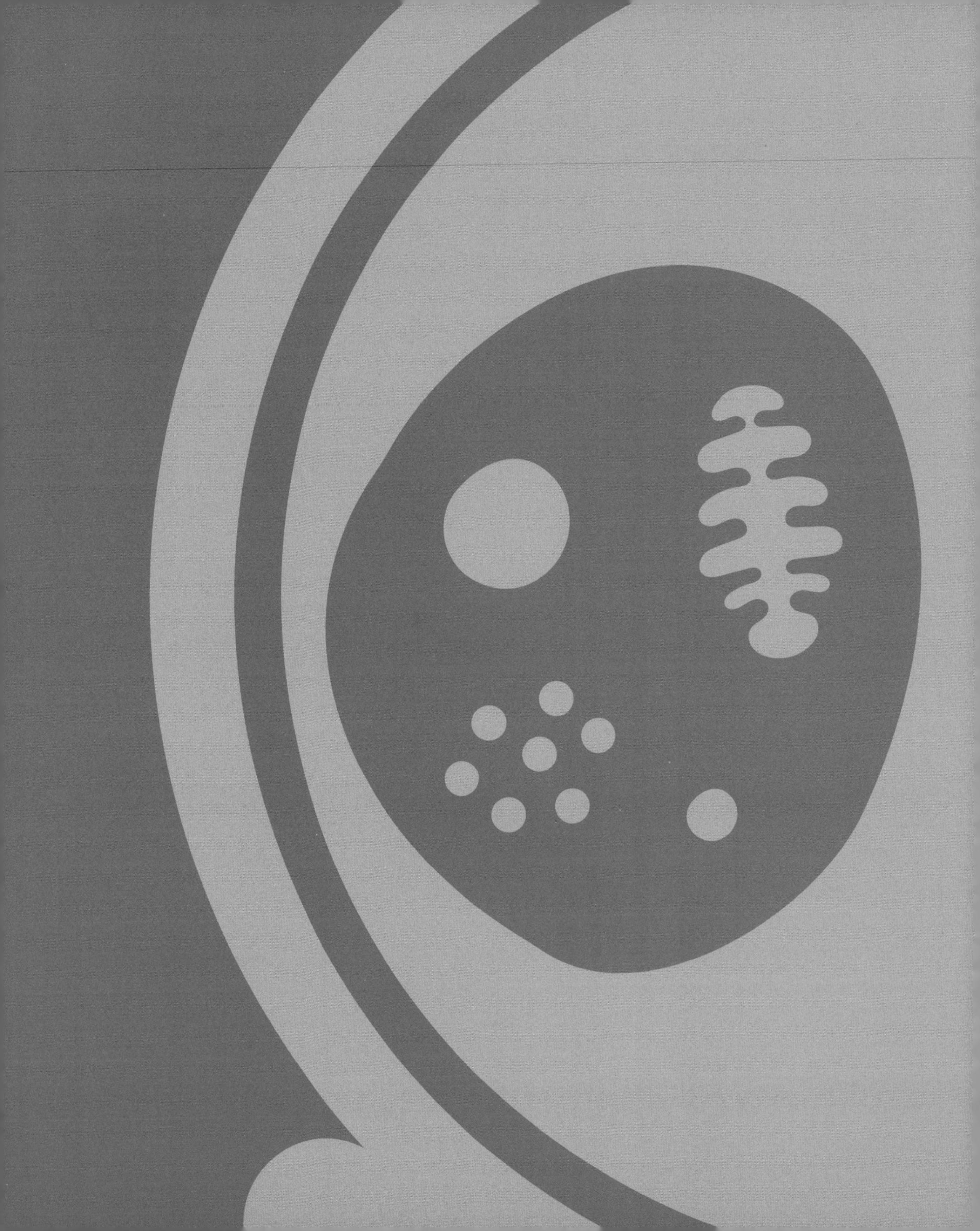

HOW CELLS WORK

Light microscope image
This image of the vascular tissue of corn (magnified 100 times) is typical of what can be seen through a light microscope. The tissues have been stained with color for extra visibility.

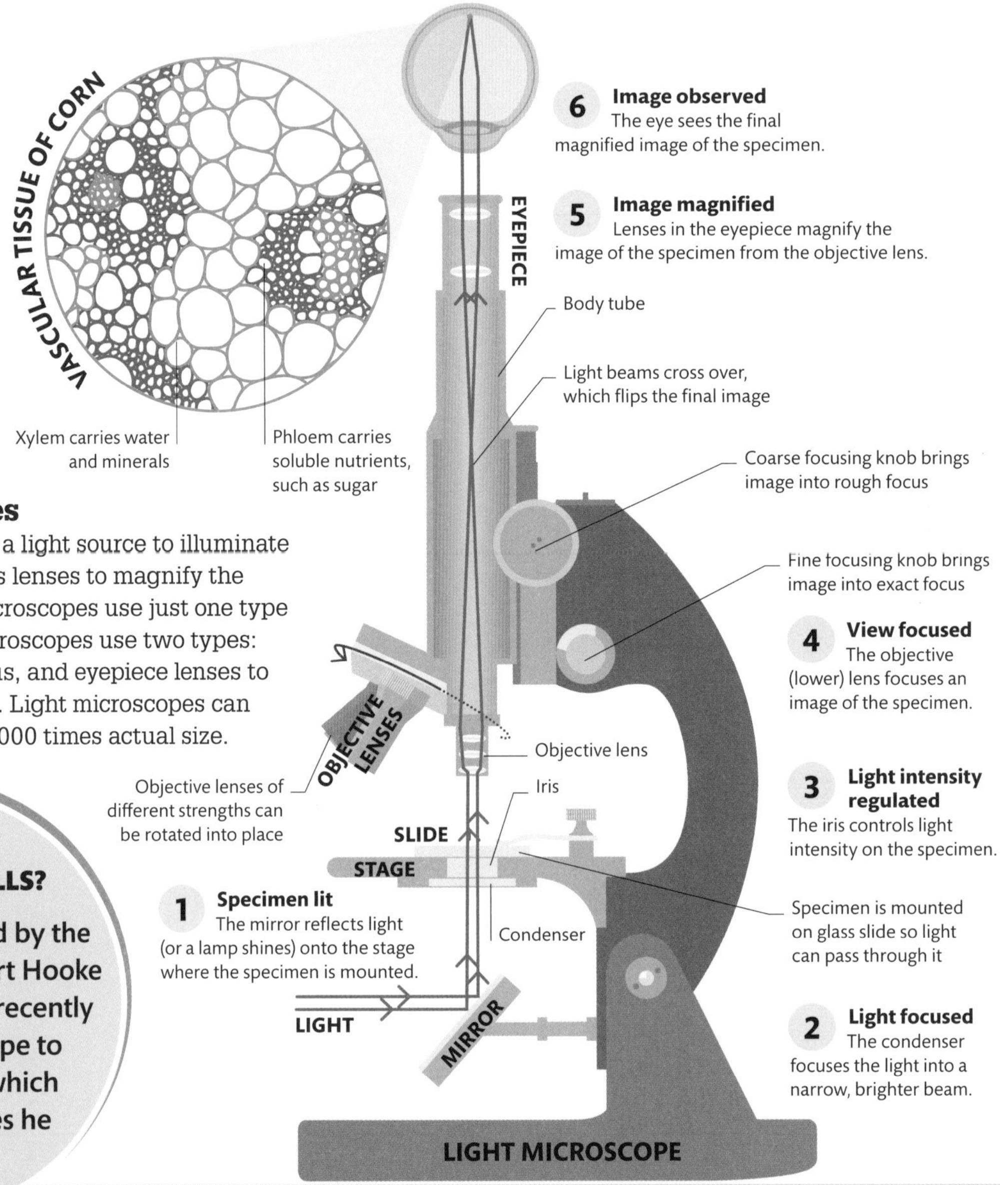
VASCULAR TISSUE OF CORN
Xylem carries water and minerals
Phloem carries soluble nutrients, such as sugar
6 Image observed
The eye sees the final magnified image of the specimen.
EYEPIECE
5 Image magnified
Lenses in the eyepiece magnify the image of the specimen from the objective lens.
Body tube
Light beams cross over, which flips the final image
Coarse focusing knob brings image into rough focus
Fine focusing knob brings image into exact focus
4 View focused
The objective (lower) lens focuses an image of the specimen.
OBJECTIVE LENSES
Objective lens
Objective lenses of different strengths can be rotated into place
Iris
3 Light intensity regulated
The iris controls light intensity on the specimen.
SLIDE
STAGE
Specimen is mounted on glass slide so light can pass through it
1 Specimen lit
The mirror reflects light (or a lamp shines) onto the stage where the specimen is mounted.
Condenser
LIGHT
MIRROR
2 Light focused
The condenser focuses the light into a narrow, brighter beam.
LIGHT MICROSCOPE

Light microscopes

A light microscope has a light source to illuminate the specimen and glass lenses to magnify the image. Simple light microscopes use just one type of lens. Compound microscopes use two types: objective lenses to focus, and eyepiece lenses to increase the resolution. Light microscopes can magnify up to about 2,000 times actual size.

WHO DISCOVERED CELLS?

Cells were discovered by the English scientist Robert Hooke in 1665. He used the recently invented microscope to examine cork, in which he saw small pores he called cells.

Studying cells

Tiny structures such as cells are studied using microscopes: devices that illuminate objects and then produce magnified images for viewing. Light (optical) microscopes illuminate specimens with beams of light and magnify them with glass lenses. Electron microscopes illuminate specimens and produce images using beams of electrons.

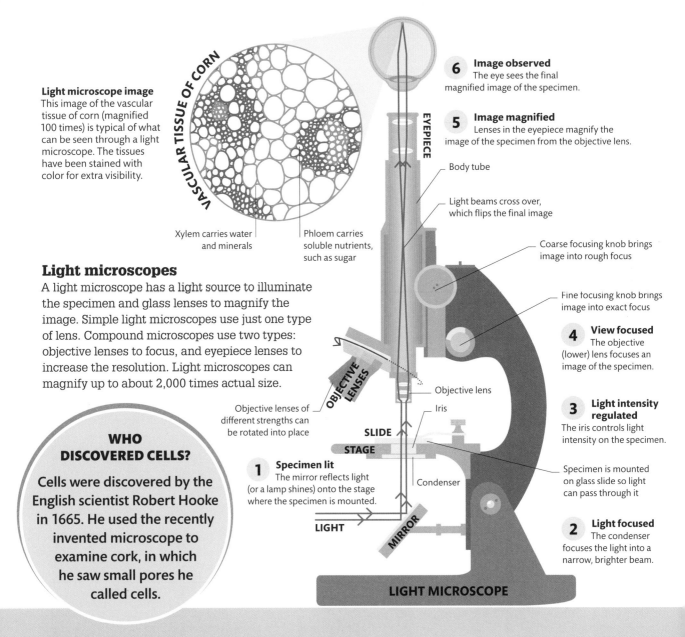

THE **LARGEST SINGLE CELL** IS THAT OF THE **ALGA *CAULERPA TAXIFOLA*,** WHICH CAN GROW TO A **LENGTH** OF ABOUT **12 IN (30 CM)**

Electron microscopes

An electron microscope makes an image using beams of electrons in a vacuum, focused with magnetic fields. Electron beams have short wavelengths, so they do not diffract and create blurs, enabling electron microscopes to magnify objects to 50 million times their actual size. There are two main types: transmission electron microscopes (as shown here), which pass electrons through a specimen; and scanning electron microscopes, which reflect electrons from the surface of a specimen.

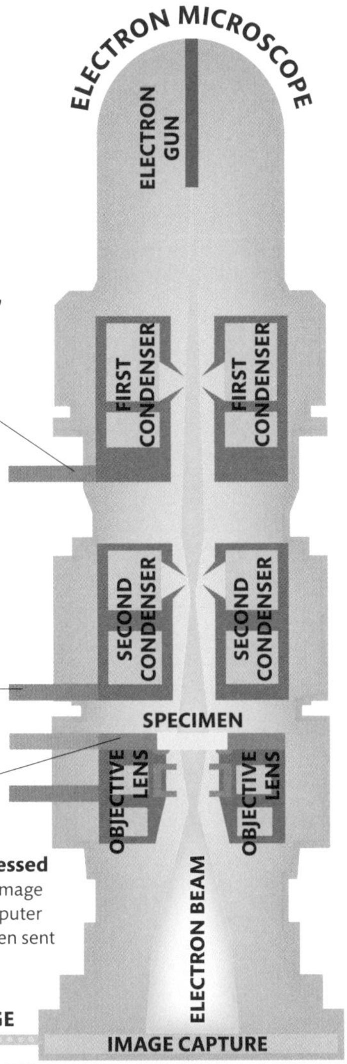
ELECTRON MICROSCOPE
ELECTRON GUN
FIRST CONDENSER
FIRST CONDENSER
Condenser aperture blocks stray electrons
SECOND CONDENSER
SECOND CONDENSER
Condenser aperture blocks stray electrons
SPECIMEN
Specimen holder and air lock
OBJECTIVE LENS
OBJECTIVE LENS
ELECTRON BEAM
IMAGE CAPTURE
DIGITIZED IMAGE

1 Electron beam generated
An electron gun fires a beam of electrons at the specimen.

2 Initial beam focus
The first condenser partially focuses the electron beam.

3 Secondary beam focus
The second condenser further focuses the electron beam.

4 Specimen
The electron beam is scattered as it passes through the specimen.

5 Image magnified
The objective lens detects electrons scattered by the specimen and magnifies the information.

6 Image capture
Information from the objective lens is captured by a digital camera and/or displayed on a monitor as an image.

7 Image processed
The digitized image may be sent to a computer for processing and then sent to a monitor screen.

8 Image displayed
The processed image of the specimen is displayed on a monitor.

TEM image
This color-enhanced transmission electron microscope (TEM) image shows a human lymphocyte (a type of white blood cell) magnified about 6,000 times.

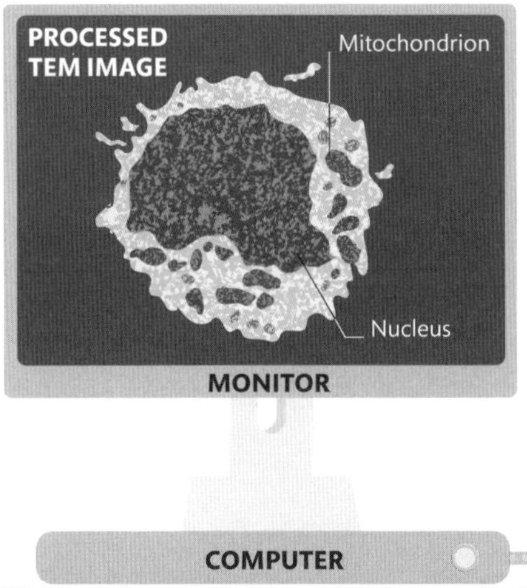
PROCESSED TEM IMAGE
Mitochondrion
Nucleus
MONITOR
COMPUTER

THE SIZES OF CELLS

The smallest things we can see with the naked eye are about 100 micrometers (microns), or 1/10 of a millimeter. Human ova (eggs) are just about visible. Other cells, and the structures within them, are visible only with a microscope.

KEY
- Visible only with electron microscope
- Visible with light microscope
- Visible with naked eye

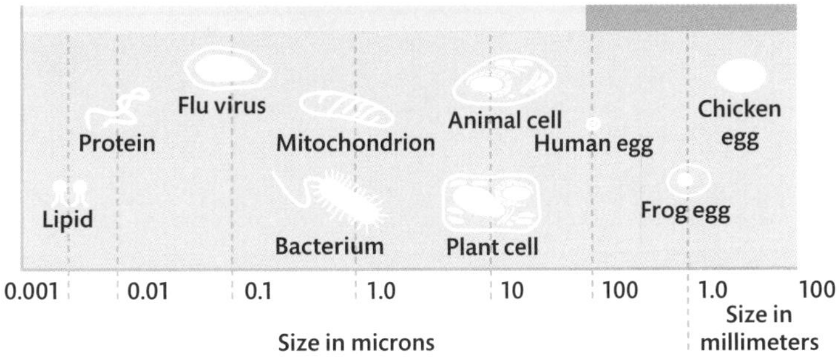
Lipid
Protein
Flu virus
Mitochondrion
Bacterium
Animal cell
Plant cell
Human egg
Frog egg
Chicken egg
0.001
0.01
0.1
1.0
10
100
1.0
100
Size in microns
Size in millimeters

Parts of the cell

The cell is the basic unit of all life forms, from single-celled microorganisms to animals, plants, and fungi. In complex organisms such as animals and plants, many of the cells contain internal structures known as organelles that perform specific functions.

Animal cells

A typical animal cell is 10–30 micrometers (millionths of a meter) in diameter. It is enclosed in a flexible plasma membrane that allows substances to move in and out of the cell. An internal “skeleton” called the cytoskeleton enables the cell to hold its shape and allows substances to be transported inside it. Most cells have a central nucleus, which contains DNA. All cells contain a watery fluid called cytoplasm; this holds subcellular bodies called organelles, stores energy in the form of glycogen (which is made from glucose), and contains enzymes and amino acids that are needed for cellular functions.

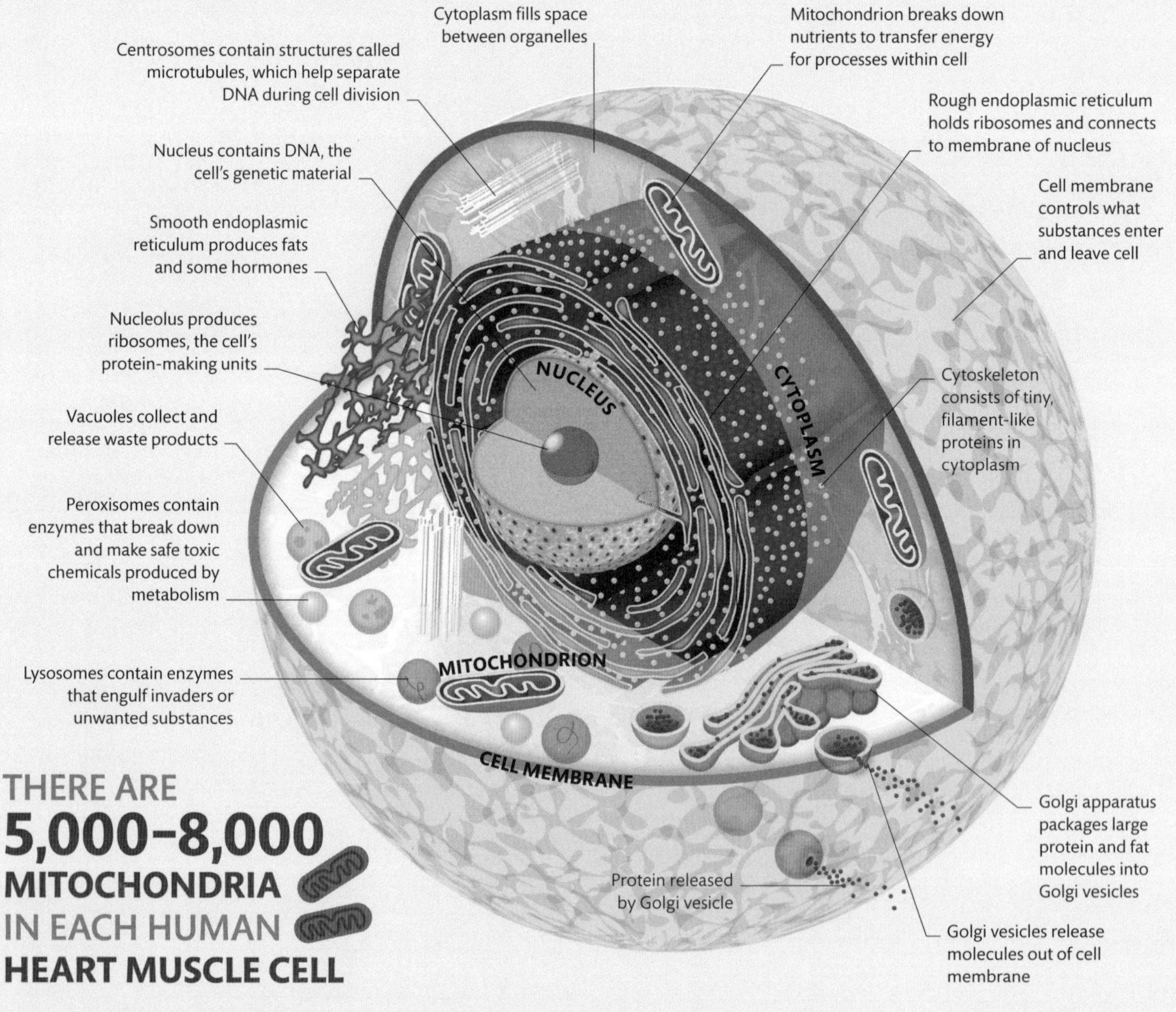
Cytoplasm fills space between organelles
Mitochondrion breaks down nutrients to transfer energy for processes within cell
Centrosomes contain structures called microtubules, which help separate DNA during cell division
Rough endoplasmic reticulum holds ribosomes and connects to membrane of nucleus
Nucleus contains DNA, the cell’s genetic material
Cell membrane controls what substances enter and leave cell
Smooth endoplasmic reticulum produces fats and some hormones
Nucleolus produces ribosomes, the cell’s protein-making units
NUCLEUS
CYTOPLASM
Cytoskeleton consists of tiny, filament-like proteins in cytoplasm
Vacuoles collect and release waste products
Peroxisomes contain enzymes that break down and make safe toxic chemicals produced by metabolism
MITOCHONDRION
Lysosomes contain enzymes that engulf invaders or unwanted substances
CELL MEMBRANE
Golgi apparatus packages large protein and fat molecules into Golgi vesicles
Protein released by Golgi vesicle
Golgi vesicles release molecules out of cell membrane

THERE ARE **5,000–8,000** **MITOCHONDRIA** IN EACH HUMAN **HEART MUSCLE CELL**

Plant cells

Like animal cells, plant cells have a nucleus and organelles such as endoplasmic reticula and mitochondria. However, plant cells tend to be larger—up to 100 micrometers across. The most obvious difference from animal cells is that plant cells have rigid cell walls, made largely of cellulose, which protect the cells and give the plant's body its overall structural integrity. Another major difference is that plant cells contain chloroplasts. These organelles contain a green pigment called chlorophyll, which converts sunlight into starch for energy in a process called photosynthesis (see pp.46–47).

CILIA AND FLAGELLA

Some cells in animals and microorganisms have hairlike extensions on their walls to move cells or substances over the cell surface, or to propel the whole organism. Cilia are groups of short "hairs" that act in waves. Flagella are longer single or paired structures that make whiplike movements.

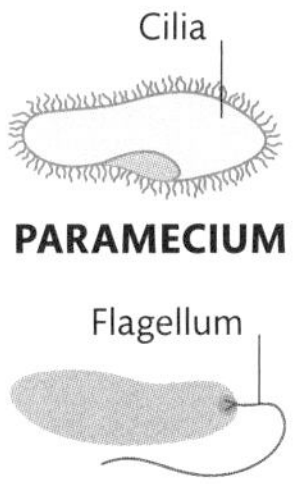
Cilia
PARAMECIUM
Flagellum
EUGLENA

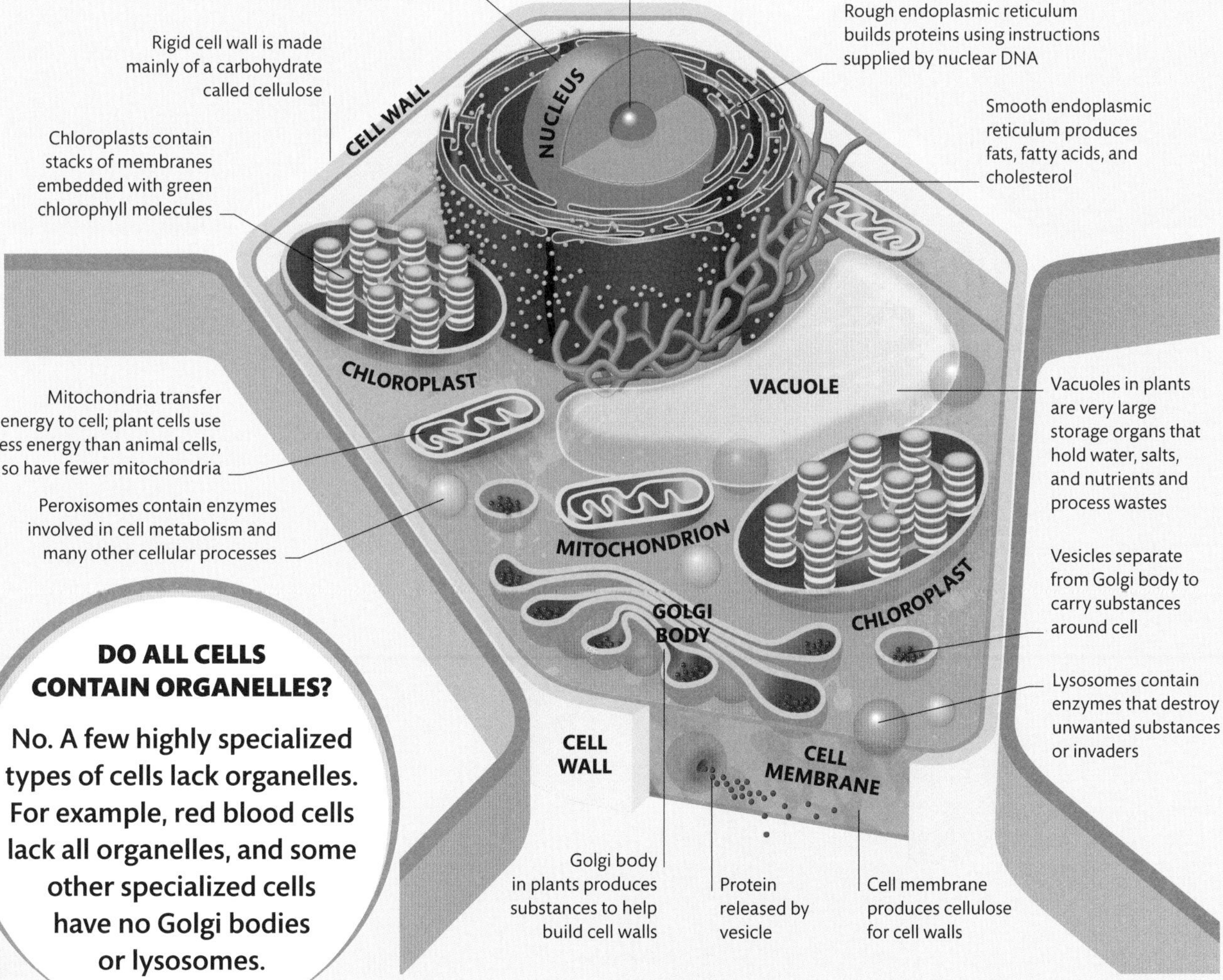
Nucleus contains DNA, the cell's genetic material
Nucleolus helps to make ribosomes
Rough endoplasmic reticulum builds proteins using instructions supplied by nuclear DNA
Rigid cell wall is made mainly of a carbohydrate called cellulose
CELL WALL
NUCLEUS
Smooth endoplasmic reticulum produces fats, fatty acids, and cholesterol
Chloroplasts contain stacks of membranes embedded with green chlorophyll molecules
CHLOROPLAST
VACUOLE
Vacuoles in plants are very large storage organs that hold water, salts, and nutrients and process wastes
Mitochondria transfer energy to cell; plant cells use less energy than animal cells, so have fewer mitochondria
Peroxisomes contain enzymes involved in cell metabolism and many other cellular processes
MITOCHONDRION
GOLGI BODY
CHLOROPLAST
Vesicles separate from Golgi body to carry substances around cell
Lysosomes contain enzymes that destroy unwanted substances or invaders
CELL WALL
CELL MEMBRANE
Golgi body in plants produces substances to help build cell walls
Protein released by vesicle
Cell membrane produces cellulose for cell walls

DO ALL CELLS CONTAIN ORGANELLES?

No. A few highly specialized types of cells lack organelles. For example, red blood cells lack all organelles, and some other specialized cells have no Golgi bodies or lysosomes.

The cell membrane

Also called the plasma membrane, the cell membrane protects the cell's internal structures. It is also semipermeable, allowing certain substances to enter and leave the cell.

PROTEINS MAKE UP HALF THE MASS OF SOME CELL MEMBRANES

Structure of the cell membrane

The membrane is a double-layered sheet of molecules called phospholipids that is less than 10 nanometers (10 billionths of a meter) thick. It also includes cholesterol, which strengthens and stabilizes the cell and regulates fluid levels. In addition, the membrane contains proteins, some of which are embedded in it and allow large molecules into and out of the cell, and some of which are attached to the outer surface.

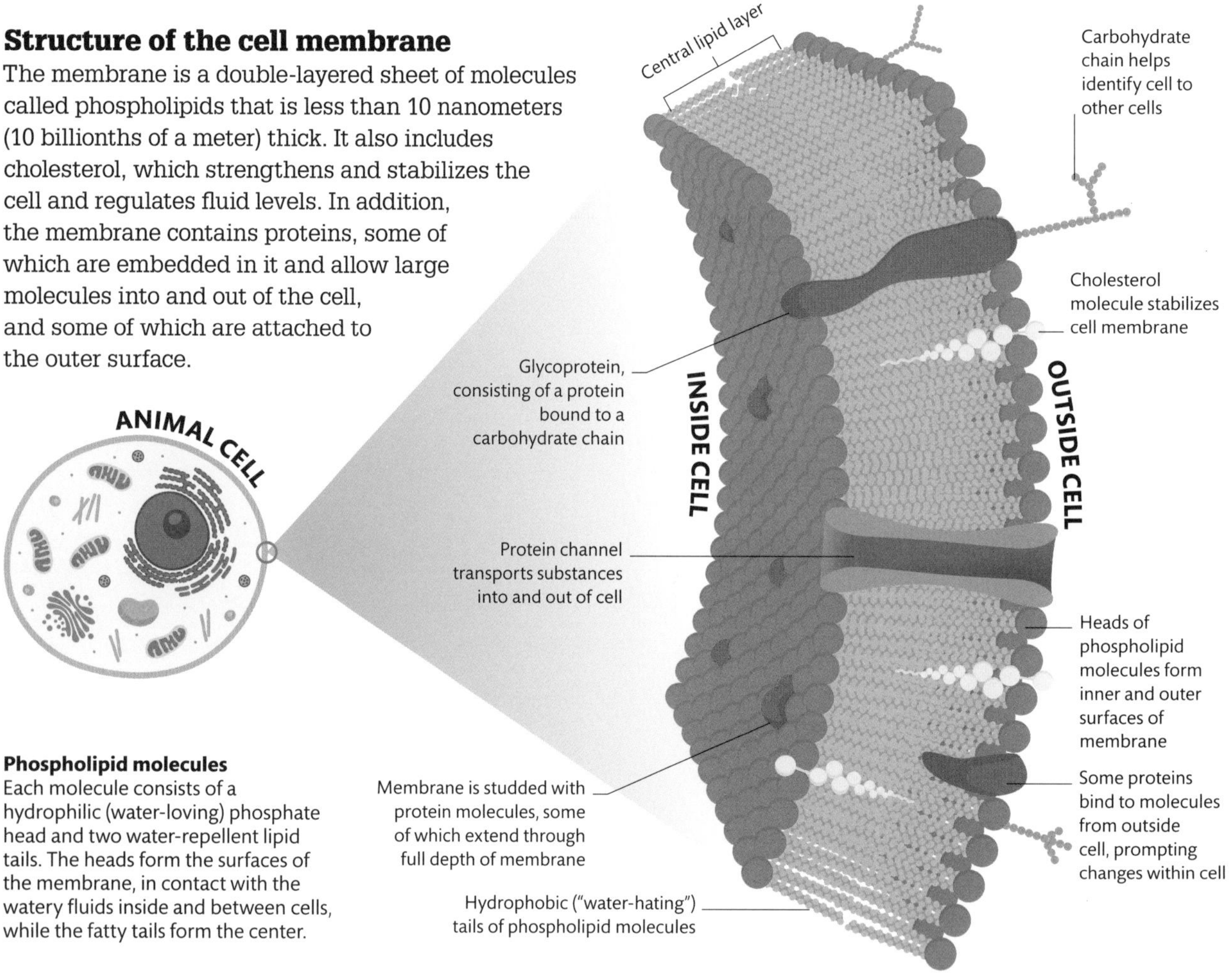
ANIMAL CELL
Central lipid layer
Carbohydrate chain helps identify cell to other cells
Cholesterol molecule stabilizes cell membrane
Glycoprotein, consisting of a protein bound to a carbohydrate chain
INSIDE CELL
OUTSIDE CELL
Protein channel transports substances into and out of cell
Heads of phospholipid molecules form inner and outer surfaces of membrane
Membrane is studded with protein molecules, some of which extend through full depth of membrane
Some proteins bind to molecules from outside cell, prompting changes within cell
Hydrophobic ("water-hating") tails of phospholipid molecules

Phospholipid molecules
Each molecule consists of a hydrophilic (water-loving) phosphate head and two water-repellent lipid tails. The heads form the surfaces of the membrane, in contact with the watery fluids inside and between cells, while the fatty tails form the center.

The cell wall

Plants, fungi, and many single-celled organisms have a wall just outside the cell membrane that supports and protects the cell. Plant cell walls are constructed from a network of cellulose fibers, interwoven with more tangled hemicellulose molecules. Pectin, a gluelike material, sticks the cell wall to its neighbors. Proteins strengthen the cell wall and aid in growth.

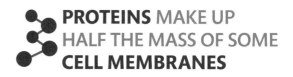

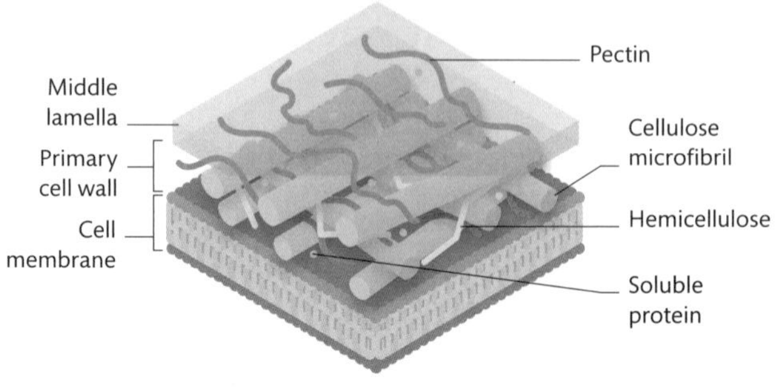
Pectin
Middle lamella
Primary cell wall
Cell membrane
Cellulose microfibril
Hemicellulose
Soluble protein

Transporting substances

Cell membranes are semipermeable. Tiny molecules such as oxygen and carbon dioxide can pass straight through. Some larger molecules, such as nutrients or electrically charged molecules (ions), can only enter or leave via channels in the cell membrane. For other large molecules, the cell creates a structure called a vesicle, in which the molecule is enclosed in a membrane, in order to take in or expel the molecule.

KEY

- Taking in (endocytosis) and processing of useful molecules
- Synthesis and processing of lipids
- Processing of proteins and other substances and elimination of unwanted substances (exocytosis)

Vesicles
Some vesicles form within a cell and merge with the cell membrane to release their contents, in a process called exocytosis. In the opposite process, endocytosis, a vesicle forms at the cell membrane to bring a substance into the cell. Some common processes are shown here.

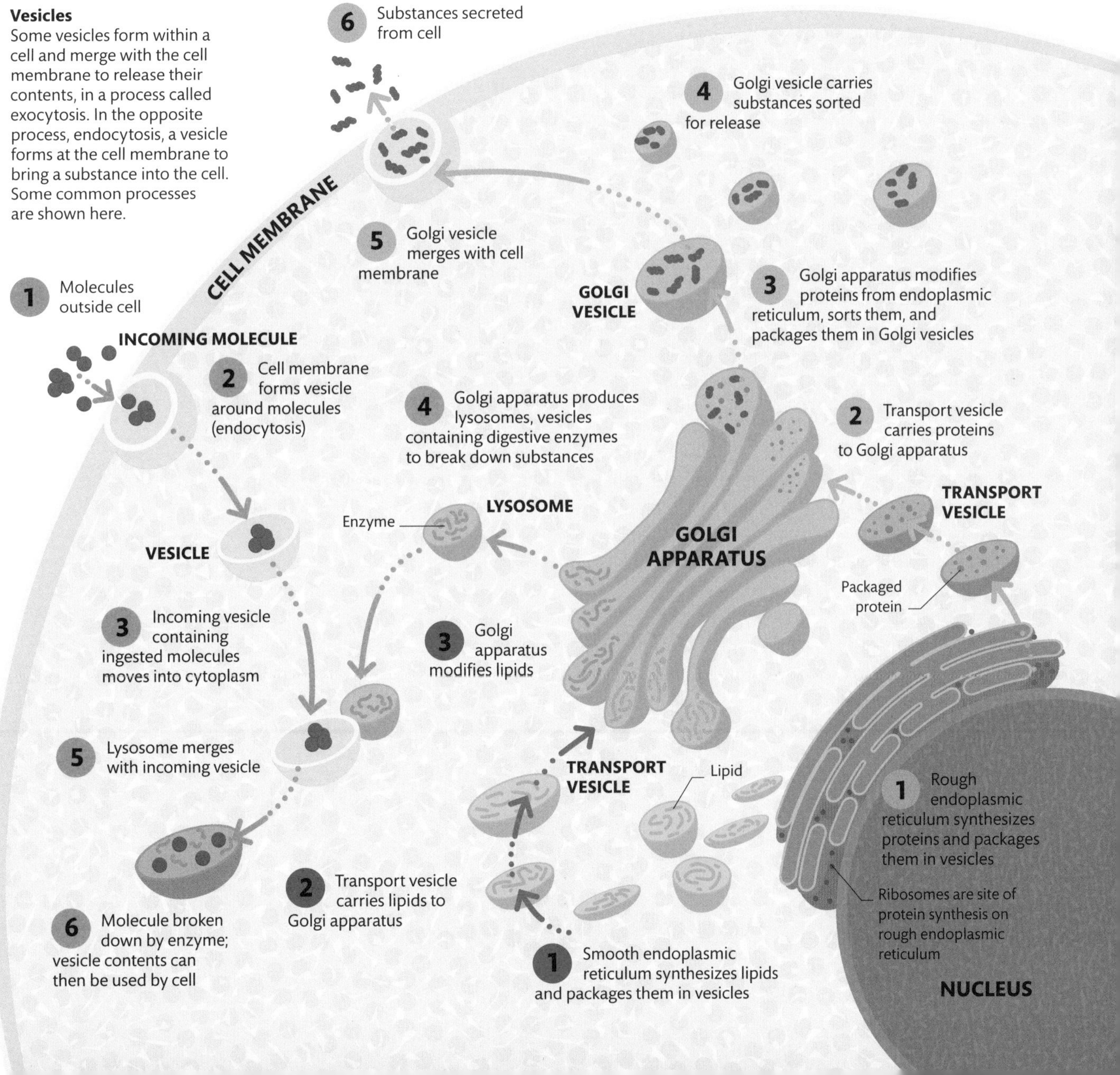
6 Substances secreted from cell
4 Golgi vesicle carries substances sorted for release
CELL MEMBRANE
5 Golgi vesicle merges with cell membrane
1 Molecules outside cell
GOLGI VESICLE
3 Golgi apparatus modifies proteins from endoplasmic reticulum, sorts them, and packages them in Golgi vesicles
INCOMING MOLECULE
2 Cell membrane forms vesicle around molecules (endocytosis)
4 Golgi apparatus produces lysosomes, vesicles containing digestive enzymes to break down substances
2 Transport vesicle carries proteins to Golgi apparatus
TRANSPORT VESICLE
LYSOSOME
Enzyme
VESICLE
GOLGI APPARATUS
Packaged protein
3 Incoming vesicle containing ingested molecules moves into cytoplasm
3 Golgi apparatus modifies lipids
5 Lysosome merges with incoming vesicle
TRANSPORT VESICLE
Lipid
1 Rough endoplasmic reticulum synthesizes proteins and packages them in vesicles
2 Transport vesicle carries lipids to Golgi apparatus
6 Molecule broken down by enzyme; vesicle contents can then be used by cell
Ribosomes are site of protein synthesis on rough endoplasmic reticulum
1 Smooth endoplasmic reticulum synthesizes lipids and packages them in vesicles
NUCLEUS

The nucleus

The nucleus is the cell's control center. It contains DNA, a long molecule that holds genetic information, which is used to make proteins that carry out the cell's development and functions.

The structure of the nucleus

The nucleus is generally the largest organelle (subcellular structure) in a cell. Its main functions are to protect the DNA and to supply information for making proteins. The nucleus is filled with a liquid called nucleoplasm; this is where the DNA is held, as long strands called chromatin. It also contains at least one nucleolus (see opposite). The nuclear surface is formed from a double layer of membranes. The outer membrane is joined to the rough endoplasmic reticulum (see p.57), which holds ribosomes for making proteins.

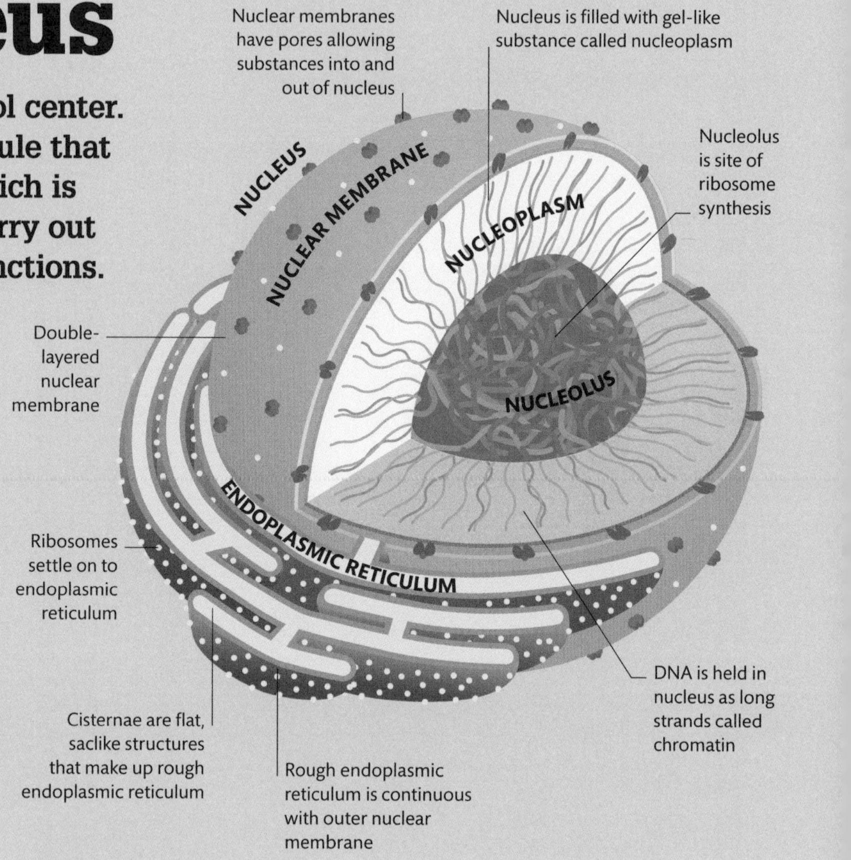

Nuclear membranes have pores allowing substances into and out of nucleus
Nucleus is filled with gel-like substance called nucleoplasm
Nucleolus is site of ribosome synthesis
NUCLEUS
NUCLEAR MEMBRANE
NUCLEOPLASM
NUCLEOLUS
Double-layered nuclear membrane
ENDOPLASMIC RETICULUM
Ribosomes settle on to endoplasmic reticulum
Cisternae are flat, saclike structures that make up rough endoplasmic reticulum
Rough endoplasmic reticulum is continuous with outer nuclear membrane
DNA is held in nucleus as long strands called chromatin

DNA and chromosomes

Each long DNA strand in a nucleus is wound around supporting protein structures called histones. Most of the time, the DNA exists in the form of chromatin. When a cell prepares to divide into two new cells, the strands are duplicated and supercoiled into compact, tightly coiled structures called chromosomes. Each chromosome is made of two identical units called chromatids; these break apart during division to provide a full set of DNA for each new cell.

The structure of DNA

The DNA molecule is a double helix with rungs like a ladder. It is coiled twice around sets of eight proteins called histones, creating a structure called a nucleosome.

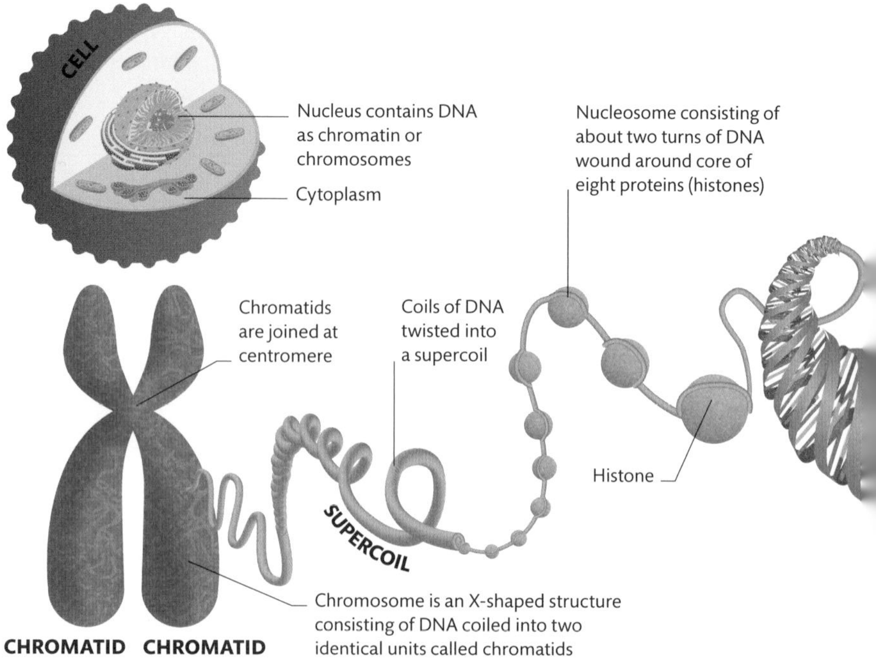

CELL
Nucleus contains DNA as chromatin or chromosomes
Cytoplasm
Nucleosome consisting of about two turns of DNA wound around core of eight proteins (histones)
Chromatids are joined at centromere
Coils of DNA twisted into a supercoil
SUPERCOIL
Histone
Chromosome is an X-shaped structure consisting of DNA coiled into two identical units called chromatids
CHROMATID
CHROMATID

Inside the nucleolus

The nucleolus is a dense region within the nucleus where the nucleoplasm becomes more gel-like. The nucleolus produces ribosomal RNA (rRNA) and proteins, which combine to make the interlocking subunits that form ribosomes. Once assembled within the nucleolus, the ribosomes pass into the cytoplasm, where they build protein molecules (see p.91).

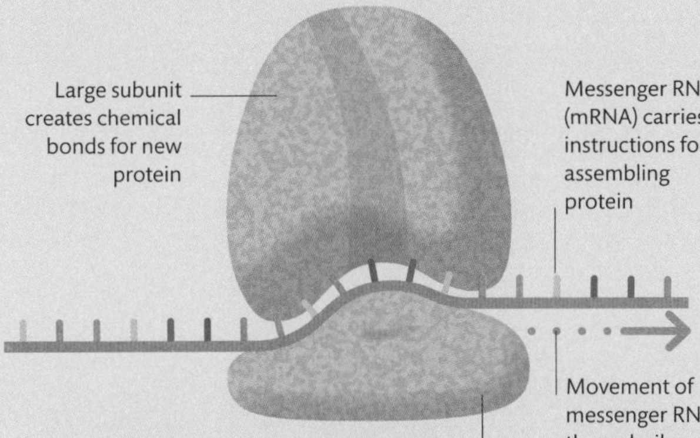

Action of ribosomes
Ribosomes lock around messenger RNA (mRNA), which carries genetic instructions from DNA in the cell's nucleus, and read the bases to create a new protein.

DO ALL CELLS HAVE A NUCLEUS?

No, only the cells of eukaryotes have a defined nucleus. In humans, mature red blood cells and cornified cells in the skin, hair, and nails also lack a nucleus.

ALL THE DNA IN A PERSON'S CELLS COULD STRETCH TO THE SUN AND BACK 16 TIMES

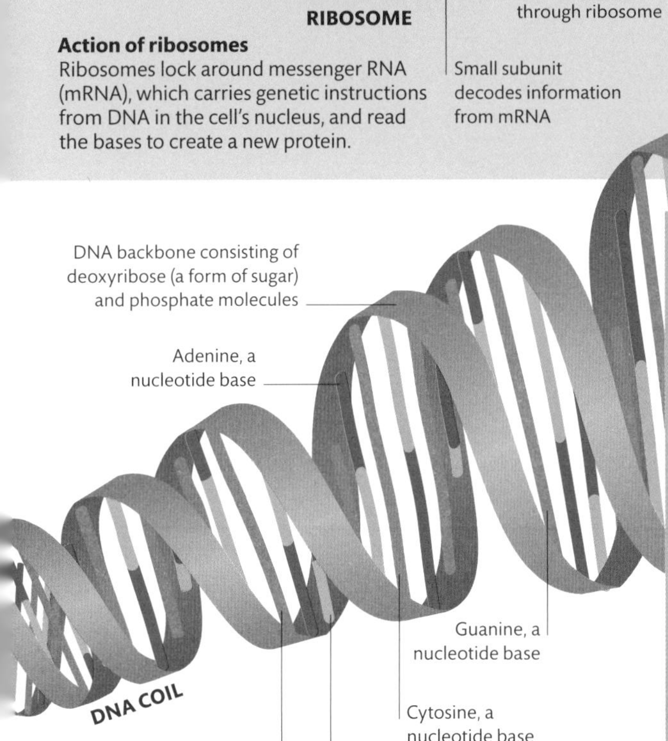

COILS AND SUPERCOILS

DNA molecules coil and coil again to form supercoils. A human cell nucleus has around 6.5 ft (2 m) of DNA; when supercoiled, the total length of 46 chromosomes is just 200 nanometers (billionths of a meter).

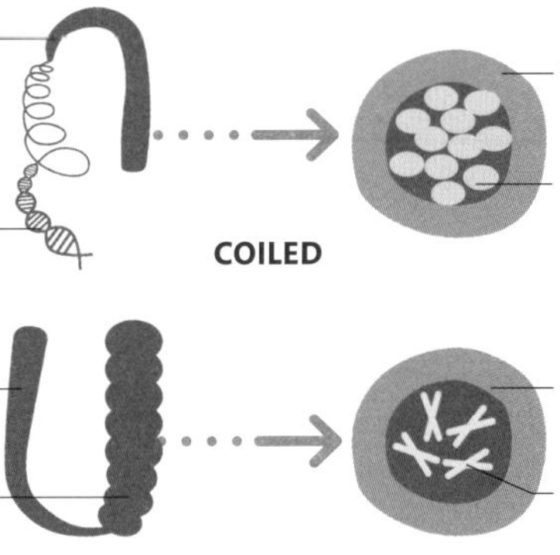

Energy factories

The energy used by animals and plants comes ultimately from sunlight. It enters a food chain by means of photosynthesis, which takes place in the chloroplasts of plants and algae, and results in the formation of sugars for fuel. This fuel is used by "powerhouse" organelles called mitochondria.

WHERE DID MITOCHONDRIA AND CHLOROPLASTS COME FROM?

These organelles evolved from ancient free-living bacteria, and now live symbiotically inside larger cells (see pp.54-55).

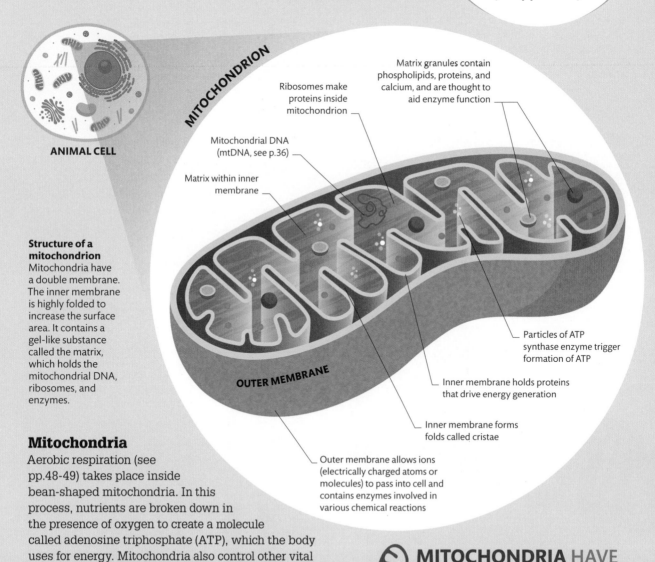

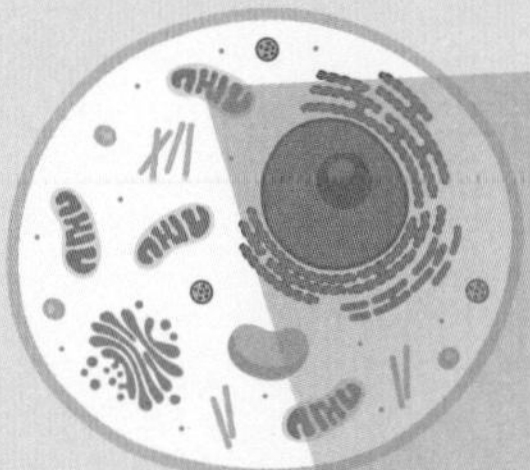
ANIMAL CELL

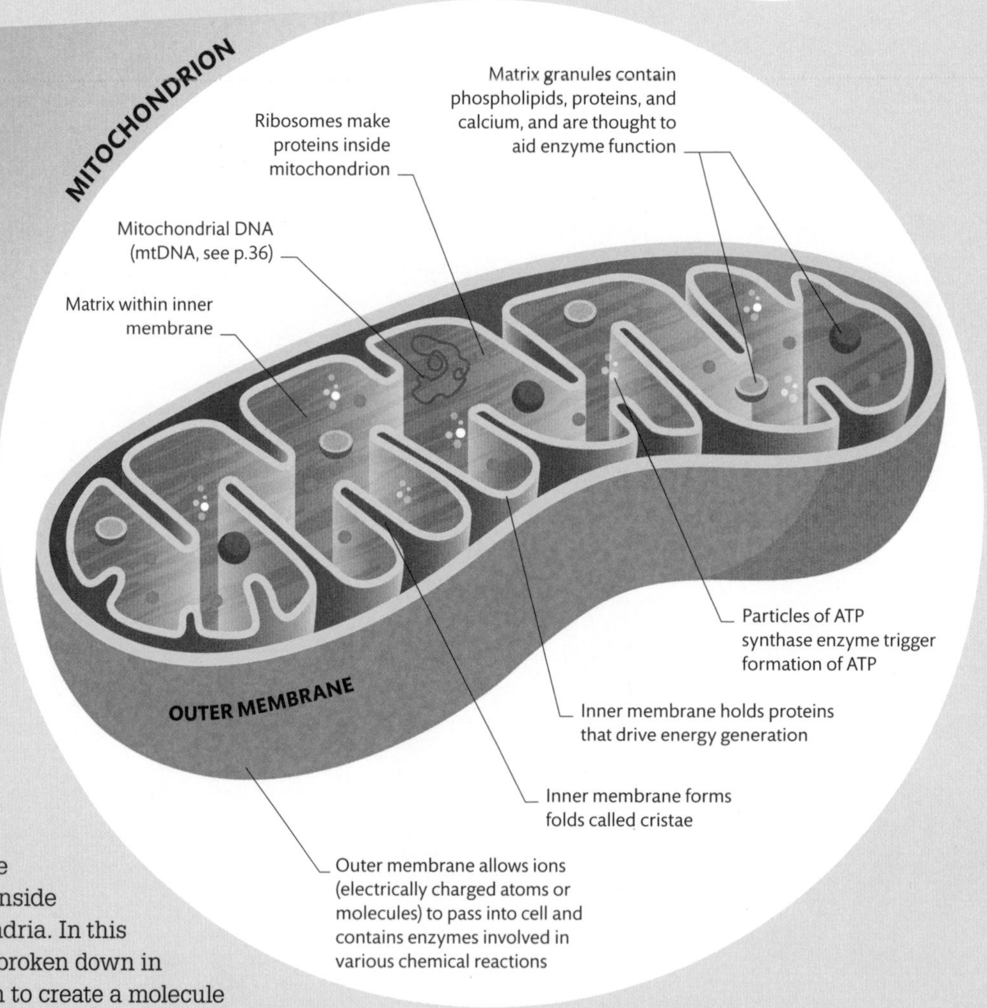
MITOCHONDRION
Ribosomes make proteins inside mitochondrion
Matrix granules contain phospholipids, proteins, and calcium, and are thought to aid enzyme function
Mitochondrial DNA (mtDNA, see p.36)
Matrix within inner membrane
Particles of ATP synthase enzyme trigger formation of ATP
OUTER MEMBRANE
Inner membrane holds proteins that drive energy generation
Inner membrane forms folds called cristae
Outer membrane allows ions (electrically charged atoms or molecules) to pass into cell and contains enzymes involved in various chemical reactions

Structure of a mitochondrion
Mitochondria have a double membrane. The inner membrane is highly folded to increase the surface area. It contains a gel-like substance called the matrix, which holds the mitochondrial DNA, ribosomes, and enzymes.

Mitochondria

Aerobic respiration (see pp.48-49) takes place inside bean-shaped mitochondria. In this process, nutrients are broken down in the presence of oxygen to create a molecule called adenosine triphosphate (ATP), which the body uses for energy. Mitochondria also control other vital processes such as cell growth, differentiation, and death. They are found in most of the cells of plants and animals, but not in organisms such as bacteria.

MITOCHONDRIA HAVE THEIR **OWN FORM** OF **DNA,** CALLED **mtDNA**

CHROMOPLASTS

These organelles contain carotenoids (orange pigments), xanthophylls (yellow pigments), and red pigments, giving flowers, fruits, and fall leaves their color. In ripening fruit, the thylakoids in chloroplasts are broken down, and carotenoids build up in crystals and in lipid structures called plastoglobules.

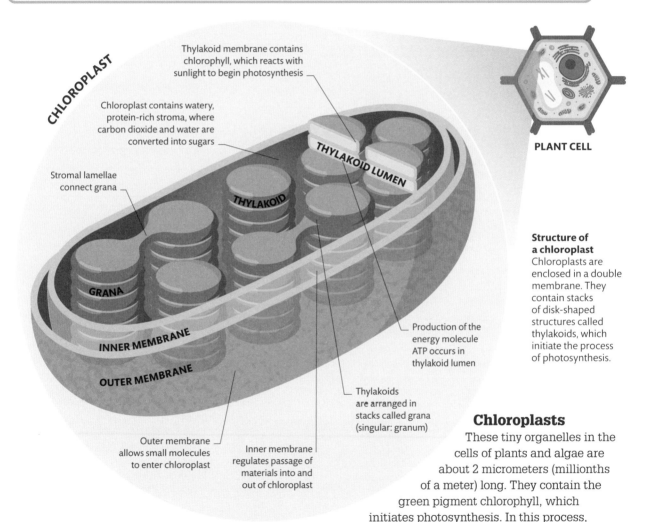

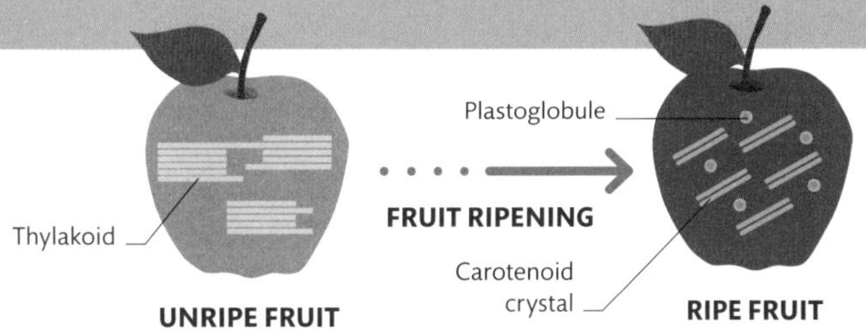
Thylakoid
UNRIPE FRUIT
FRUIT RIPENING
Plastoglobule
Carotenoid crystal
RIPE FRUIT

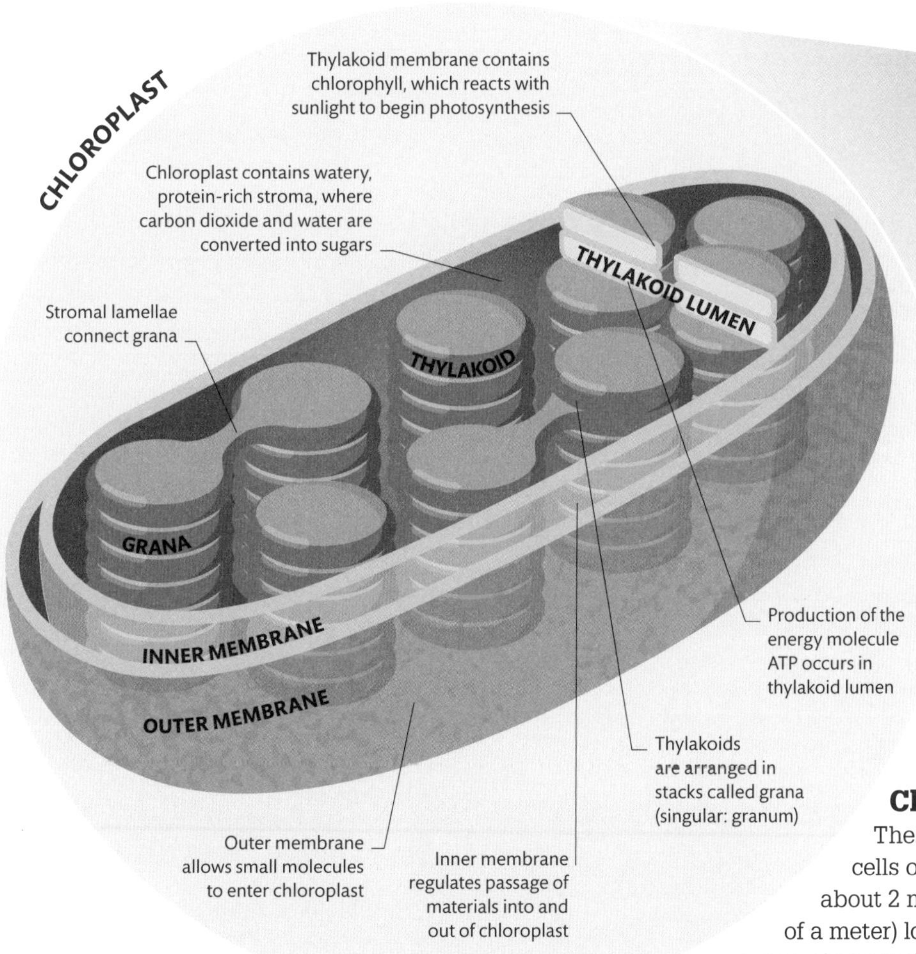
CHLOROPLAST
Thylakoid membrane contains chlorophyll, which reacts with sunlight to begin photosynthesis
Chloroplast contains watery, protein-rich stroma, where carbon dioxide and water are converted into sugars
Stromal lamellae connect grana
THYLAKOID LUMEN
THYLAKOID
GRANA
INNER MEMBRANE
OUTER MEMBRANE
Production of the energy molecule ATP occurs in thylakoid lumen
Thylakoids are arranged in stacks called grana (singular: granum)
Outer membrane allows small molecules to enter chloroplast
Inner membrane regulates passage of materials into and out of chloroplast

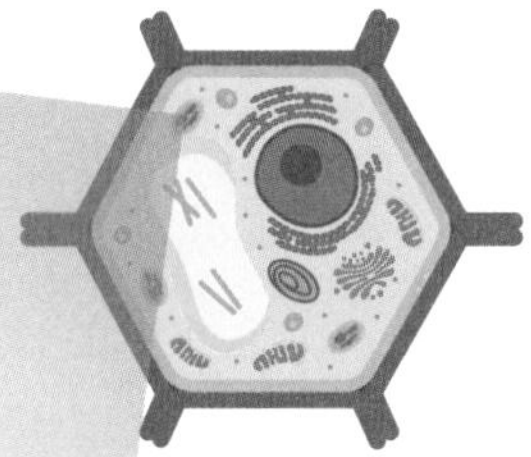
PLANT CELL

Structure of a chloroplast
Chloroplasts are enclosed in a double membrane. They contain stacks of disk-shaped structures called thylakoids, which initiate the process of photosynthesis.

Chloroplasts

These tiny organelles in the cells of plants and algae are about 2 micrometers (millionths of a meter) long. They contain the green pigment chlorophyll, which initiates photosynthesis. In this process, chloroplasts capture energy from sunlight and use it to convert carbon dioxide and water into glucose. During this process, a molecule called adenosine triphosphate (ATP) is produced, providing energy to drive processes inside the cell. Oxygen is also released in this process.

EACH **PHOTOSYNTHETIC PLANT CELL** CONTAINS **50–60 CHLOROPLASTS**

Cytoskeleton and vacuoles

A cell's cytoplasm contains an immensely complex network of proteins called the cytoskeleton, which maintains the cell's internal structure. Within the cytoplasm of many types of cell is a large, baglike organelle called a vacuole.

WHAT IS THE DIFFERENCE BETWEEN VACUOLES AND VESICLES?

While both are saclike structures inside a cell, vesicles are temporary transporters. Vacuoles are larger and tend to remain as distinct organelles for longer.

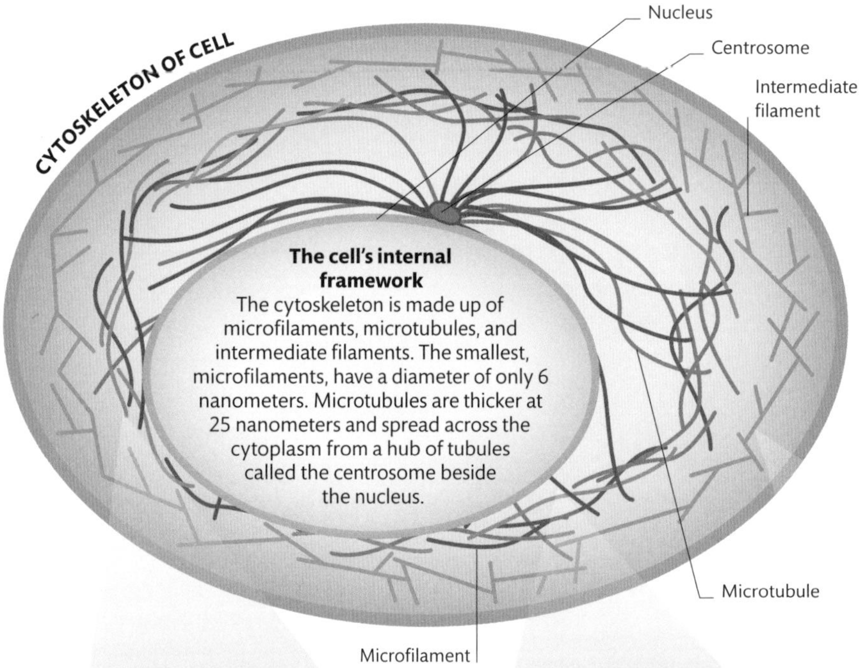

CYTOSKELETON OF CELL
Nucleus
Centrosome
Intermediate filament
The cell's internal framework
The cytoskeleton is made up of microfilaments, microtubules, and intermediate filaments. The smallest, microfilaments, have a diameter of only 6 nanometers. Microtubules are thicker at 25 nanometers and spread across the cytoplasm from a hub of tubules called the centrosome beside the nucleus.
Microtubule
Microfilament

The cytoskeleton

Made up of ultra-thin filaments and tubules, the cytoskeleton is constantly adapting to the needs of the cell. Its main role is to maintain structural integrity. In animal cells, which lack a rigid cell wall, the combined cytoskeletons of millions of cells play an important role in giving shape to entire tissues and organs. Secondary roles of the cytoskeleton involve shifting material within the cell and deforming the cell membrane to aid locomotion or feeding. It also plays a significant role in cell division (see pp.68–69).

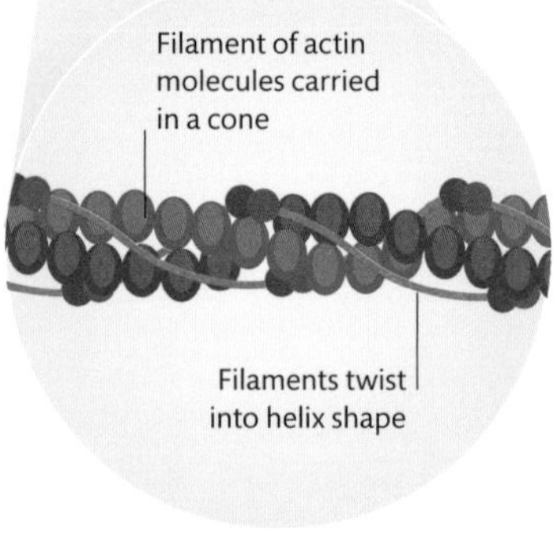

Filament of actin molecules carried in a cone
Filaments twist into helix shape

Microfilament
Actin molecules chained together in spirals form microfilaments. Actin is an active protein and determines cell shape and allows the cell surface to move, enabling cell motion.

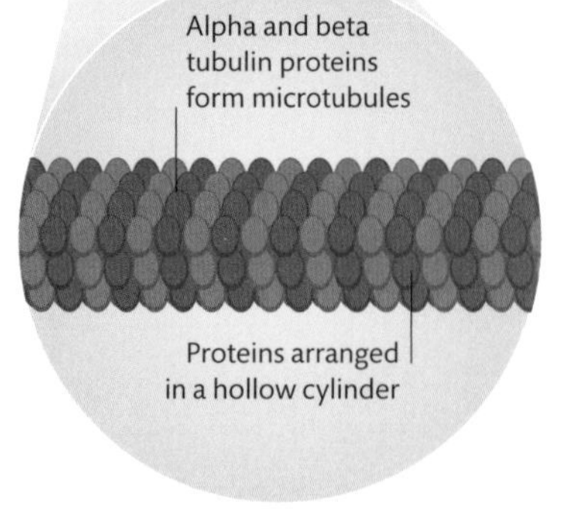

Alpha and beta tubulin proteins form microtubules
Proteins arranged in a hollow cylinder

Microtubules
Tube-shaped threads called microtubules are made from coils of proteins known as tubulins. The tubes hold the organelles and cell membrane in place.

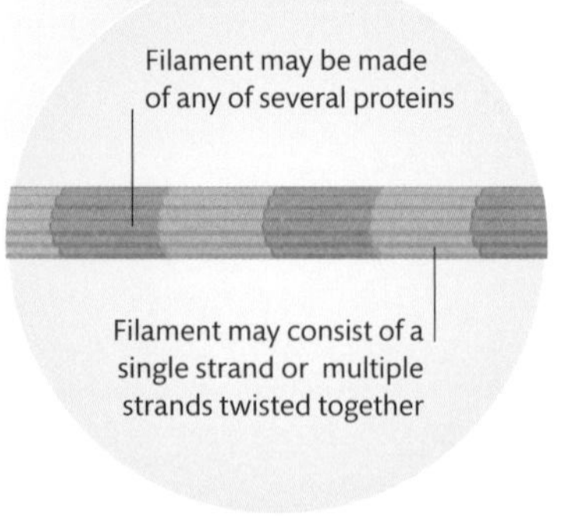

Filament may be made of any of several proteins
Filament may consist of a single strand or multiple strands twisted together

Intermediate filaments
The intermediate filaments are made from a range of proteins. They act as a sturdy structural support for microtubules and are less able to grow or spread through the cell.

Vacuoles

A vacuole is a large, relatively simple organelle enclosed by a membrane. Vacuoles often have a storage role, holding water, salts, food, or waste materials, and they are also structurally useful in helping maintain cell pressure. Vacuoles are seen in animal cells, although they are generally small and hard to differentiate from the vesicles used to transport materials. In plants, fungi, and single-celled life forms, however, vacuoles are an intrinsic part of the cell, removing excess water, accumulating gas, holding ions or nutrients, or acting as floats.

TYPES OF VACUOLES

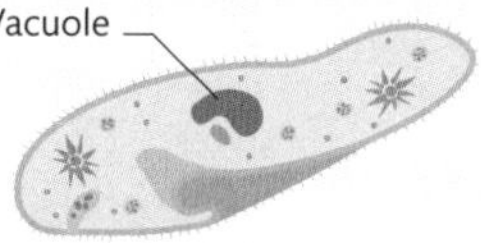
Vacuole

CONTRACTILE

Contractile vacuoles are found in freshwater protists (see pp.126–27). Water floods into the cell via osmosis and threatens to burst the cell. The excess water is sequestered into the vacuole, which then contracts, pumping water back out of the cell.

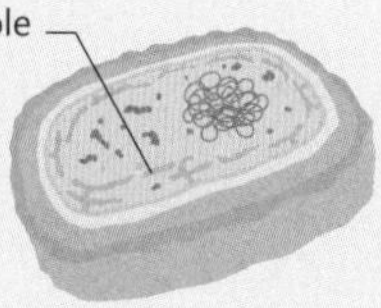
Vacuole

GAS

Cyanobacteria, which are types of photosynthetic bacteria, and various archaea have gas vacuoles, which hold tiny bubbles of air. The gas vacuoles act as floats, so the organism can control its buoyancy and move up and down in the water column.

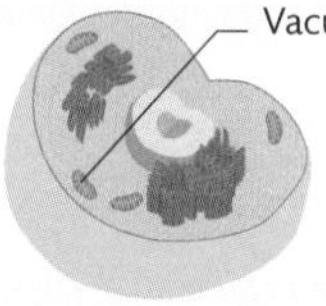
Vacuole

FOOD

Food-storage vacuoles perform the role of a stomach for many protists. Nutritious substances are collected inside the vacuole and mixed with digestive enzymes, which break the substances into simpler materials.

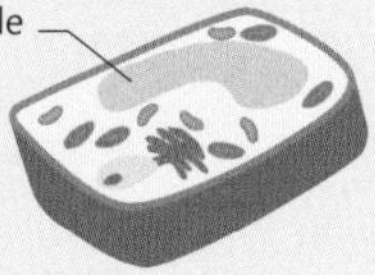
Vacuole

CENTRAL

Plant cells have big central vacuoles. They may have strands of cytoplasm tunneling through and can take up 80 percent of the cell volume. The vacuole stores ions to regulate the acidity of the cytoplasm and water to maintain the cell's internal pressure.

MICROTUBULES MAY BE AS **LONG** AS **0.0002 IN** (0.05 MM)

THE CENTROSOME

Many microtubules of the cytoskeleton, including the spindle used in cell division (see pp.68–69), grow from a central region near the nucleus called the centrosome. The centrosome is normally made from two cylindrical bundles of tubulin proteins called centrioles. These centrioles are constructed of tubulin proteins that connect to form a pipelike structure. The centrioles are an important microtubule organizing center (MTOC), a site from where cytoskeletal structures develop. There are also many other smaller MTOCs dotted around the cell.

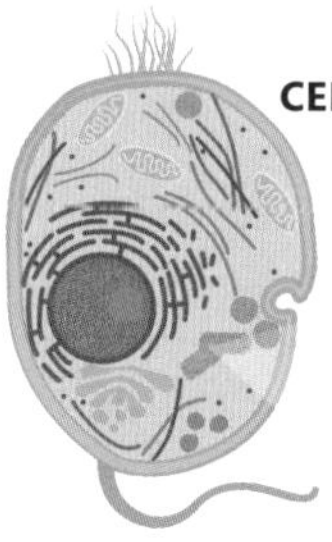
CELL

Mother and daughter
Centrioles are in pairs. The mother centriole originates from the mother cell; the daughter centriole is made after cell division is complete.

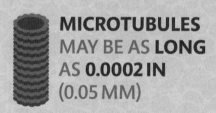

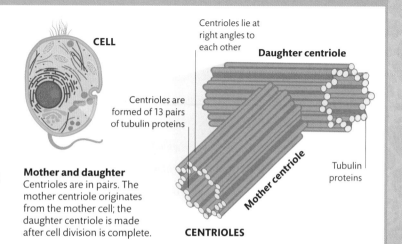

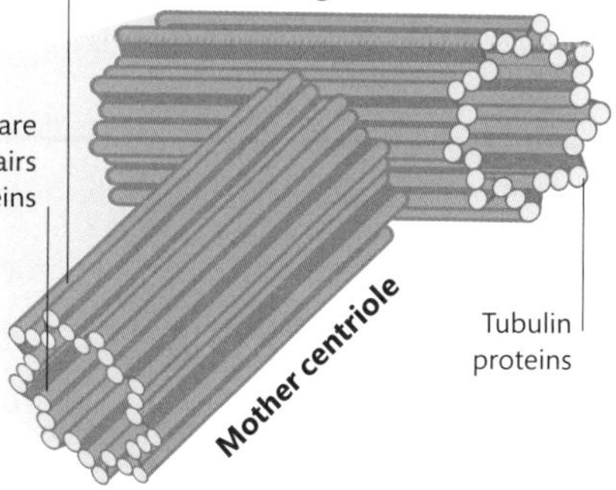
Centrioles lie at right angles to each other
Daughter centriole
Centrioles are formed of 13 pairs of tubulin proteins
Mother centriole
Tubulin proteins

CENTRIOLES

Cell transport

Cells rely on a supply of materials entering through their membranes and being moved between organelles. To do this, cells use a range of passive and active transport systems.

Diffusion

When a substance is free to move, it will tend to drift from areas where there is a high concentration to a place where there is not; this process is called diffusion. As an entirely passive process, it occurs without an input of energy, although it will happen faster when the material and its medium are warmer. Diffusion always occurs in the same direction along a concentration gradient with molecules moving from high to low concentrations. It can occur in a mixture of gases, such as air, or in more complex solutions like cytoplasm, which contain many dissolved substances.

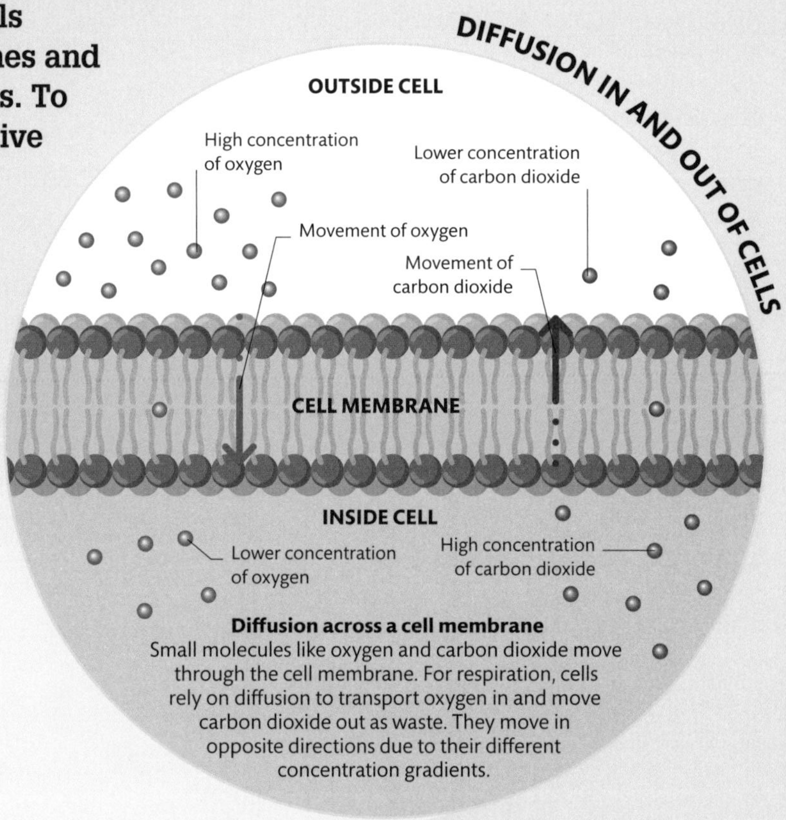
DIFFUSION IN AND OUT OF CELLS
OUTSIDE CELL
High concentration of oxygen
Lower concentration of carbon dioxide
Movement of oxygen
Movement of carbon dioxide
CELL MEMBRANE
INSIDE CELL
Lower concentration of oxygen
High concentration of carbon dioxide

Diffusion across a cell membrane
Small molecules like oxygen and carbon dioxide move through the cell membrane. For respiration, cells rely on diffusion to transport oxygen in and move carbon dioxide out as waste. They move in opposite directions due to their different concentration gradients.

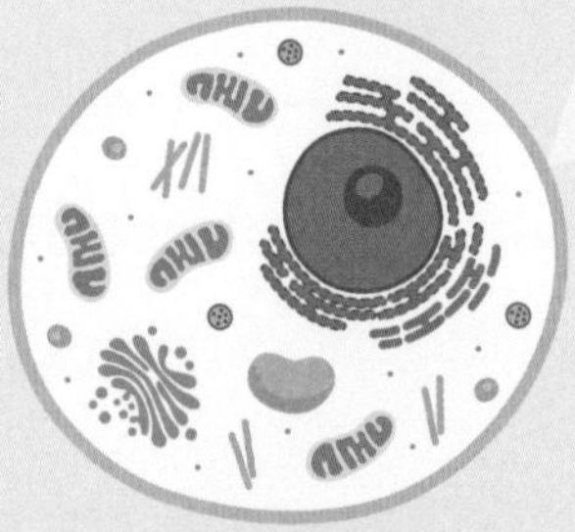

ANIMAL CELL

Diffusion within cells
Larger molecules, such as fats and amino acids, cannot diffuse through the cell membrane and instead need to be actively transported in and out of the cell. However, within the cell's cytoplasm, these molecules will diffuse from high-concentration areas to areas of low concentration until they become evenly dispersed inside the cell.

SKIN WRINKLES IN WATER PARTLY DUE TO AN INFLUX OF WATER BY OSMOSIS

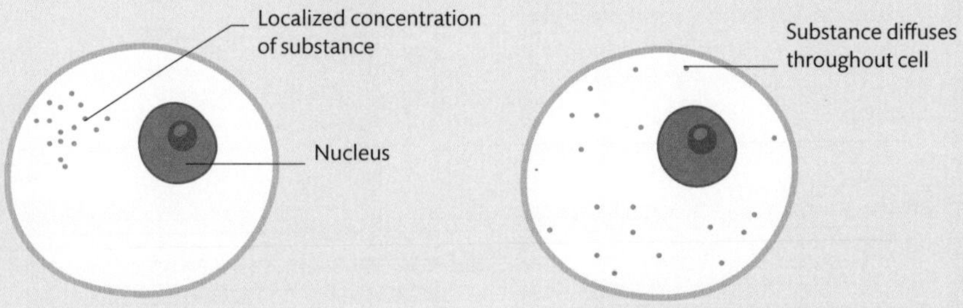
Localized concentration of substance
Nucleus
Substance diffuses throughout cell

1 Before diffusion
A substance is produced in one part of the cell, creating a high-concentration area. The molecules begin to move randomly in all directions and gradually spread out.

2 After diffusion
Once the molecules are evenly dispersed, they continue to make random movements that ensure the concentration in the cytoplasm remains even.

Osmosis

A form of diffusion called osmosis occurs when water moves from an area of high concentration to an area of low concentration. Substances dissolved in water, such as salt, are also blocked from moving by the cell membrane. This has the result of equalizing the concentration of substances either side of the membrane. When there is not enough water content, a body is dehydrated and fluids outside its cells become more highly concentrated; osmosis therefore makes water flow out of cells.

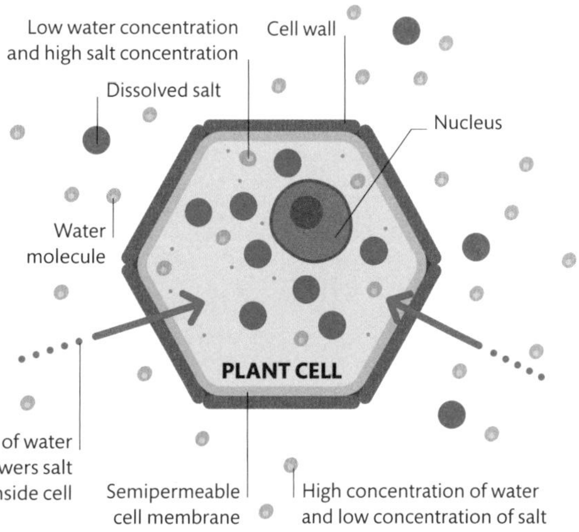
Low water concentration and high salt concentration
Cell wall
Dissolved salt
Nucleus
Water molecule
PLANT CELL
Movement of water into cell lowers salt concentration inside cell
Semipermeable cell membrane
High concentration of water and low concentration of salt

Osmotic pressure
Plant cells rely on osmosis to push water into their cells, creating a high internal pressure. If this pressure drops, the cells shrink and become flaccid, causing the plant to wilt and shrivel (see p.149).

Active transport

A cell must expend energy to transport some substances in and out. Known as active transport, this moves substances from low to high concentrations, against the gradient. One form of active transport in animal cells brings glucose into the intestine. Energy is used to move the glucose inward, to an area of high concentration, to maximize absorption.

Carrier proteins
Machinelike carrier proteins in membranes work like pumps or pores in active transport. Energy inputs flex the protein shape so it can pump molecules across the membrane.

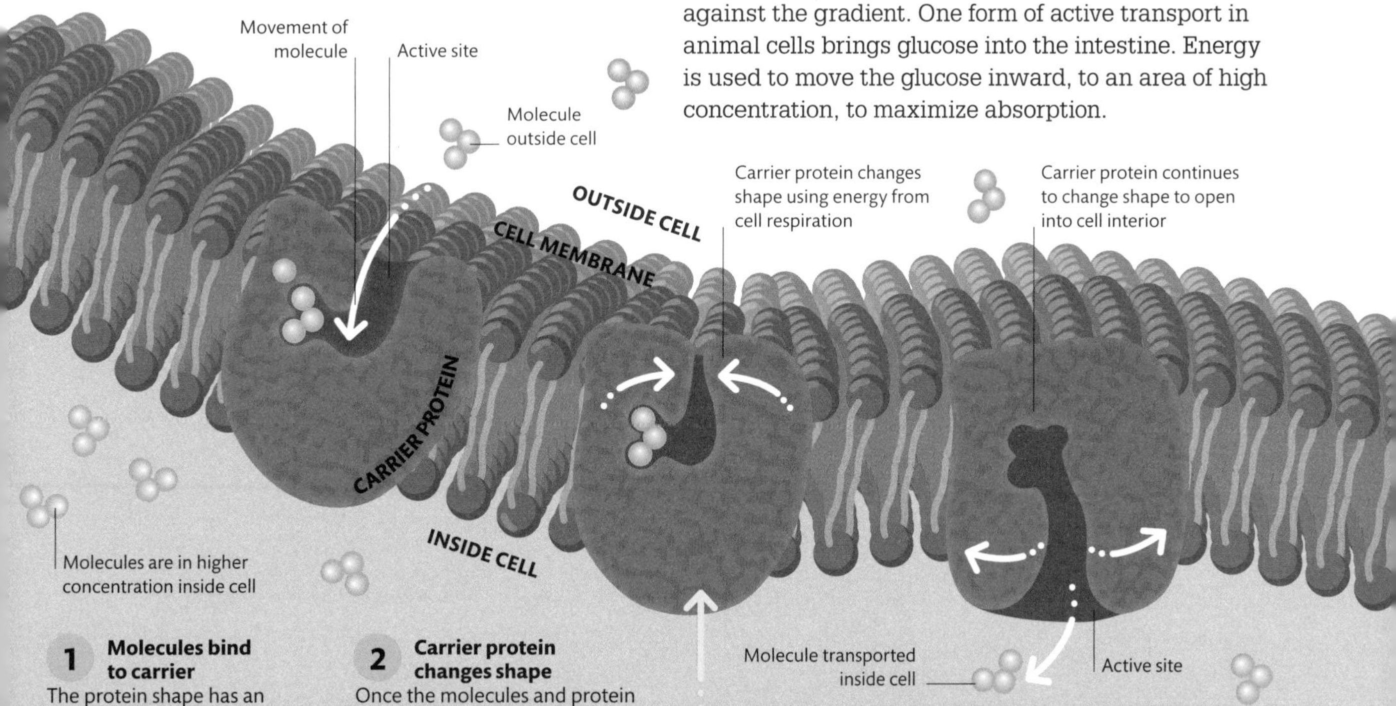
Movement of molecule
Active site
Molecule outside cell
OUTSIDE CELL
CELL MEMBRANE
CARRIER PROTEIN
Carrier protein changes shape using energy from cell respiration
Carrier protein continues to change shape to open into cell interior
INSIDE CELL
Molecules are in higher concentration inside cell
Molecule transported inside cell
Active site
Energy from cell respiration

1 Molecules bind to carrier
The protein shape has an active site on the outside of the membrane where the molecules can bond to it. That initial bond is passive, requiring no energy.

2 Carrier protein changes shape
Once the molecules and protein form a temporary link, energy supplied by the cell is used to alter the shape of the protein. This process is driven by the presence of the molecules.

3 Molecule enters cell
Changed by the addition of energy and molecules, the active site shifts to the other side of the membrane. This shape change allows the molecules to be released into the cytoplasm.

Cell motion

Many cells are embedded in body tissues or are carried in fluids such as blood. Some kinds of cell, however, have mobile parts or can even move themselves independently.

Flagella

A flagellum (plural: flagella) is a long, whiplike tail that extends from one end of a cell. Flagella are found on many bacteria, protists, and other microorganisms; they also exist on the sex cells of animals and of some plants. In these organisms, the flagellum is made from a bundle of microtubules that slide against each other to create wavelike propulsive motion. Bacterial flagella are stiffer and screw-shaped and create motion by rotating like a propeller. Some organisms, such as the protist *Euglena*, use both types of motion.

HOW LONG ARE FLAGELLA?

Flagella on most single-celled organisms are typically up to about 20 microns (about one-thousandth of an inch) long.

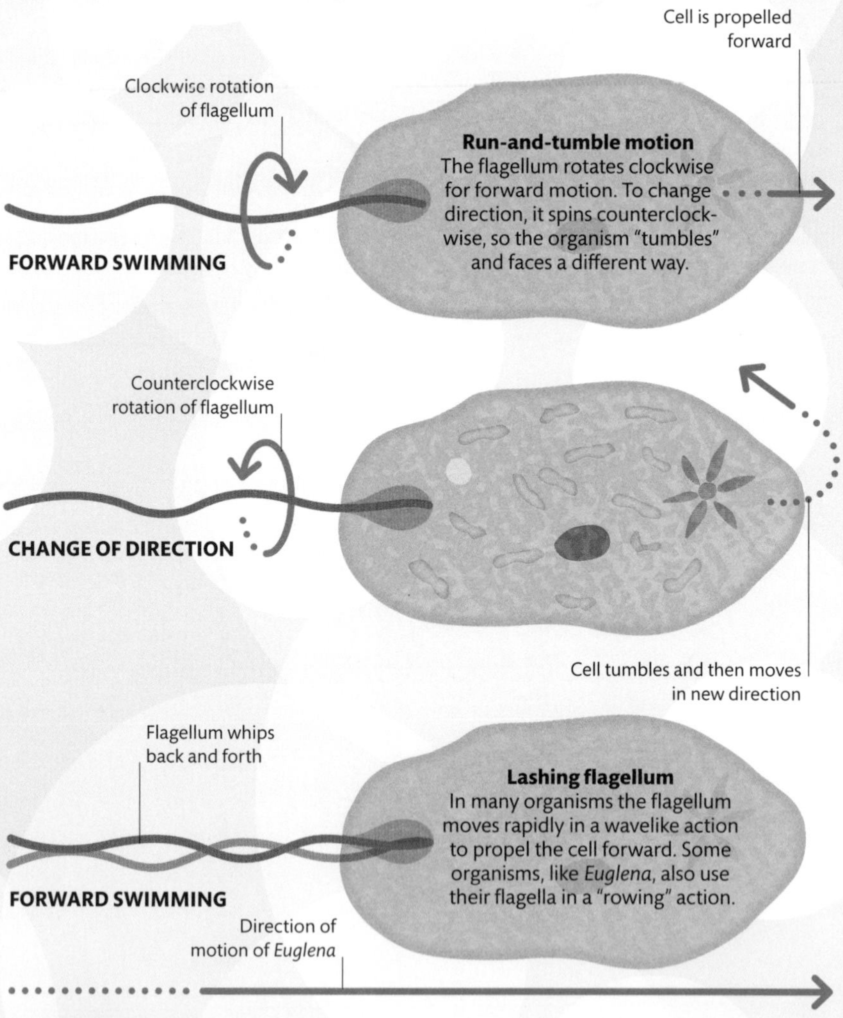

Cell is propelled forward
Clockwise rotation of flagellum
Run-and-tumble motion
The flagellum rotates clockwise for forward motion. To change direction, it spins counterclockwise, so the organism "tumbles" and faces a different way.
FORWARD SWIMMING
Counterclockwise rotation of flagellum
CHANGE OF DIRECTION
Cell tumbles and then moves in new direction
Flagellum whips back and forth
Lashing flagellum
In many organisms the flagellum moves rapidly in a wavelike action to propel the cell forward. Some organisms, like Euglena, also use their flagella in a "rowing" action.
FORWARD SWIMMING
Direction of motion of Euglena

CELL MOTION IN THE HUMAN BODY

Movement with flagella and cilia and by amoeboid motion can all be seen in the human body. Sperm cells (male sex cells) are the only body cells to have flagella. Cells with cilia are widespread in the linings of the lungs and airways. White blood cells use amoeboid motion to move through blood and infected tissues.

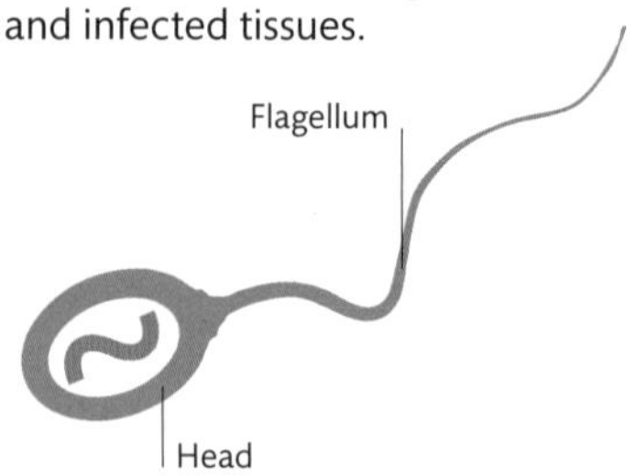

Flagellum
Head

HUMAN SPERM CELL

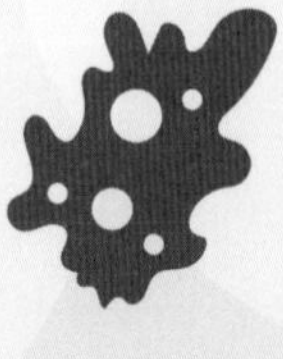

AMOEBAS CAN MOVE AT SPEEDS OF ABOUT 0.1–0.2 IN (2–5 MM) PER MINUTE

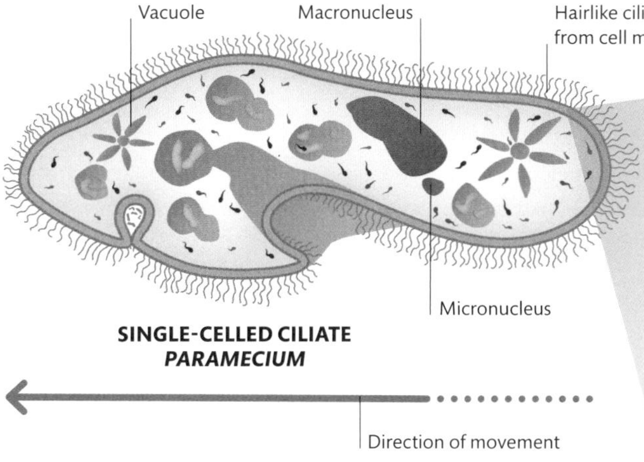
Vacuole
Macronucleus
Hairlike cilia projecting from cell membrane
Micronucleus
Direction of movement

SINGLE-CELLED CILIATE *PARAMECIUM*

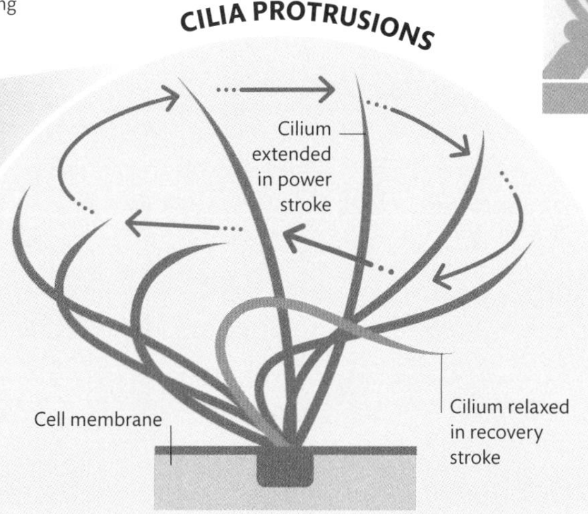
CILIA PROTRUSIONS
Cilium extended in power stroke
Cell membrane
Cilium relaxed in recovery stroke

Ciliary movement
Groups of cilia move back and forth in a coordinated rhythm so that each cilium makes a power stroke in the right direction at the right time. Together, they create a wavelike force that pushes on surrounding objects or fluids.

Cilia

A cilium (plural: cilia) is a short, hairlike protrusion from a cell membrane. Cilia work together in large numbers, moving back and forth. In some organisms, known as ciliates, the cilia move the entire cell with concerted wafting. In others, the cilia direct a current of water around the cell, which brings with it food or oxygen. Nonmobile cells with cilia are also found in some tissues in animals and plants; these act to push material along internal tubes and vessels.

Pseudopodia

Some cells can move by extending limblike projections of membrane called pseudopodia (meaning "false feet"). The cell moves in the direction of the pseudopodia, which later shrink back into the body of the cell. Organisms such as amoebas move in this way, so this is sometimes called amoeboid motion. Pseudopodia can also be sent out in several directions at once to engulf nearby food particles.

Amoeboid motion
The changes in the cell's shape are controlled by a microscopic network of protein fibers in the cytoskeleton (the network of filaments that gives the cell its structure).

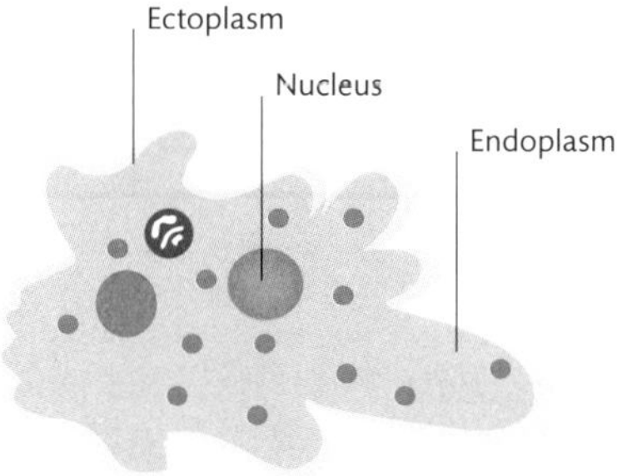
Ectoplasm
Nucleus
Endoplasm

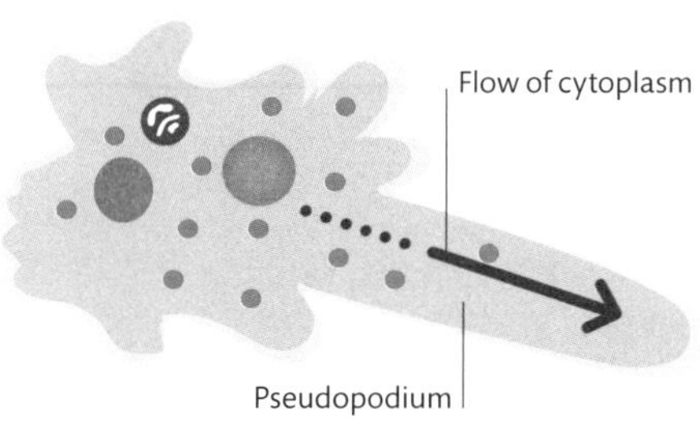
Flow of cytoplasm
Pseudopodium

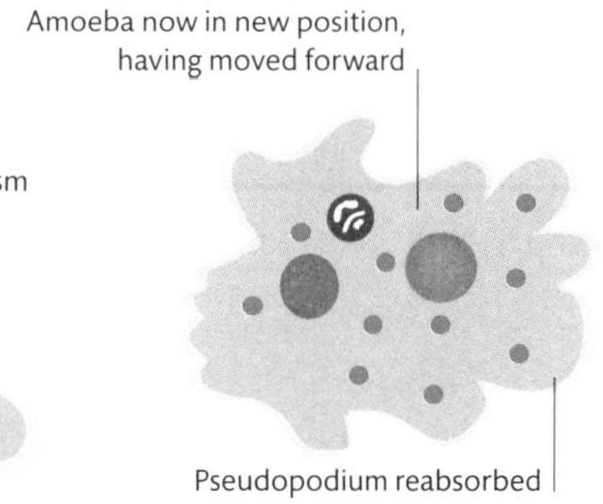
Amoeba now in new position, having moved forward
Pseudopodium reabsorbed

1 Structure of amoeba
The cytoplasm (internal material) of an amoeba's body comprises an outer layer (ectoplasm) formed from a semisolid gel, and an inner layer (endoplasm) of fluid.

2 Pseudopodium forms
Filaments in the cytoskeleton extend, causing a pseudopodium to form. Some of the gel turns to fluid and flows into the pseudopodium, carrying the cell contents.

3 Cell pulled forward
The flow of the cytoplasm pulls the whole cell forward, and as a result the pseudopodium is reabsorbed, with some of the fluid turning back into gel.

Cell division

In all life forms, cells reproduce by dividing into two new daughter cells. In bacteria and other simple organisms, this happens by binary fission. In eukaryotes—organisms the cells of which have a nucleus—it happens by mitosis.

HOW FAST CAN BACTERIA REPRODUCE?

In ideal conditions, some bacteria can undergo binary fission about every 20 minutes.

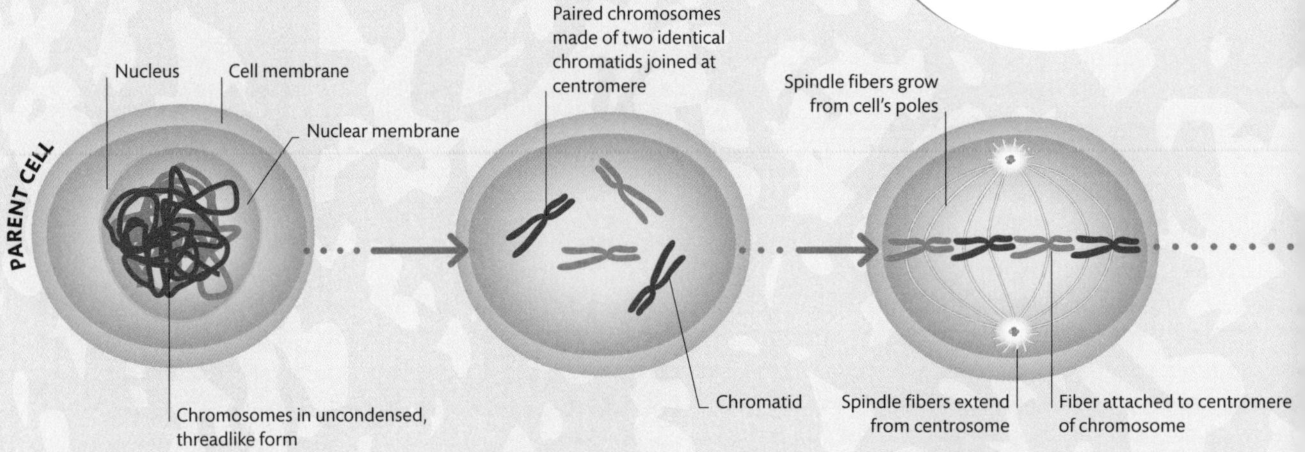

Nucleus
Cell membrane
Nuclear membrane
PARENT CELL
Chromosomes in uncondensed, threadlike form
Paired chromosomes made of two identical chromatids joined at centromere
Chromatid
Spindle fibers grow from cell's poles
Spindle fibers extend from centrosome
Fiber attached to centromere of chromosome

1 Interphase In the cell's nonreproducing phase, when division is not happening, the cell's genetic material is contained in long strands of chromatin, a complex formed from DNA wound around proteins called histones.

2 Prophase As interphase ends, each chromatin strand duplicates itself. The strands coil more tightly into X shapes, made from pairs of duplicate chromosomes, or chromatids. The nuclear membrane begins to break down.

3 Metaphase The chromosomes, now released from the nucleus, line up at the center of the cell. Spindle fibers grow from opposite ends (poles) of the cell and attach to proteins at the centromere of each chromosome.

Mitosis

The body cells of eukaryotic life forms divide by mitosis to create two identical daughter cells, each containing a copy of the parent cell's genetic material (DNA). Strictly speaking, mitosis is the splitting of pairs of chromosomes (the structures that contain DNA) in the cell nucleus to produce a full set of DNA in each daughter cell. In the process, the DNA in the nucleus is duplicated and the cell contents dispersed equally. The term mitosis is derived from the Greek word *mitos*, meaning "warp thread," due to the threadlike fibers that form to separate the cell's chromosomes.

THE CELL CYCLE

Mitosis is just a small part of the cell cycle that a cell will undergo in its lifetime. The longest part of the cell cycle is interphase, which begins once the daughter cells separate. During interphase, the cell will grow larger and make copies of the organelles (the separate structures inside the cell, such as ribosomes and mitochondria). Just before division, the DNA is duplicated into chromatids, and the final interphase step is to organize the cell's contents so they can be divided once mitosis begins.

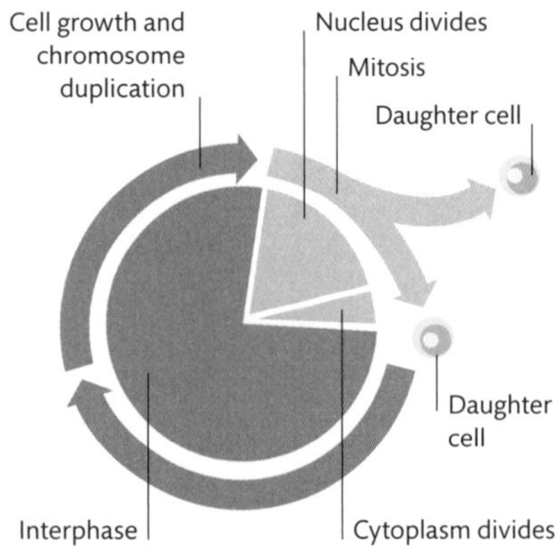

Cell growth and chromosome duplication
Nucleus divides
Mitosis
Daughter cell
Daughter cell
Interphase
Cytoplasm divides

STAGES OF THE CELL CYCLE

THE **HUMAN BODY** UNDERGOES ABOUT **10 QUADRILLION CELL DIVISIONS** IN A LIFETIME

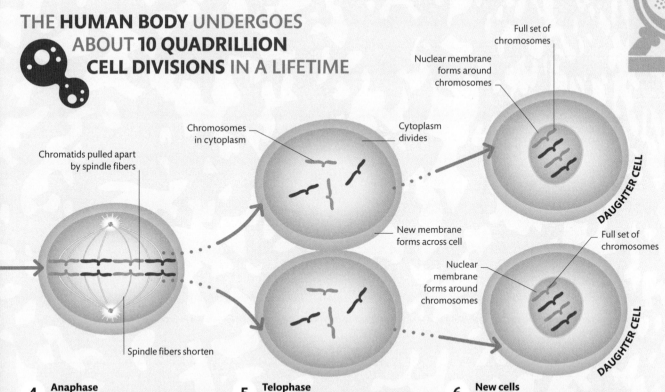

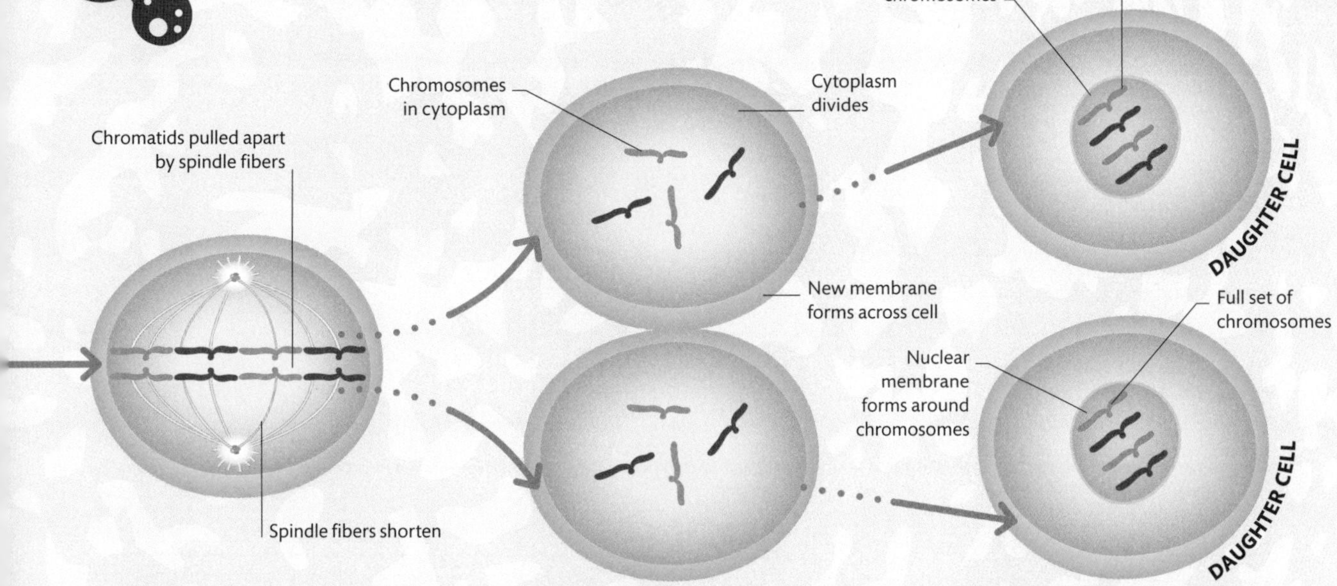
Full set of chromosomes
Nuclear membrane forms around chromosomes
Chromosomes in cytoplasm
Cytoplasm divides
Chromatids pulled apart by spindle fibers
DAUGHTER CELL
New membrane forms across cell
Full set of chromosomes
Nuclear membrane forms around chromosomes
DAUGHTER CELL
Spindle fibers shorten

4 Anaphase
The spindle fibers contract and pull the chromatids of each chromosome apart. As the fibers contract further, the new individual chromosomes are pulled toward the poles of the cell.

5 Telophase
New membranes form around each new set of chromosomes. The cell material (cytoplasm) itself divides, in a process called cytokinesis. A cell membrane forms across the center of the cell, dividing it into two parts.

6 New cells
Two new identical daughter cells form, each with a full set of chromosomes. The chromosomes will then revert to their threadlike form of chromatin. The new cells will then be in their own interphase.

Binary fission

The cells of prokaryotes—bacteria and archaea—are much smaller than those of eukaryotes and do not have nuclei or organelles, so they divide using a different, simpler process called binary fission. Like mitosis, the process results in one parent cell producing two daughters that each carry the same genetic material.

Binary fission in bacteria
In most bacteria, division involves the cell growing longer as the contents increase, before dividing into two daughter cells. Alternatively, some bacteria divide by budding a daughter cell off the main cell.

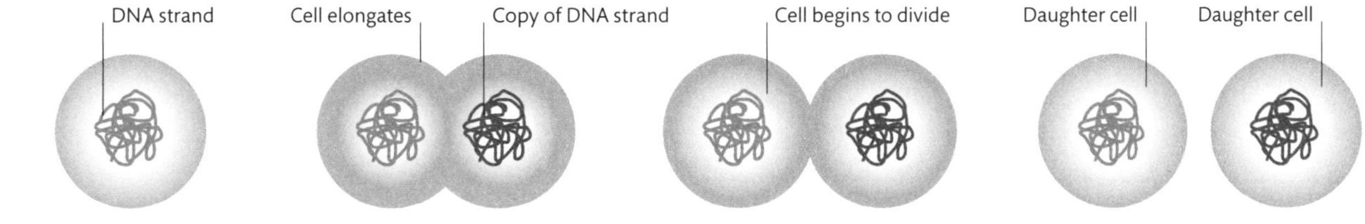
DNA strand
Cell elongates
Copy of DNA strand
Cell begins to divide
Daughter cell
Daughter cell

1 Parent bacterium
The genetic material in the cell consists of a single chromosome, forming one circular strand of DNA.

2 DNA duplicates
The DNA strand is duplicated. The cell enlarges, and the copies move toward the ends.

3 Cytoplasm divides
A new membrane and wall form at the center of the cell, dividing the cytoplasm and sets of DNA.

4 Daughter cells separate
The two daughter cells separate. Each is genetically identical to the parent, and they are soon ready to divide in turn.

Nerve cells

Also called neurons, nerve cells are the basic units of the nervous system. They carry information within the brain and from the brain to all parts of the body.

Structure of a nerve cell

A neuron has four parts, each of which performs a different function. The axon generates and carries a nerve signal, or impulse, along its length. The cell body processes the signal. The axon terminal transmits the signal to the next neuron. A dendrite receives the signal from a neighboring neuron. Most axons are insulated with a myelin sheath, which increases the speed of nerve signals, although some are unmyelinated and transmit signals more slowly.

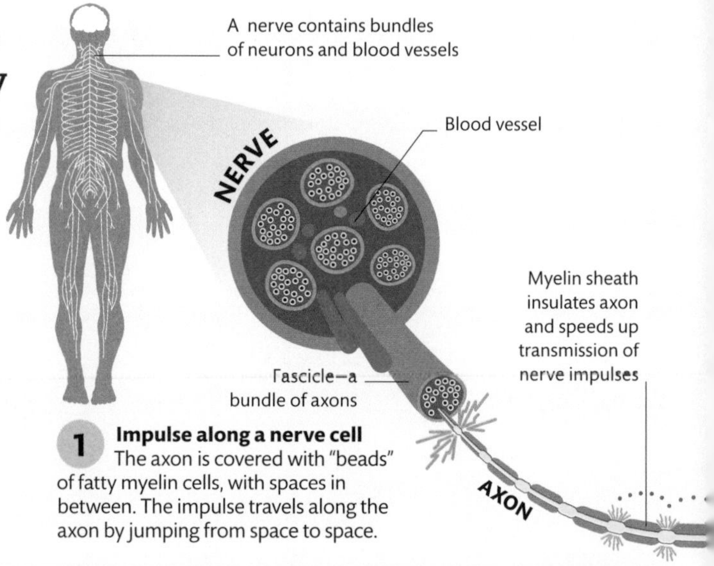

A nerve contains bundles of neurons and blood vessels
NERVE
Blood vessel
Myelin sheath insulates axon and speeds up transmission of nerve impulses
Fascicle—a bundle of axons
AXON

1 Impulse along a nerve cell The axon is covered with "beads" of fatty myelin cells, with spaces in between. The impulse travels along the axon by jumping from space to space.

Nerve signals

Neurons communicate with each other using electrochemical signals. When a signal is strong enough, it passes along the axon. On reaching the axon terminal, chemicals called neurotransmitters are released into the synapse, the small gap between neurons. The neurotransmitters bind to receptors on the next neuron, causing the nerve signal to be passed on.

500,000,000
THE NUMBER OF **NEURONS** IN THE WALL OF THE **HUMAN INTESTINE**

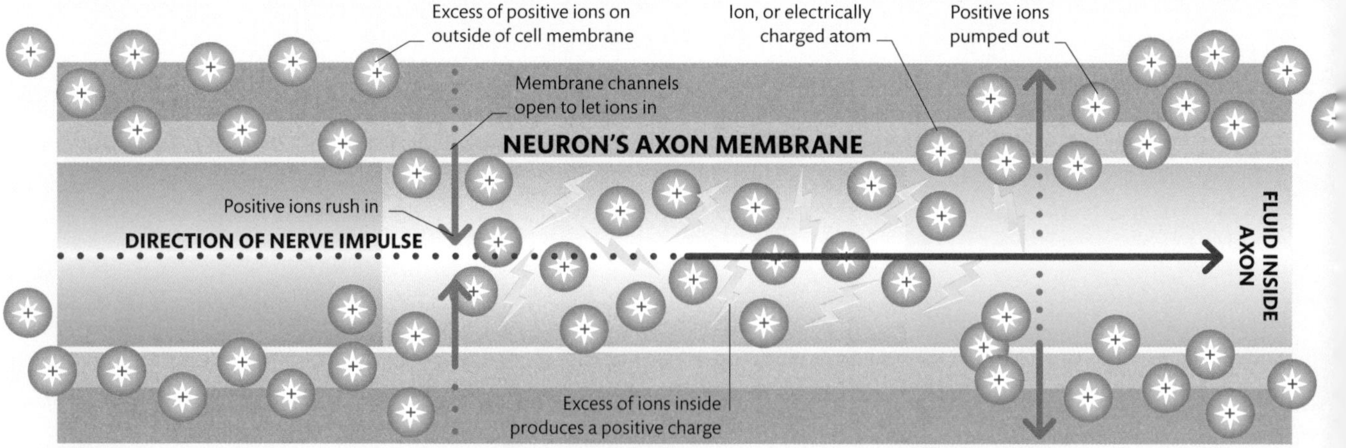

Excess of positive ions on outside of cell membrane
Ion, or electrically charged atom
Positive ions pumped out
Membrane channels open to let ions in
NEURON'S AXON MEMBRANE
Positive ions rush in
DIRECTION OF NERVE IMPULSE
FLUID INSIDE AXON
Excess of ions inside produces a positive charge

1 Resting potential When at rest, a neuron has more positive ions outside its membrane than inside. This difference in polarization, or electrical potential, across the membrane is called the resting potential.

2 Depolarization As a result of chemical changes from the cell body, positive ions enter the cell through the membrane. This influx of positive ions reverses the polarization of the axon, making the outside negative.

3 Repolarization Depolarization of part of the axon causes the adjacent section to undergo the same process. The cell pumps out positive ions, which repolarizes the membrane back to its resting potential.

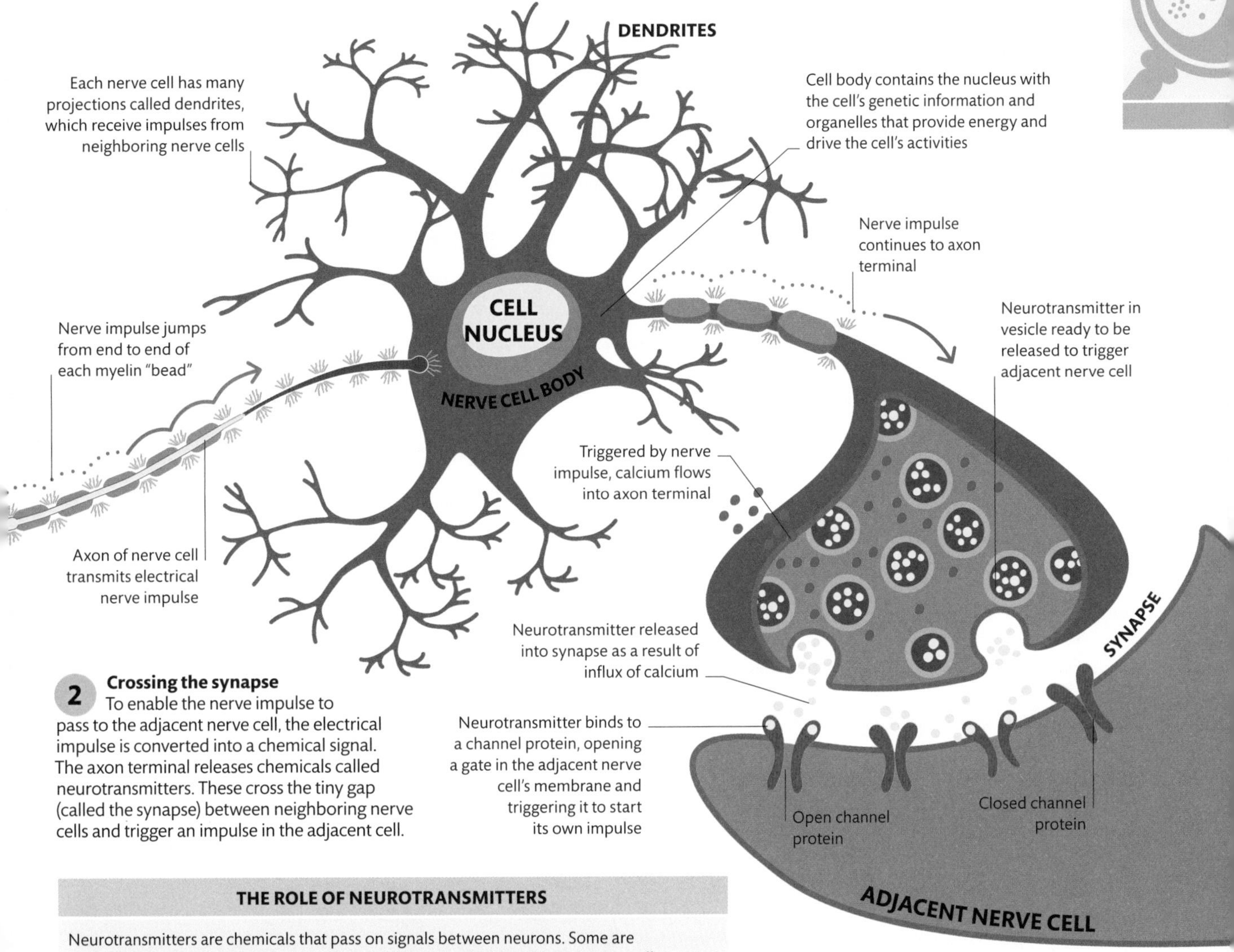
DENDRITES
Each nerve cell has many projections called dendrites, which receive impulses from neighboring nerve cells
Cell body contains the nucleus with the cell's genetic information and organelles that provide energy and drive the cell's activities
Nerve impulse continues to axon terminal
CELL NUCLEUS
NERVE CELL BODY
Nerve impulse jumps from end to end of each myelin "bead"
Neurotransmitter in vesicle ready to be released to trigger adjacent nerve cell
Triggered by nerve impulse, calcium flows into axon terminal
Axon of nerve cell transmits electrical nerve impulse
Neurotransmitter released into synapse as a result of influx of calcium
SYNAPSE
Neurotransmitter binds to a channel protein, opening a gate in the adjacent nerve cell's membrane and triggering it to start its own impulse
Open channel protein
Closed channel protein
ADJACENT NERVE CELL

2 **Crossing the synapse**
To enable the nerve impulse to pass to the adjacent nerve cell, the electrical impulse is converted into a chemical signal. The axon terminal releases chemicals called neurotransmitters. These cross the tiny gap (called the synapse) between neighboring nerve cells and trigger an impulse in the adjacent cell.

THE ROLE OF NEUROTRANSMITTERS

Neurotransmitters are chemicals that pass on signals between neurons. Some are excitatory—they help continue transmission of the nerve signal to the next nerve cell. Inhibitory neurotransmitters have the opposite effect. For example, serotonin has an inhibitory effect, helping reduce anxiety and to regulate sleep and hunger.

NEUROTRANSMITTER	USUAL EFFECT
Acetylcholine	Mostly excitatory
Gamma-aminobutyric acid (GABA)	Inhibitory
Glutamate	Excitatory
Dopamine	Excitatory and inhibitory
Norepinephrine	Mostly excitatory
Serotonin	Inhibitory
Histamine	Excitatory

HOW FAST DO NERVE SIGNALS TRAVEL?

Different types of signals travel at different speeds: pain signals travel at about 2 ft/sec (0.6 m/sec), while touch signals travel at up to 400 ft/sec (120 m/sec).

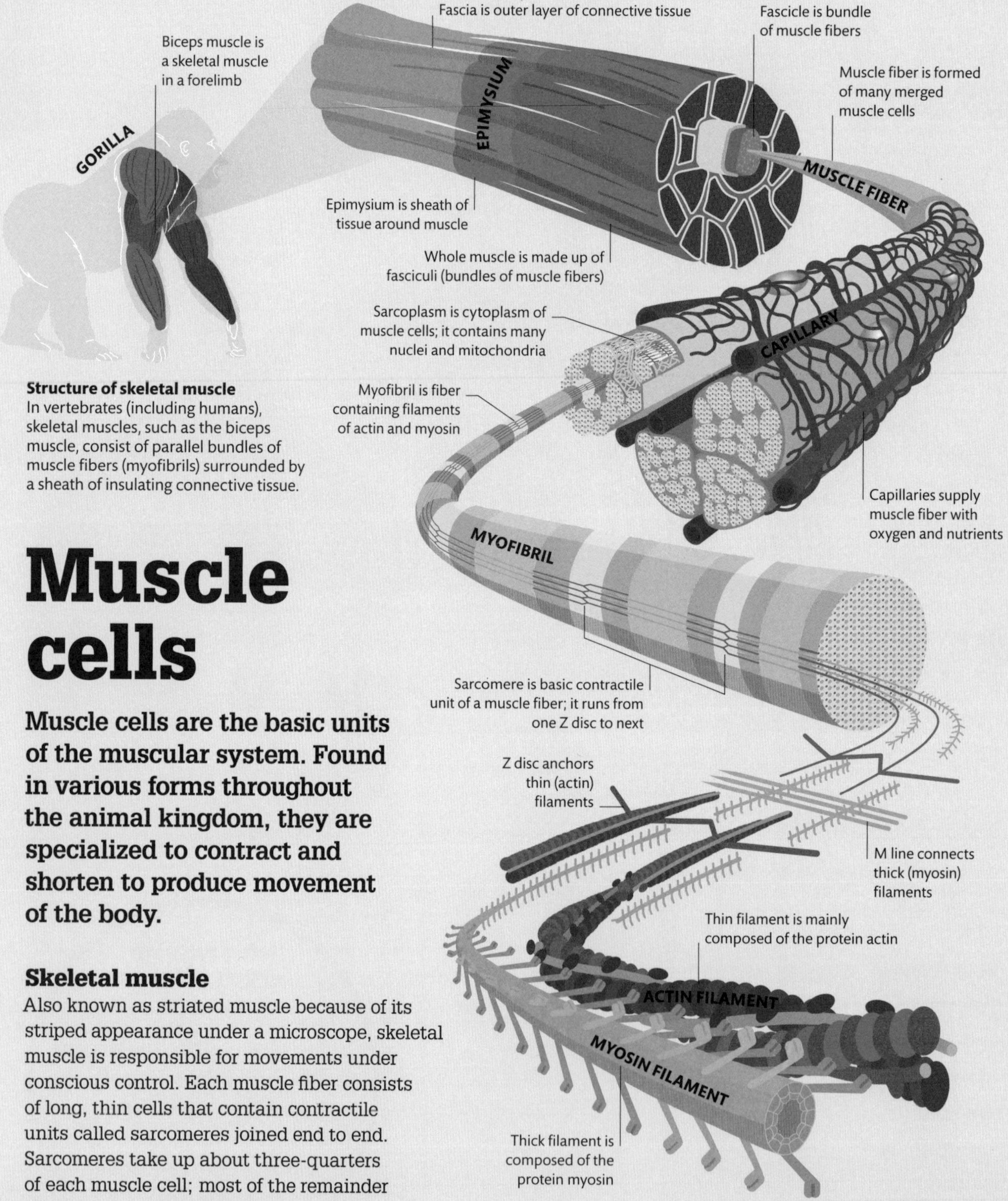
Fascia is outer layer of connective tissue
Fascicle is bundle of muscle fibers
Biceps muscle is a skeletal muscle in a forelimb
Muscle fiber is formed of many merged muscle cells
GORILLA
EPIMYSIUM
MUSCLE FIBER
Epimysium is sheath of tissue around muscle
Whole muscle is made up of fasciculi (bundles of muscle fibers)
Sarcoplasm is cytoplasm of muscle cells; it contains many nuclei and mitochondria
CAPILLARY
Myofibril is fiber containing filaments of actin and myosin
Capillaries supply muscle fiber with oxygen and nutrients
MYOFIBRIL
Sarcomere is basic contractile unit of a muscle fiber; it runs from one Z disc to next
Z disc anchors thin (actin) filaments
M line connects thick (myosin) filaments
Thin filament is mainly composed of the protein actin
ACTIN FILAMENT
MYOSIN FILAMENT
Thick filament is composed of the protein myosin

Structure of skeletal muscle
In vertebrates (including humans), skeletal muscles, such as the biceps muscle, consist of parallel bundles of muscle fibers (myofibrils) surrounded by a sheath of insulating connective tissue.

Muscle cells

Muscle cells are the basic units of the muscular system. Found in various forms throughout the animal kingdom, they are specialized to contract and shorten to produce movement of the body.

Skeletal muscle

Also known as striated muscle because of its striped appearance under a microscope, skeletal muscle is responsible for movements under conscious control. Each muscle fiber consists of long, thin cells that contain contractile units called sarcomeres joined end to end. Sarcomeres take up about three-quarters of each muscle cell; most of the remainder consists of mitochondria that provide energy in the form of ATP molecules (see pp.60–61).

Muscle contraction

The sarcomeres transform chemical energy in the form of ATP into the mechanical work of muscle contraction. Within each sarcomere, thin actin filaments slide over the thick myosin filaments, causing the sarcomeres to contract and shorten. Sarcomeres are joined end to end, so their simultaneous contraction causes contraction of the entire muscle.

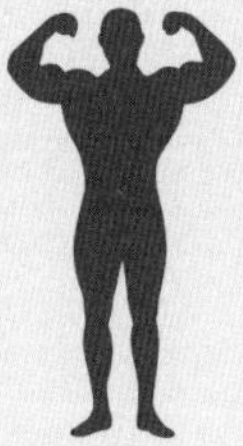

IN HUMANS, MUSCLE CONTRACTIONS PRODUCE ABOUT 85% OF BODY HEAT

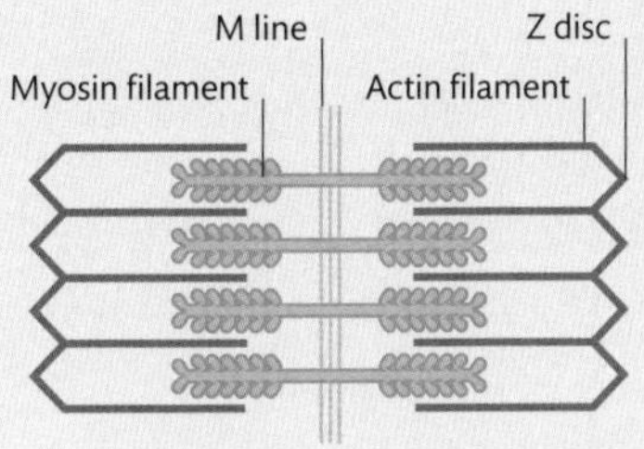
M line
Z disc
Myosin filament
Actin filament

1 Sarcomere of relaxed muscle
In a relaxed muscle, myosin heads are not attached to the thin actin filaments. At this point, the distance between the Z discs is at its greatest.

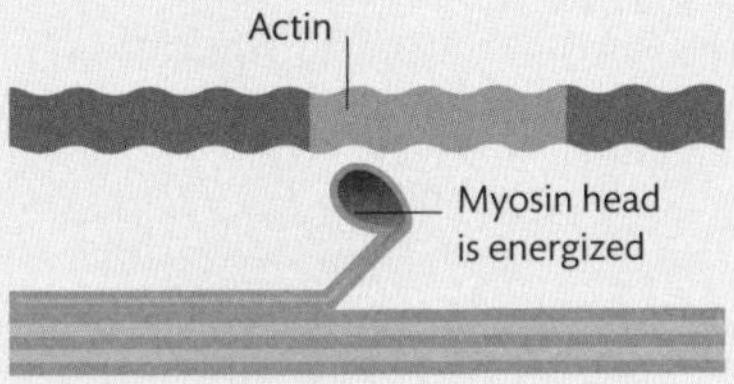
Actin
Myosin head is energized

2 Myosin energized
The myosin head is energized by ATP, produced in mitochondria from sugars and oxygen, and the actin is ready to attach to the myosin filaments.

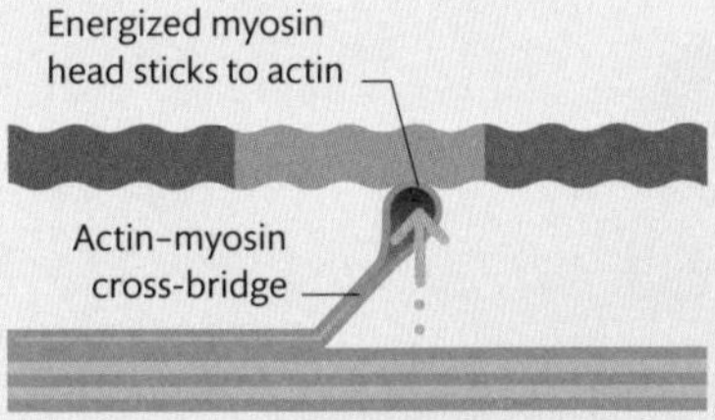
Energized myosin head sticks to actin
Actin–myosin cross-bridge

3 Myosin head sticks to actin
The energized myosin head attaches to a binding site on the actin filament, forming an actin–myosin cross-bridge between the filaments.

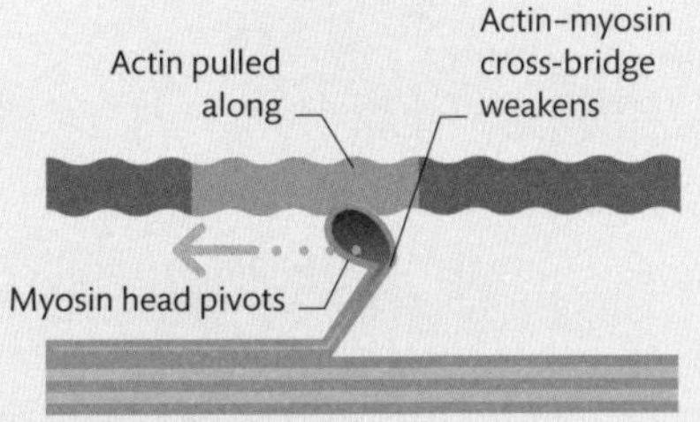
Actin pulled along
Actin–myosin cross-bridge weakens
Myosin head pivots

4 Head pivots
The myosin head releases energy and pivots. As a result, the actin filament moves along. The cross-bridge between the filaments becomes weaker.

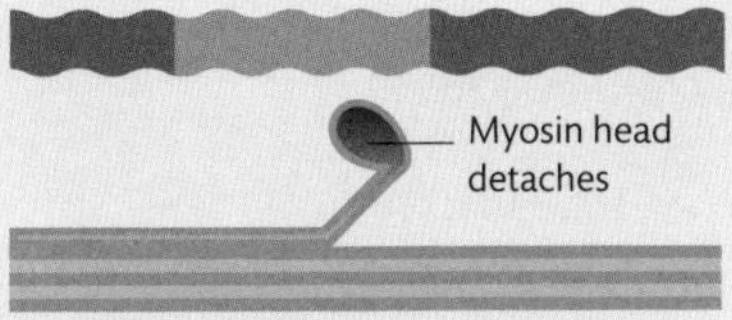
Myosin head detaches

5 Re-energizing
The cross-bridge releases, and the myosin is re-energized. Myosin is energized, sticks to actin, releases energy, and pivots many times during a single contraction.

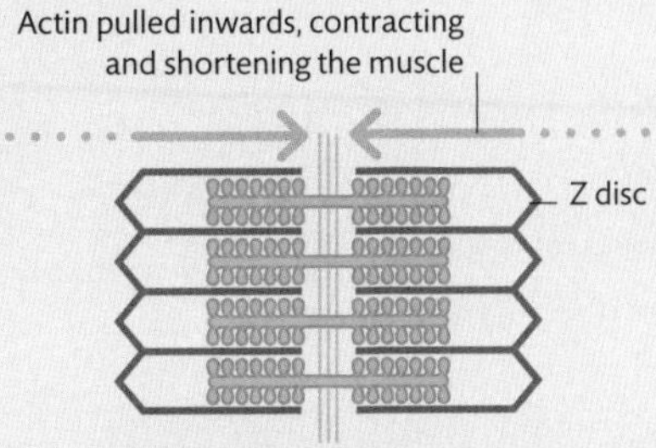
Actin pulled inwards, contracting and shortening the muscle
Z disc

6 Sarcomere of contracted muscle
In a contracted muscle, the actin has been pulled inward. When a muscle is contracted, the Z discs are closer together and the muscle is shorter.

WHAT CAUSES MUSCLE FATIGUE AND SORENESS?

Various factors cause muscle discomfort after exercise, such as inflammation of the muscle tissue. However, the buildup of lactic acid is not a cause.

MUSCLE TYPES

There are three main types of muscle in vertebrates. Striated muscle is under voluntary control; smooth and cardiac muscle are not under conscious control.

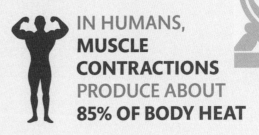

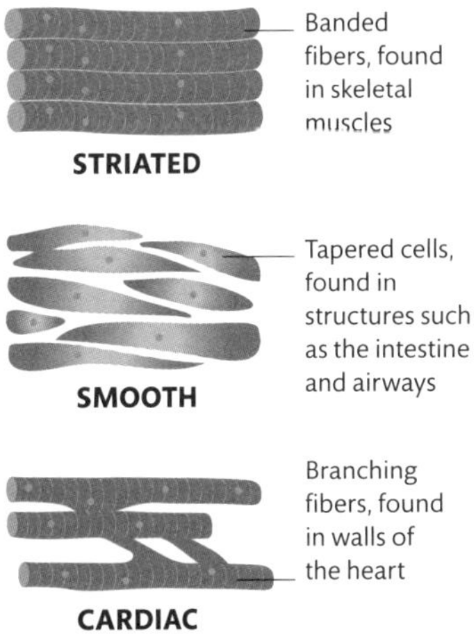
Banded fibers, found in skeletal muscles
STRIATED
Tapered cells, found in structures such as the intestine and airways
SMOOTH
Branching fibers, found in walls of the heart
CARDIAC

WHITE BLOOD CELLS

There are several types of white blood cell, each with specific functions. Some engulf and destroy invading organisms or foreign substances. Others produce proteins called antibodies to fight disease; some of these cells retain a memory of the disease in case of reinfection.

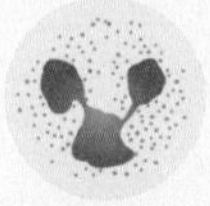

Neutrophils
Making up 50–70 percent of white blood cells, these are attracted to sites of inflammation, where they engulf and destroy infectious organisms.

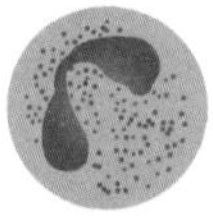

Eosinophils
Eosinophils are found in tissues and fight parasitic infestations and certain infections. High levels occur with allergic reactions and some autoimmune diseases.

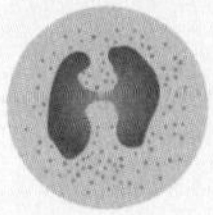

Basophils
Found in tissues, these trigger inflammatory reactions to allergens and parasites, and promote blood flow, so the body can expel unwanted substances.

Monocytes
The largest type of white blood cell, monocytes circulate in the blood and travel to sites of inflammation, to digest bacteria in a process called phagocytosis.

Natural killer cells
These cells detect and destroy other body cells that carry abnormal proteins, such as cancerous cells and those that are infected by viruses.

B cells
Plasma cells secrete antibodies to fight specific infections. Memory B cells retain "memory" of an infection, to enable a rapid response to reinfection.

T cells
Cytotoxic T cells directly attack infected body cells. Helper T cells activate B cells. Memory T cells remember infection to enable a rapid response to re-infection.

The composition of blood
Blood cells make up just under half the volume of blood. Red blood cells form 45 percent of the volume, while white blood cells and platelets account for 1 percent.

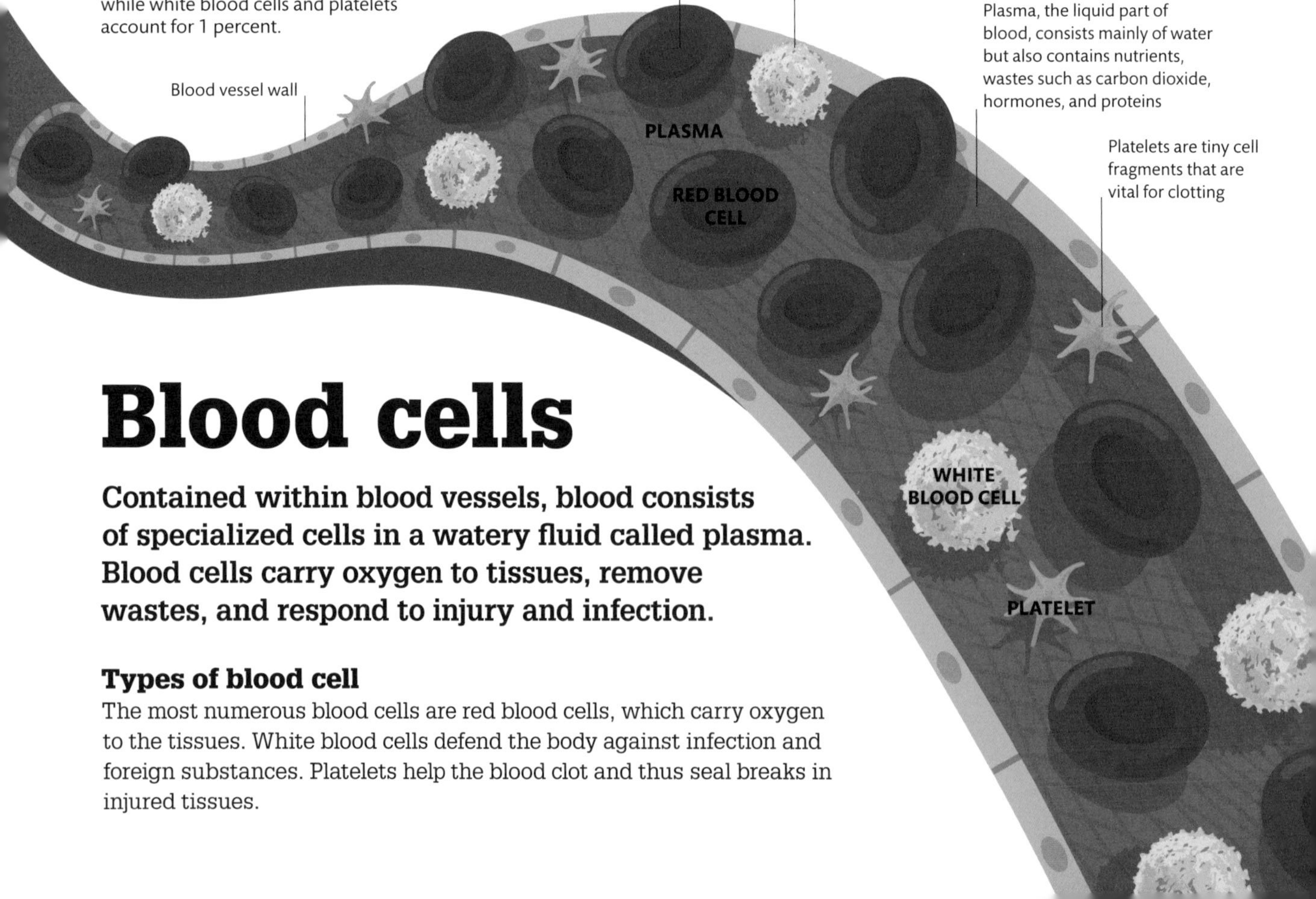
Red blood cells carry oxygen to body tissues
White blood cells, also known as leucocytes, are key components of the immune system
Plasma, the liquid part of blood, consists mainly of water but also contains nutrients, wastes such as carbon dioxide, hormones, and proteins
Platelets are tiny cell fragments that are vital for clotting
Blood vessel wall
PLASMA
RED BLOOD CELL
WHITE BLOOD CELL
PLATELET

Blood cells

Contained within blood vessels, blood consists of specialized cells in a watery fluid called plasma. Blood cells carry oxygen to tissues, remove wastes, and respond to injury and infection.

Types of blood cell

The most numerous blood cells are red blood cells, which carry oxygen to the tissues. White blood cells defend the body against infection and foreign substances. Platelets help the blood clot and thus seal breaks in injured tissues.

OXYHEMOGLOBIN

Heme molecule

Iron atom

Globin chain of protein

Oxygen

2 Travel to tissues
The red blood cells carry oxyhemoglobin in the bloodstream to all of the body tissues.

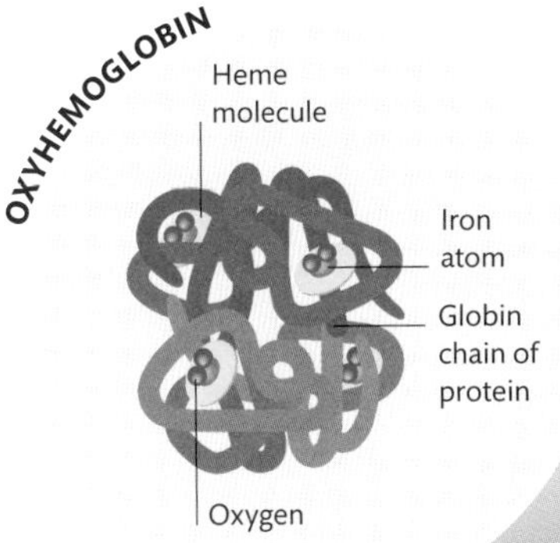

Red blood cells

Each red blood cell (or erythrocyte) contains millions of molecules of hemoglobin, an iron-rich protein. The cells are concave on both sides, giving a large surface area for absorbing oxygen or carbon dioxide, and they have a flexible "skeleton" so they can fit through even the smallest blood vessels.

3 Oxygen released
When oxygenated red blood cells reach tissues with low oxygen, oxyhemoglobin gives up its oxygen to become deoxyhemoglobin.

1 Oxygenation
Oxygen in the lungs diffuses through blood vessel walls into the blood cells. It binds to hemoglobin, forming oxyhemoglobin.

BLOOD FLOW

OXYGEN

CARBON DIOXIDE

LUNGS

Transporting gases
Red blood cells transport oxygen from the lungs to body tissues, where they release it. The cells can also absorb some carbon dioxide and return it to the lungs for it to be exhaled.

OXYGEN

CARBON DIOXIDE

BODY TISSUES

6 Carbon dioxide released
When the blood reaches the lungs, carbon dioxide in the plasma and red blood cells is released to be breathed out.

BLOOD VESSEL

Heme molecule

DEOXYHEMOGLOBIN

5 Travel back to lungs
Deoxygenated blood flows back to the lungs, carrying carbon dioxide in the blood plasma and the red blood cells.

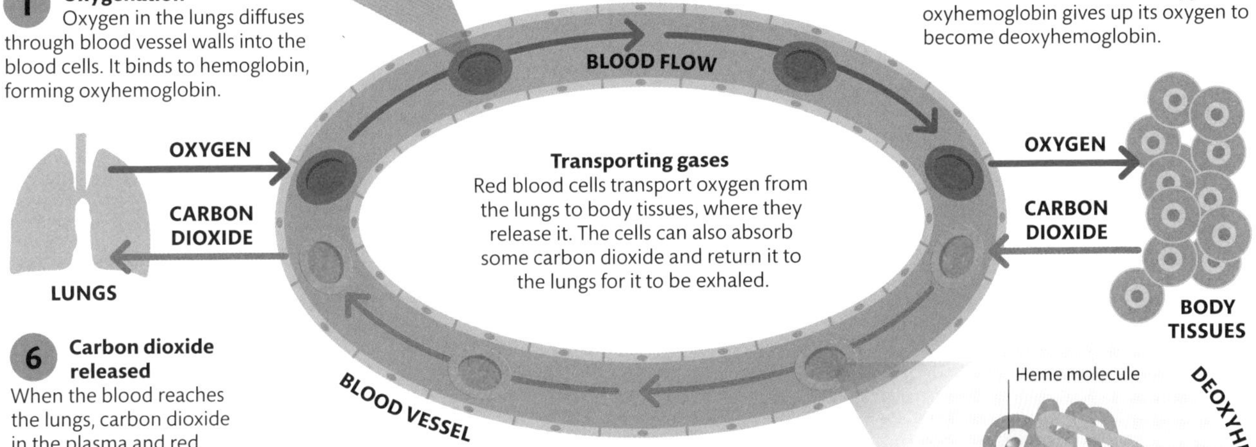

4 Carbon dioxide taken up
Carbon dioxide from body tissues passes into the deoxygenated blood. Most of the carbon dioxide dissolves in the blood plasma, but some combines with hemoglobin in red blood cells.

Globin chain of protein

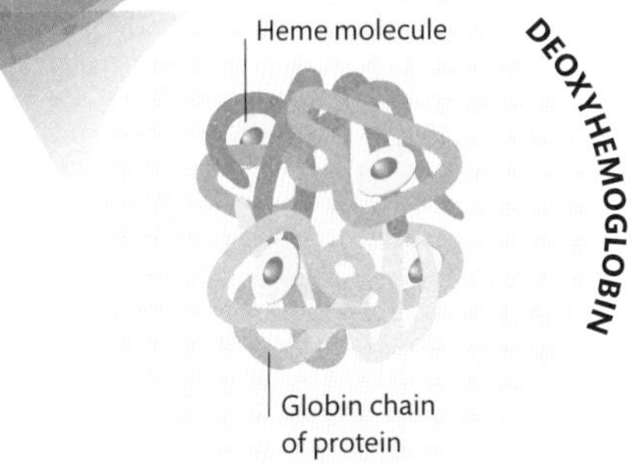

HOW LONG DO BLOOD CELLS LIVE?

In humans, the average lifespan of a red blood cell is 120 days. White blood cells live from minutes to hours, depending on their type and whether they are fighting infection.

THE COLORS OF BLOOD

Blood in different species contains different pigments that bind to oxygen, making the blood various colors. All blood looks brighter when it is carrying oxygen and darker or colorless when deoxygenated.

Red
In humans and most animals, birds, and fish, blood is red due to the iron in hemoglobin.

Blue
Some mollusks, crustaceans, and spiders have blue blood due to copper-based hemocyanin.

Green
The blood of some worms and leeches contains chlorocruorin, another iron-based pigment.

Violet
The iron-based pigment of some marine worms is hemerythrin, making their blood violet.

Tissues to organisms

The tiniest organisms consist of just one cell. In more complex life forms, such as plants and animals, collections of cells form tissues, organs, and systems to carry out the functions of life.

Animal tissues and organs

In animals, groups of cells form tissues to carry out specific functions. For example, epithelial tissues form the skin and the linings of hollow organs, and connective tissues link structures such as bones and muscles. Groups of different tissues form organs; for example, the heart includes muscle tissue to pump blood and nerve tissue to stimulate muscle contractions. Groups of organs, in turn, form systems to carry out the functions enabling the animal to live.

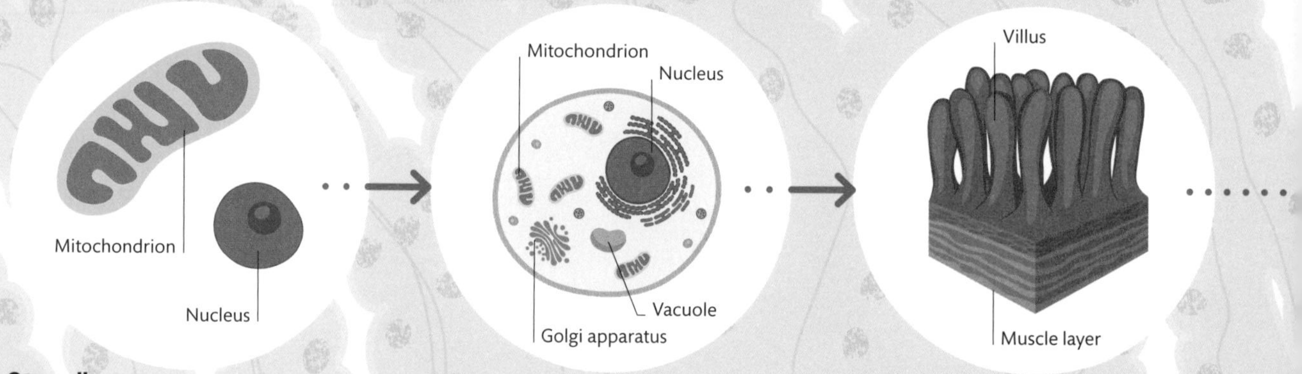

Organelle
Organelles perform specific functions within a cell. For example, the nucleus stores genetic information, while mitochondria produce chemical energy.

Cell
Cells are specialized to carry out specific functions. Some, such as blood cells, travel freely around the body; others, such as muscle cells, come together in tissues.

Tissue
Cells with similar structures and functions form tissues. The nutrient-absorbing cells of epithelial tissues form projections, called villi, to line the small intestine.

Plant tissues and organs

The most complex plants, known as vascular plants, have tissues and organs formed from various cells. One major cell type is parenchyma, which carries out photosynthesis, water and sugar storage, and exchange of oxygen and carbon dioxide. Other cell types include sclerenchyma and collenchyma, which give the plant structure and stability; epidermal cells, forming the surfaces of leaves and stems; tracheids, found in the xylem (water-conducting system); and sieve-tube members, in the phloem (sugar-carrying system).

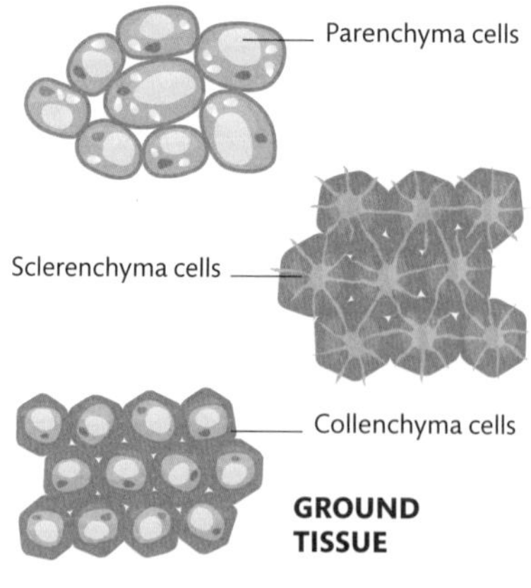

GROUND TISSUE

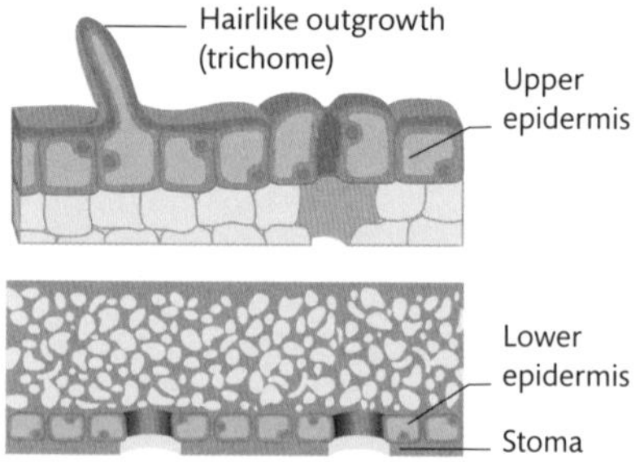

DERMAL TISSUE

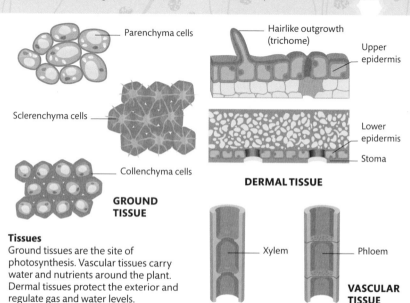

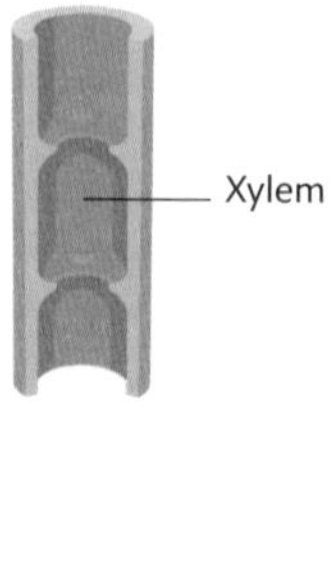

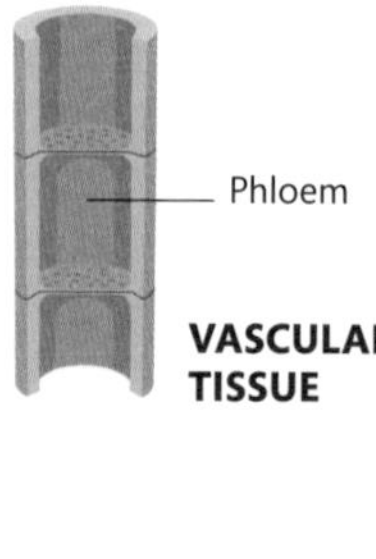

VASCULAR TISSUE

Tissues
Ground tissues are the site of photosynthesis. Vascular tissues carry water and nutrients around the plant. Dermal tissues protect the exterior and regulate gas and water levels.

WHAT IS THE LARGEST ORGAN IN ANIMALS?

The skin is the largest organ in vertebrates, making up about 12–25% of body mass, depending on the species. In humans, it makes up about 15% of body mass.

THE FIRST ORGANISMS TO DEVELOP TISSUES WERE SPONGES, ABOUT 600 MILLION YEARS AGO

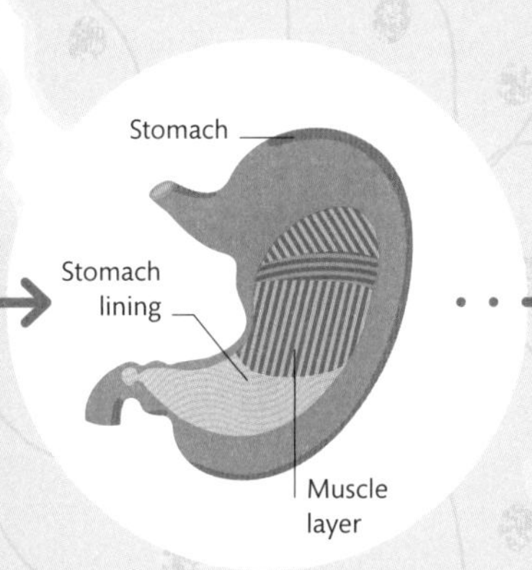
Stomach
Stomach lining
Muscle layer

Organ
Groups of tissues working together form an organ. In the human stomach, layers of muscle churn food, while glands in the lining secrete enzymes, to break down food.

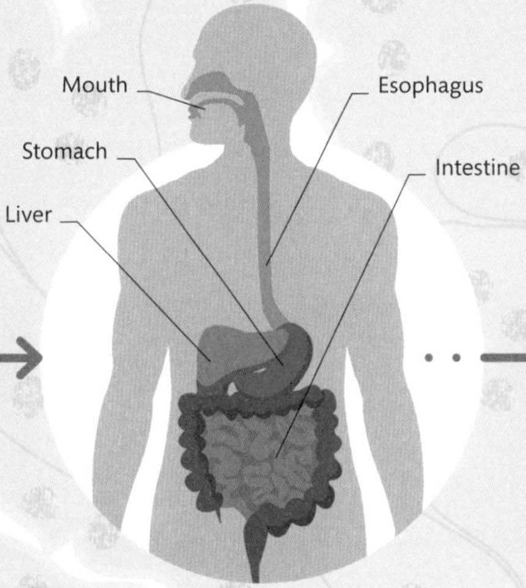
Mouth
Esophagus
Stomach
Intestine
Liver

Organ system
Groups of organs doing similar jobs form organ systems. The digestive system has organs for swallowing, moving food, absorbing nutrients, and eliminating waste.

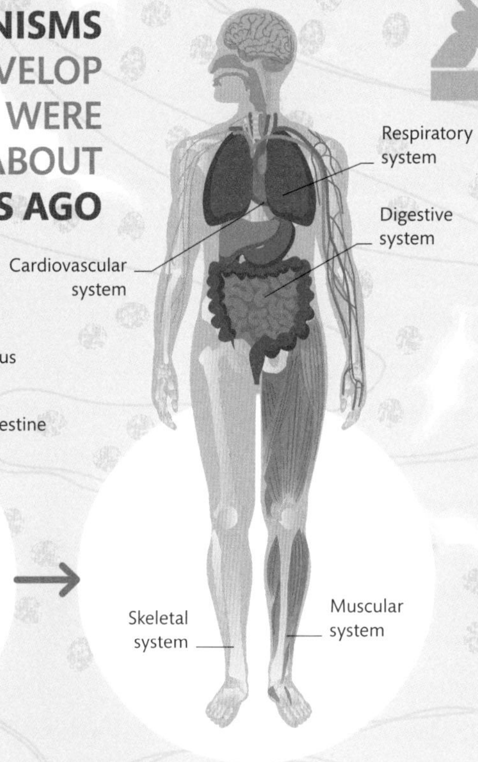
Respiratory system
Digestive system
Cardiovascular system
Skeletal system
Muscular system

Organism
Organ systems work together to fulfill all the needs of an organism, such as respiration, digestion of food, transport of blood, and support or movement of the body.

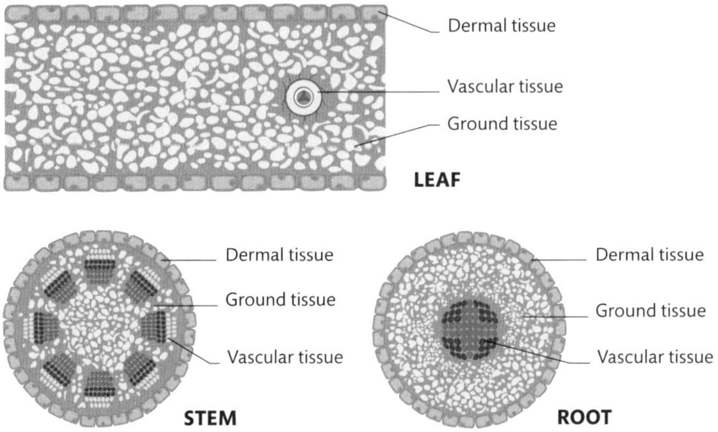
Dermal tissue
Vascular tissue
Ground tissue
LEAF
Dermal tissue
Ground tissue
Vascular tissue
STEM
Dermal tissue
Ground tissue
Vascular tissue
ROOT

Organs
A plant's organs are the leaves, which turn energy from sunlight into fuel; flowers and seeds; stems, to support the plant; and roots, to anchor the plant and draw nutrients from the soil.

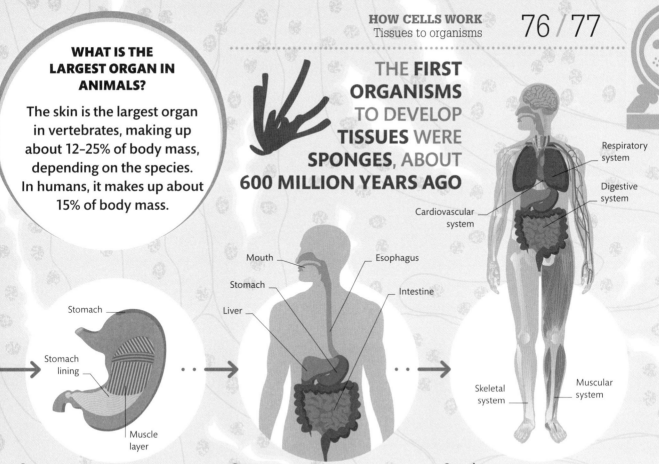

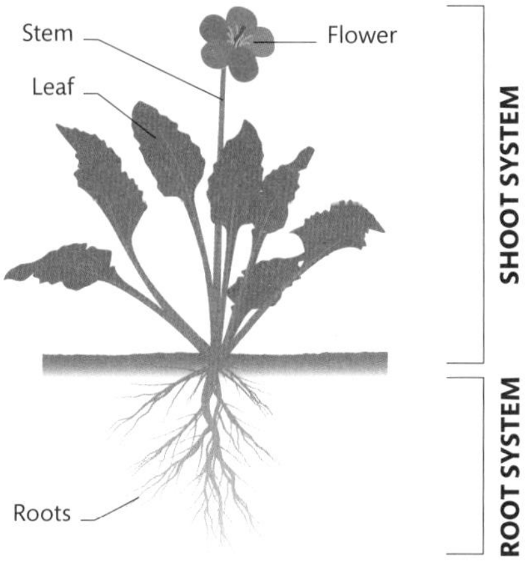
Stem
Flower
Leaf
SHOOT SYSTEM
ROOT SYSTEM
Roots

Organ systems
Plants contain several organ systems. The leaves, stem, fruits, and flowers form the shoot system, while the different types of roots underground constitute the root system.

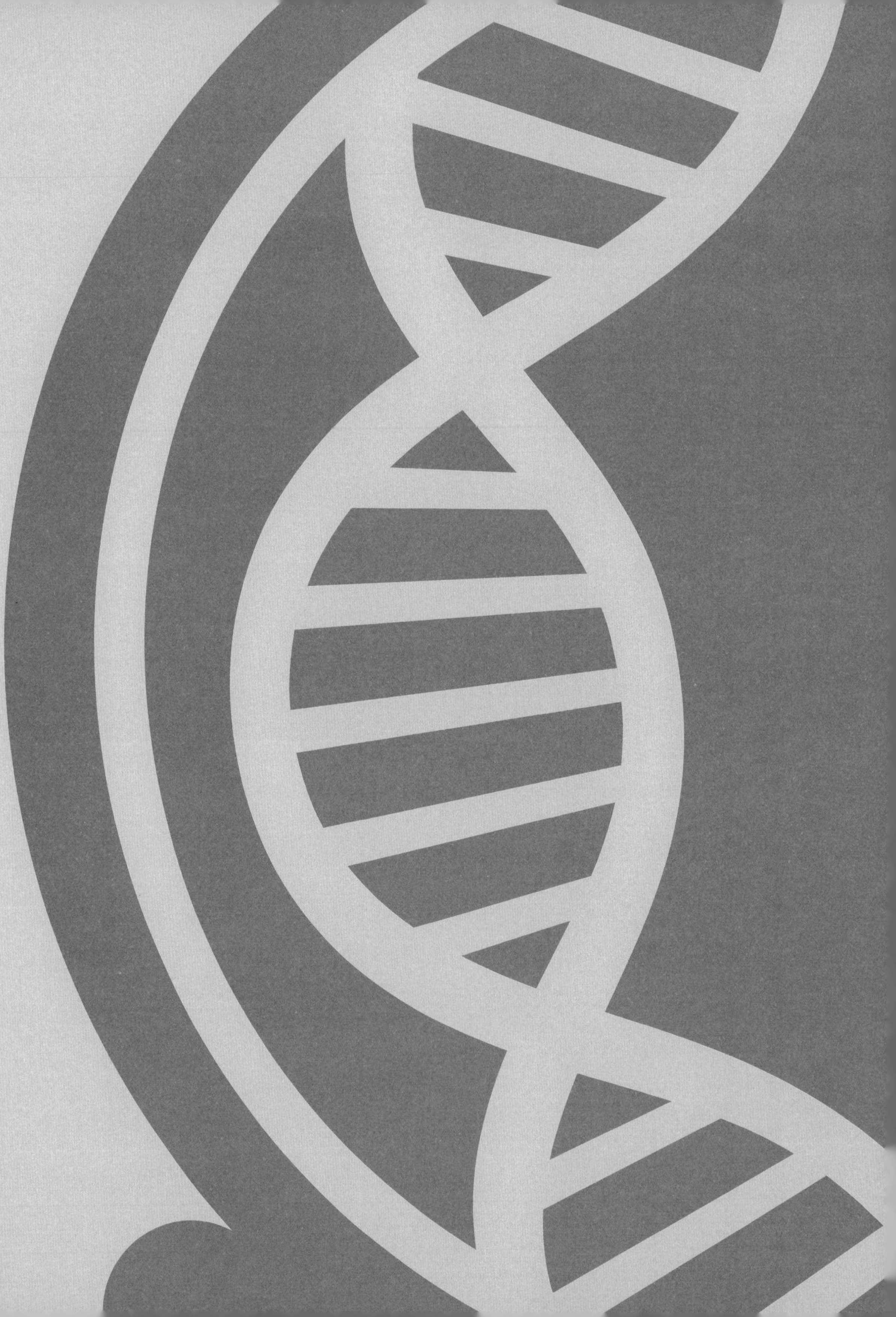

REPRODUCTION AND GENETICS

Asexual reproduction

Some organisms reproduce without mating, producing offspring that are identical to their parent and each other. This is asexual reproduction. The offspring may rapidly fill a new habitat, but genetic uniformity can leave them unable to adapt to changing conditions.

A **FLATWORM** CAN BE **FRAGMENTED** INTO **279 PIECES** THAT BECOME **NEW WORMS**

Parthenogenesis

Some organisms (mostly invertebrates) can produce eggs that do not need to be fertilized by sperm in order to develop into offspring. This type of asexual reproduction—parthenogenesis—can even feature alongside sexual reproduction in the same organism.

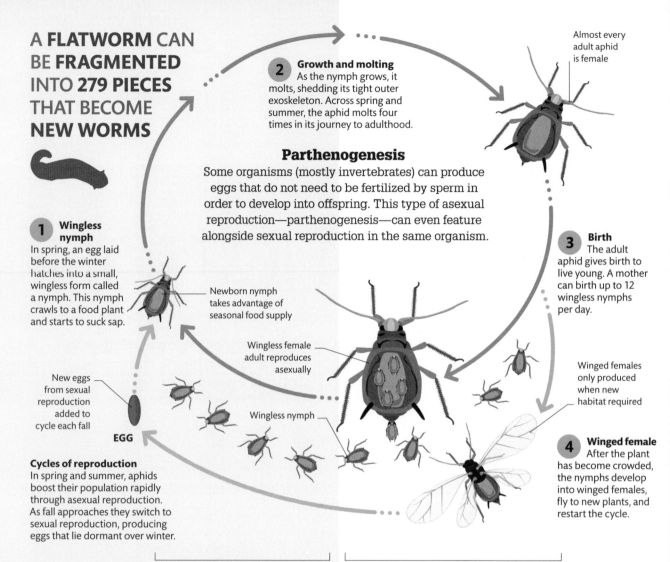

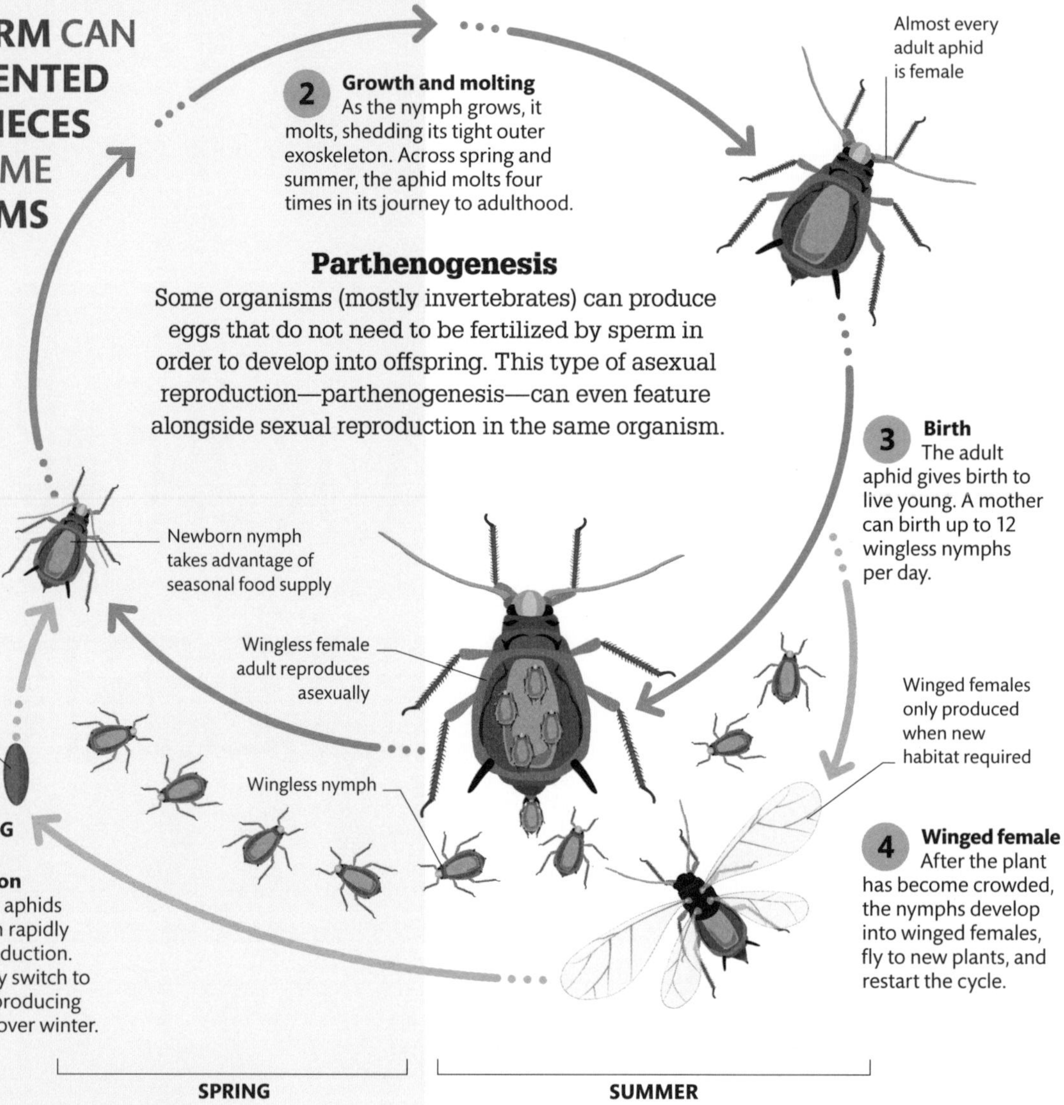

1 Wingless nymph
In spring, an egg laid before the winter hatches into a small, wingless form called a nymph. This nymph crawls to a food plant and starts to suck sap.
Newborn nymph takes advantage of seasonal food supply
2 Growth and molting
As the nymph grows, it molts, shedding its tight outer exoskeleton. Across spring and summer, the aphid molts four times in its journey to adulthood.
Almost every adult aphid is female
3 Birth
The adult aphid gives birth to live young. A mother can birth up to 12 wingless nymphs per day.
Wingless female adult reproduces asexually
Winged females only produced when new habitat required
Wingless nymph
4 Winged female
After the plant has become crowded, the nymphs develop into winged females, fly to new plants, and restart the cycle.
New eggs from sexual reproduction added to cycle each fall
EGG
Cycles of reproduction
In spring and summer, aphids boost their population rapidly through asexual reproduction. As fall approaches they switch to sexual reproduction, producing eggs that lie dormant over winter.
SPRING
SUMMER

IS PARTHENOGENESIS COMMON IN VERTEBRATES?

No—it is rare. A few species of amphibians, reptiles, and fish can reproduce asexually. However, no species of birds or mammals can.

Budding

One of the simplest methods of asexual reproduction is when part of the parent's body breaks off and develops into a full-sized and independent individual. This process, known as budding, is used by the simplest animals and by single-celled organisms, including yeasts. Budding requires no complex cellular changes, such as creating eggs, to facilitate the growth of the new individual.

Immortal organism
Hydras are relatives of jellyfish and anemones that grow on the seabed. They can reproduce by budding. This means that the hydras growing today are parts of an original body that has been growing for millions of years.

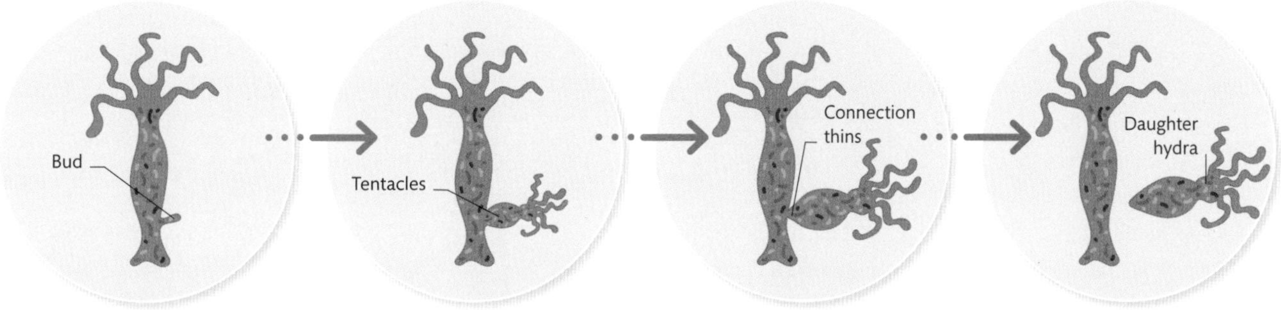

1 Bud grows When a bud first begins to form, it grows from the body wall of the adult. A new bud can grow every two days.

2 Bud develops traits Developing tentacles and other features, the bud grows, resembling a small version of the adult.

3 Bud matures Growing further still, the bud then begins to break off from the older body. This process is called cleavage.

4 Bud separates The bud breaks free from the parent and floats away. Then it settles on a solid surface and grows to full size.

Vegetative reproduction

Plants can spread asexually through a process called vegetative reproduction. This is when growths such as rhizomes and stolons (two types of stem) emerge from the parent plant. Seeking out new places to grow, these stems give rise to new daughter plants.

New stems
Rhizomes and stolons are similar but not the same. Rhizomes grow beneath the ground, while stolons develop above it. Additionally, rhizomes develop from roots, while stolons usually grow out from the parent's main stem.

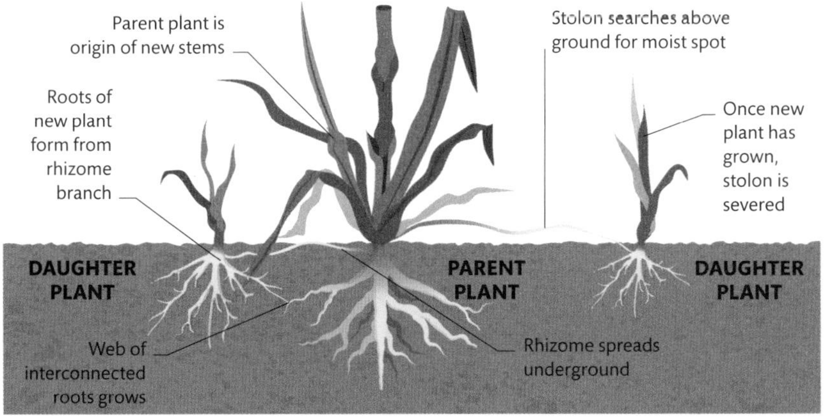
Parent plant is origin of new stems
Roots of new plant form from rhizome branch
Stolon searches above ground for moist spot
Once new plant has grown, stolon is severed
DAUGHTER PLANT
PARENT PLANT
DAUGHTER PLANT
Web of interconnected roots grows
Rhizome spreads underground

FRAGMENTATION

Some animals, such as flatworms, can multiply after their body is divided or fragmented. Each body part grows into a new fully formed body. This allows an individual to survive being injured or dismembered.

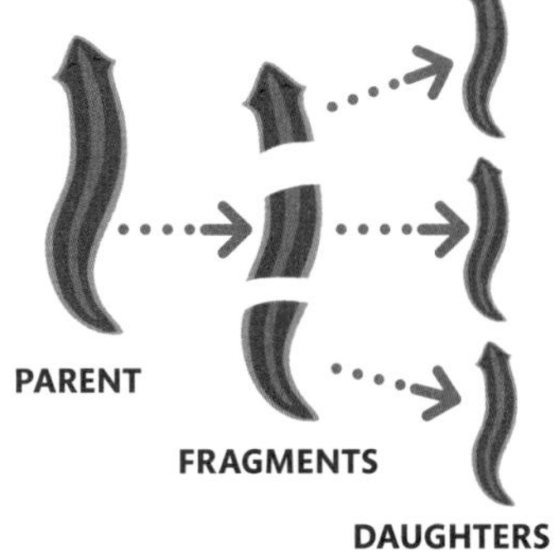
PARENT
FRAGMENTS
DAUGHTERS

Meiosis

The gametes (sex cells) used in sexual reproduction are created by a cell division process called meiosis. Meiosis creates the gametes that fuse during fertilization to form a new organism.

Processes of division

Meiosis uses the same spindle apparatus to divide the contents of the cell as mitosis (see pp.68–69). However, unlike mitosis, meiosis consists of two division processes that turn one parent cell into four daughter cells. In humans, each of these cells has a half set of 23 chromosomes, so that when an egg and sperm fuse, the full set of 46 is achieved.

1 Interphase
Meiosis begins with a single diploid cell (see opposite). Before division, the DNA in the nucleus (chromatin) duplicates, as does the centrosome.

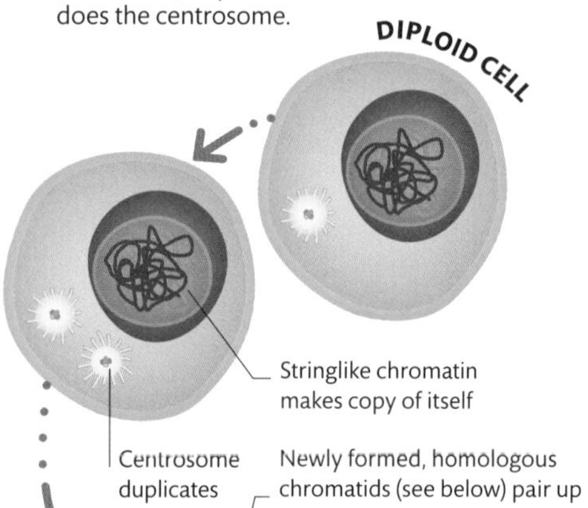

DIPLOID CELL
Stringlike chromatin makes copy of itself
Centrosome duplicates
Newly formed, homologous chromatids (see below) pair up

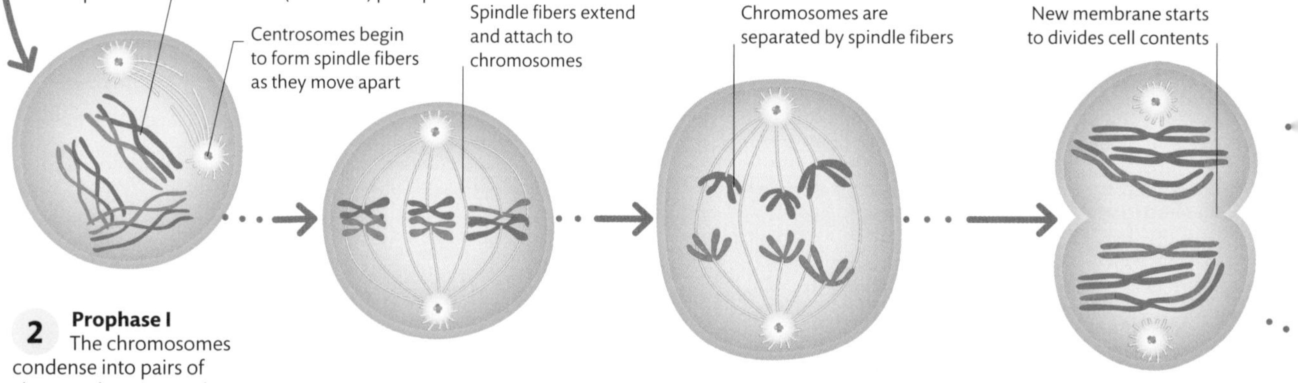

Centrosomes begin to form spindle fibers as they move apart
Spindle fibers extend and attach to chromosomes
Chromosomes are separated by spindle fibers
New membrane starts to divides cell contents

2 Prophase I
The chromosomes condense into pairs of chromatids connected at a midpoint. Adjacent chromatids from each homologous pair can exchange sections of DNA during crossing over (see below).

3 Metaphase I
Crossing over continues. A spindle apparatus forms at either end of the cell. The chromosome pairs line up along the equator of the cell.

4 Anaphase I
The spindle fibers pull one chromosome (made up of two connected chromatids) from each pair to opposite ends of the cells.

5 Telophase I
A new cell membrane forms across the middle of the cell, dividing the cytoplasm in two. This creates two haploid cells (see opposite).

Crossing over

Homologous chromosomes carry the same genes at the same fixed points. Each chromosome is made up of two identical chromatids. Neighboring chromatids are able to swap some sections in a process called crossing over. This process begins in each prophase stage and continues into the next metaphase stage. Crossing over maximizes the genetic diversity of any future offspring by producing four unique chromatids.

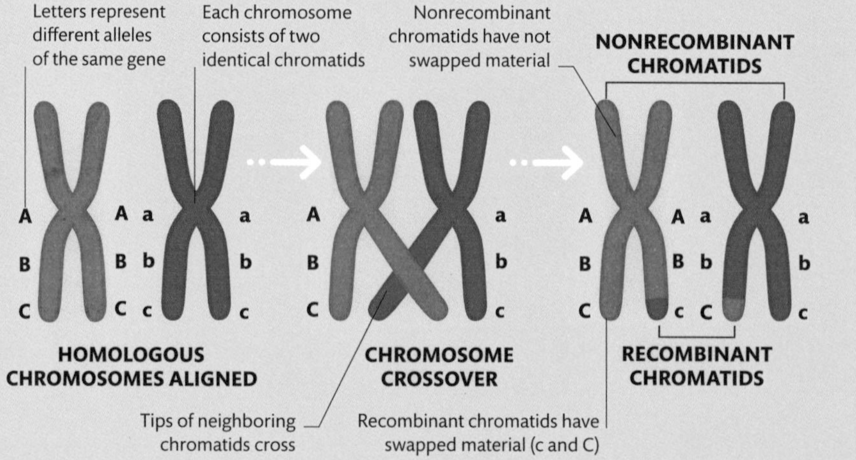

Letters represent different alleles of the same gene
Each chromosome consists of two identical chromatids
Nonrecombinant chromatids have not swapped material
NONRECOMBINANT CHROMATIDS
A B C
A B C
a b c
a b c
A B C
a b c
A B C
A B c
a b C
a b c
HOMOLOGOUS CHROMOSOMES ALIGNED
CHROMOSOME CROSSOVER
RECOMBINANT CHROMATIDS
Tips of neighboring chromatids cross
Recombinant chromatids have swapped material (c and C)

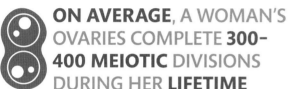

ON AVERAGE, A WOMAN'S OVARIES COMPLETE 300–400 MEIOTIC DIVISIONS DURING HER LIFETIME

10 Four daughter cells
After the second division, there are four daughter cells, each of which have a unique set of DNA.

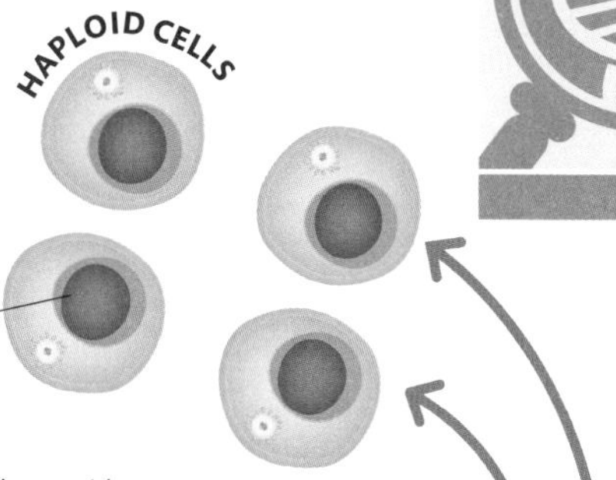
HAPLOID CELLS
Nucleus of each cell contains 23 single, genetically unique chromosomes

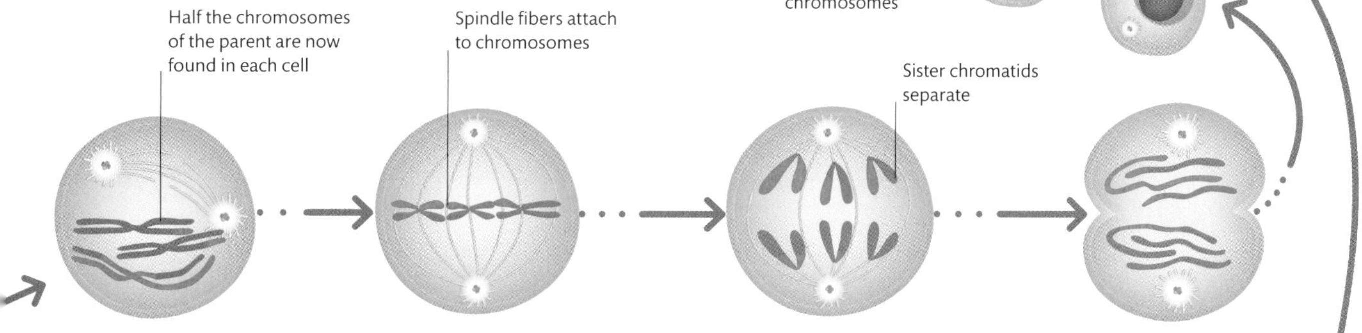
Half the chromosomes of the parent are now found in each cell
Spindle fibers attach to chromosomes
Sister chromatids separate

6 Prophase II
Both daughter cells can now undergo a second division in which each previous step is repeated. This starts with the new centrosomes moving apart.

7 Metaphase II
The spindle fibers connect to the homologous chromosomes that are lining up in the middle of the cell.

8 Anaphase II
The spindle fibers pull the chromosome apart, and the chromatids separate and move to opposite ends of the cell.

9 Telophase II
The cell membrane develops across the middle of the cell, dividing the chromosome and creating a pair of new cells.

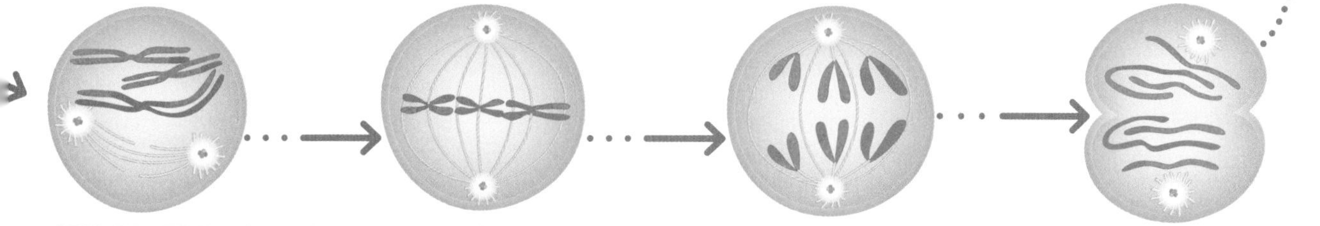

CELL SPLITS INTO TWO AND DIVIDES AGAIN

DOES MEIOSIS OCCUR IN ALL LIVING THINGS?

In multicellular organisms (such as plants, animals, and some fungi), yes, but not in many single-celled organisms (such as bacteria and archaea).

DIPLOID AND HAPLOID CELLS

Gametes, or sex cells, are haploid, which means they contain half the full set of chromosomes carried by diploid body cells. Diploid cells carry two copies of each chromosome—one provided by each of the parents. The chromosomes in a diploid cell exist in homologous pairs—that is, the DNA in each chromosome holds a version of the same genes. During meiosis, the homologous pairs are always separated.

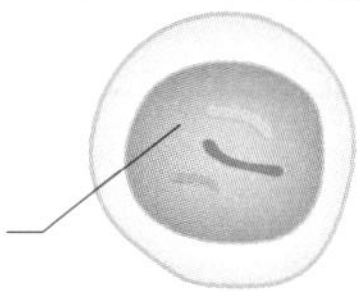
One copy of each chromosome in cells born from meiosis

HAPLOID CELL

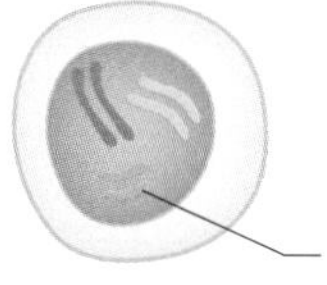
Two copies of each chromosome in cells that begin meiosis

DIPLOID CELL

Sexual reproduction

Most complex organisms reproduce sexually. This involves two parents combining their genetic information and forming a new organism. Offspring from sexual reproduction are genetically unique. This genetic variety within a species increases the odds of each new generation's survival.

Merging sex cells

For both plants and animals alike, sexual reproduction requires each parent to provide one sex cell (gamete). In order to create a new organism, the male gamete and female gamete need to meet and the nuclei of the cells then need to fuse. This is known as fertilization (see opposite). The fertilized cell, or zygote, then divides and develops into a new organism.

A **QUEEN BEE** CAN **LAY 1,500** EGGS IN A **SINGLE DAY**

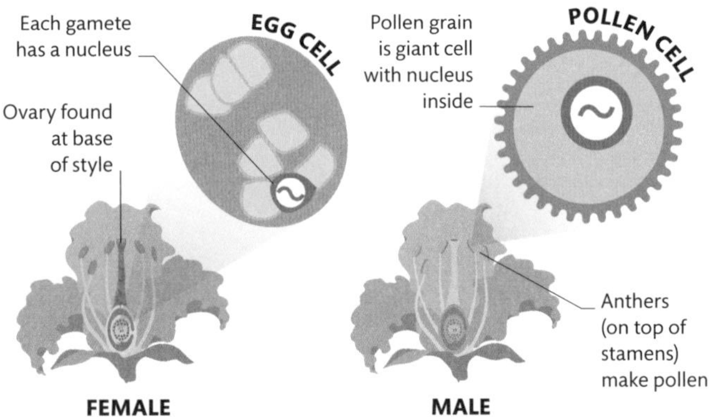

Each gamete has a nucleus
EGG CELL
Ovary found at base of style
Pollen grain is giant cell with nucleus inside
POLLEN CELL
Anthers (on top of stamens) make pollen
FEMALE
MALE

Sexual reproduction in flowering plants
The male gamete of a flowering plant is contained inside a pollen grain, which must be transferred by wind, water, or an animal from the anther, one of the male parts of the flower, to the female section, or ovary, of another flower.

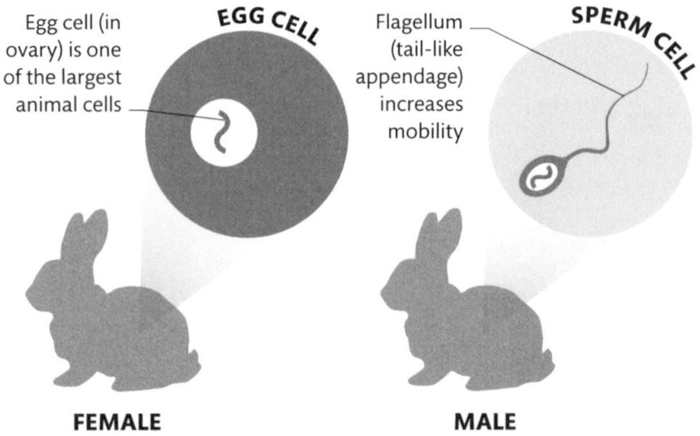

Egg cell (in ovary) is one of the largest animal cells
EGG CELL
Flagellum (tail-like appendage) increases mobility
SPERM CELL
FEMALE
MALE

Sexual reproduction in animals
An animal sperm is a highly mobile cell that easily swims to egg cells. Animal egg cells may be ejected from the body for the sperm to fertilize in large numbers, or fertilization can occur internally through sexual intercourse.

ALTERNATION OF GENERATION

Many simple plants cycle through two distinct phases – the haploid phase, where they reproduce sexually by releasing gametes, and the diploid phase, where they reproduce asexually by releasing spores. A plant alternating between a haploid phase and a diploid phase undergoes meiosis (see pp.82–83) and mitosis (see pp.68–69).

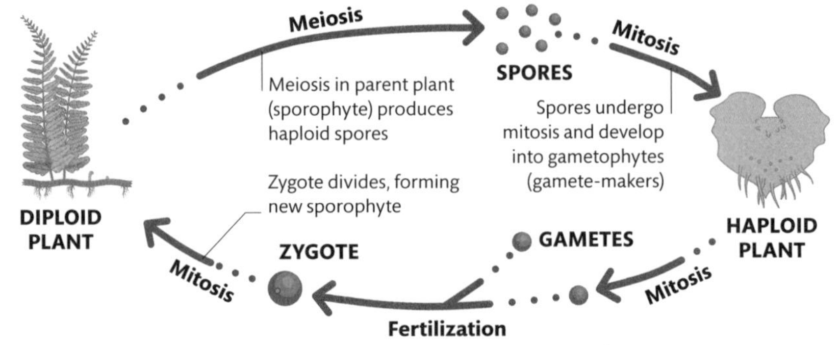

Meiosis
SPORES
Mitosis
Meiosis in parent plant (sporophyte) produces haploid spores
Spores undergo mitosis and develop into gametophytes (gamete-makers)
DIPLOID PLANT
HAPLOID PLANT
Zygote divides, forming new sporophyte
ZYGOTE
GAMETES
Mitosis
Mitosis
Fertilization

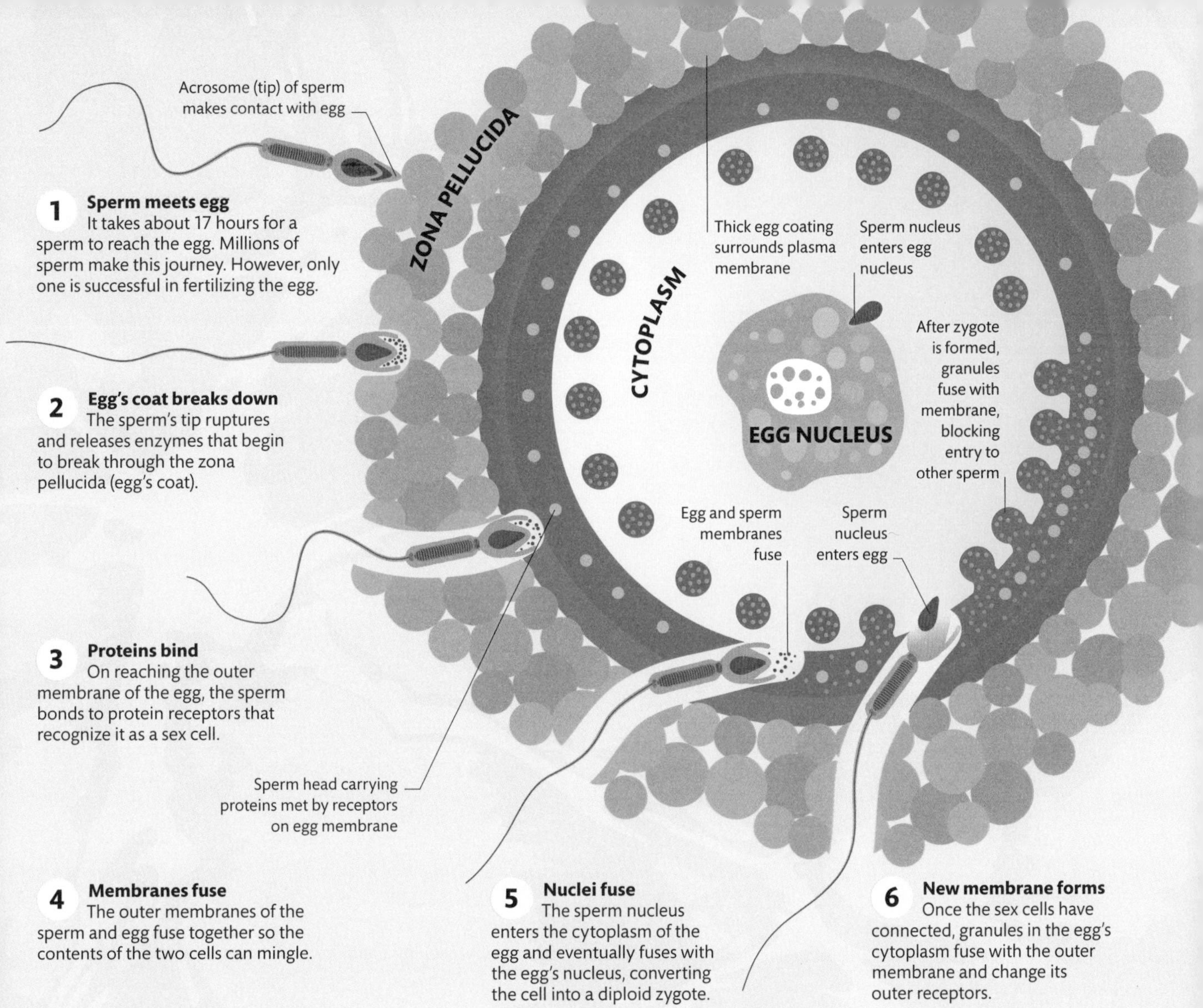
Acrosome (tip) of sperm makes contact with egg
ZONA PELLUCIDA
CYTOPLASM
Thick egg coating surrounds plasma membrane
Sperm nucleus enters egg nucleus
EGG NUCLEUS
After zygote is formed, granules fuse with membrane, blocking entry to other sperm
Egg and sperm membranes fuse
Sperm nucleus enters egg
Sperm head carrying proteins met by receptors on egg membrane
1 Sperm meets egg
It takes about 17 hours for a sperm to reach the egg. Millions of sperm make this journey. However, only one is successful in fertilizing the egg.
2 Egg's coat breaks down
The sperm's tip ruptures and releases enzymes that begin to break through the zona pellucida (egg's coat).
3 Proteins bind
On reaching the outer membrane of the egg, the sperm bonds to protein receptors that recognize it as a sex cell.
4 Membranes fuse
The outer membranes of the sperm and egg fuse together so the contents of the two cells can mingle.
5 Nuclei fuse
The sperm nucleus enters the cytoplasm of the egg and eventually fuses with the egg's nucleus, converting the cell into a diploid zygote.
6 New membrane forms
Once the sex cells have connected, granules in the egg's cytoplasm fuse with the outer membrane and change its outer receptors.

How human gametes fuse
Sperm cells swim through the uterus to find the egg, which travels down the oviduct from the ovary. The process by which a human sperm and egg combine is typical of almost all sexually reproducing animals.

HOW LARGE ARE SEX CELLS?

Each human sperm cell is about 0.002 in (0.005 mm) across—too small to see without a microscope. However, human egg cells are 20 times bigger and can be seen with the naked eye.

Fertilization

Sexual reproduction hinges on fertilization, which is the process by which the sperm and egg combine. When this occurs, the nuclei of the two sex cells merge, forming a zygote. The two haploid sex cells, each with one set of unpaired chromosomes, become one diploid cell with two sets of chromosomes. The sperm and egg each have half the genetic material needed by the zygote, which is the first body cell of the offspring. In flowering plants, fertilization is facilitated by pollination (see p.151).

Modes of animal reproduction

If an animal can raise the chances of its offspring surviving, the offspring are more likely to reproduce again. Animals have evolved a range of reproduction strategies, investing their resources in various ways that help boost these survival chances.

Live-bearing animals

Some animals invest time and energy ensuring that offspring reach an advanced level of development before they are born. Offspring grow within the parent's body where, safe from attack and often provided with nutrients, they stay until they reach a stage that they can live more independently and have a greater chance of survival.

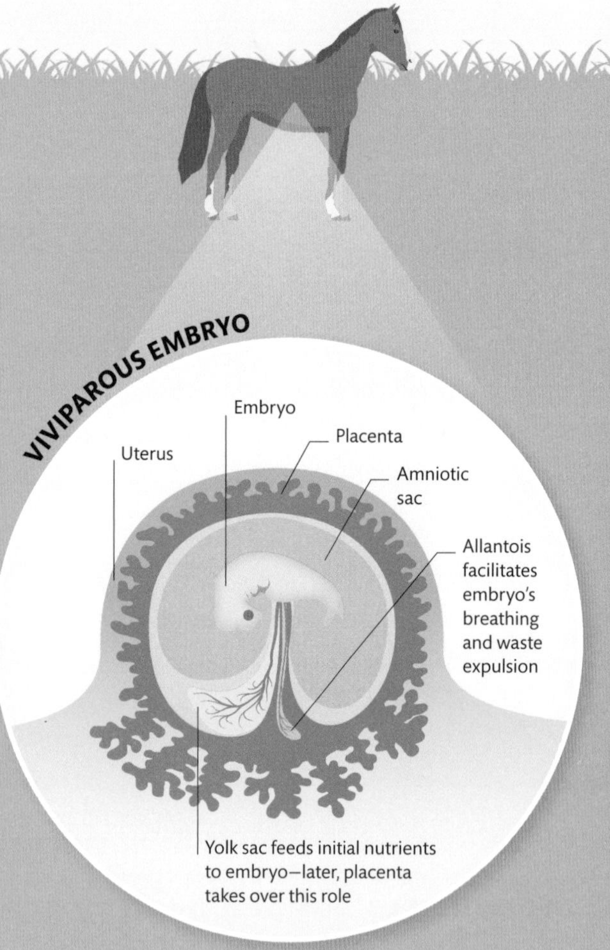

VIVIPAROUS EMBRYO
Embryo
Uterus
Placenta
Amniotic sac
Allantois facilitates embryo's breathing and waste expulsion
Yolk sac feeds initial nutrients to embryo—later, placenta takes over this role

Viviparous animals
Live-bearing animals that produce offspring within the parent's body—and not in an egg—are termed viviparous. Placental mammals, such as horses, are viviparous, as are scorpions, aphids, and velvet worms.

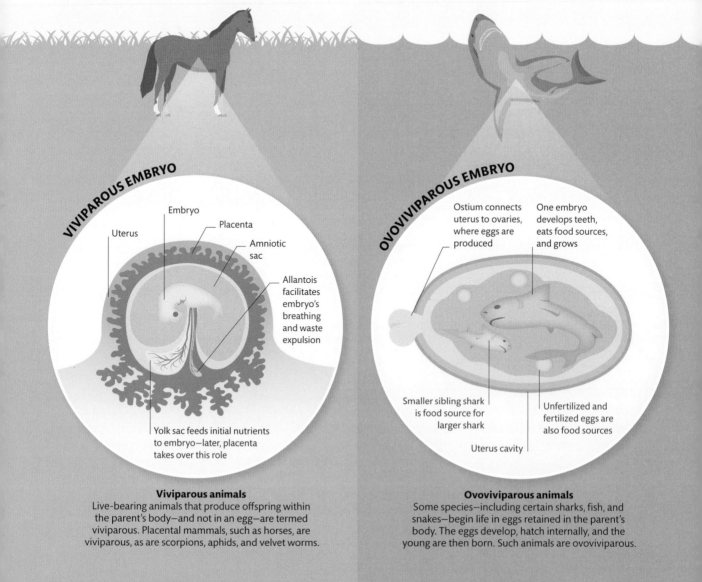

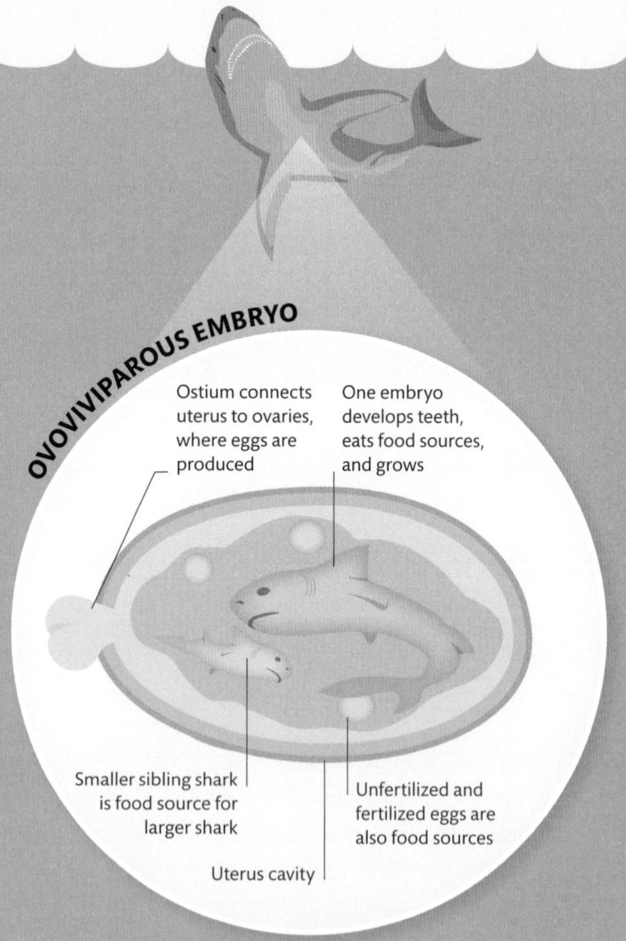

OVOVIVIPAROUS EMBRYO
Ostium connects uterus to ovaries, where eggs are produced
One embryo develops teeth, eats food sources, and grows
Smaller sibling shark is food source for larger shark
Unfertilized and fertilized eggs are also food sources
Uterus cavity

Ovoviviparous animals
Some species—including certain sharks, fish, and snakes—begin life in eggs retained in the parent's body. The eggs develop, hatch internally, and the young are then born. Such animals are ovoviviparous.

Egg-laying animals

Animals that lay eggs are able to produce large numbers of young quickly. The eggs of animals that reproduce in this way are made up, at minimum, of a protective gel coating around the embryo and a small supply of nutrients in a yolk sac. Some have an extra, amniotic layer, which effectively waterproofs the egg. For example, reptiles and birds produce eggs with a calcium carbonate shell that is impermeable to water but still allows air through. Such animals often protect their eggs by building a safe nest or by brooding (keeping them warm—for example, by sitting on them).

WHY DO SOME EGGS HAVE SHELLS?

Eggshells have many functions. Helping protect the egg against water, damage, and infection, they also regulate gas and water exchange and provide calcium for the growing embryo.

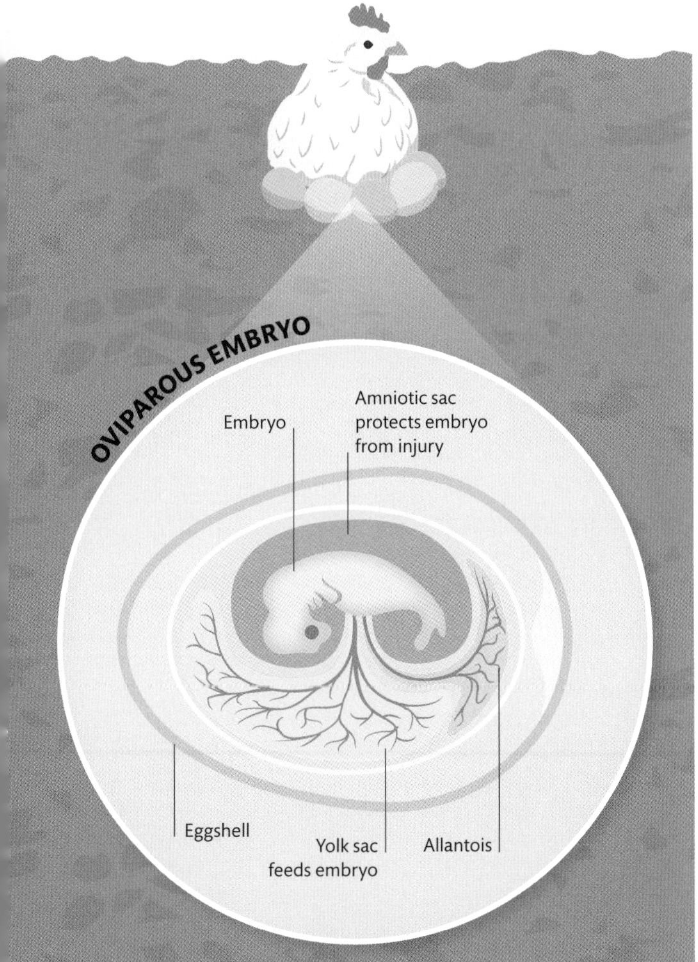
OVIPAROUS EMBRYO
Embryo
Amniotic sac protects embryo from injury
Eggshell
Yolk sac feeds embryo
Allantois

Oviparous animals
In oviparous (egg-laying) animals, such as birds and most fish, the eggs hatch outside the parent's body. Viviparous animals evolved from oviparous animals, which explains why the internal structures of their eggs are similar.

THE **OCEAN SUNFISH** CAN PRODUCE **300 MILLION EGGS AT ONCE**—MORE THAN **ANY OTHER ANIMAL**

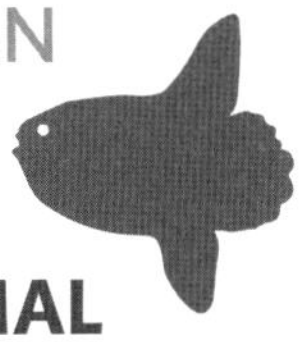

EVO-DEVO

Evolutionary development biology, or evo-devo, is the cell-by-cell comparison of how offspring develop in different organisms, and how these processes evolved. Related species develop in a similar way as embryos. However, the point at which they diverge shows us how close the relationship is. For example, humans and fish are both vertebrates, so their embryos are similar in the early stages of development.

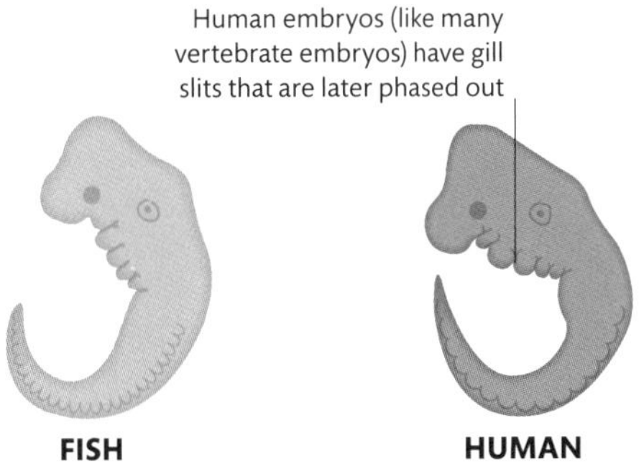
Human embryos (like many vertebrate embryos) have gill slits that are later phased out
FISH
HUMAN

Stem cells

Organisms are mostly made of cells that specialize in specific tasks. However, from embryo to adulthood, a small bank of unspecialized cells called stem cells retain the ability to develop into other cells.

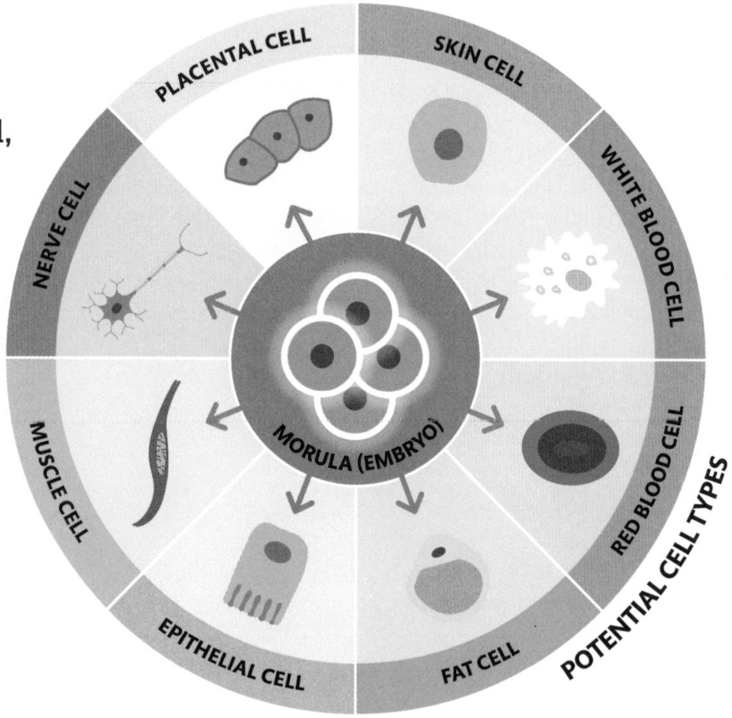

Earliest embryo
At its earliest stage, an embryo is a tiny, solid body of cells called a morula. Its cells are totipotent, meaning they can differentiate into any type of cell and form any part of the embryo. In most mammals, the morula includes the membrane that forms the placenta.

Types of stem cell

An animal embryo begins life as a ball of unspecialized cells. To go from this state to a fully grown organism, these stem cells must specialize into various types of cell through a process called differentiation. During differentiation, cells become less versatile as they are committed to certain roles. As an organism develops, its stem cells decline in number and potency (ability to specialize).

SALAMANDERS' MANY STEM CELLS LETS THEM REGROW **ORGANS** AND **WHOLE LIMBS**

STEM CELL THERAPY

The developmental potential of stem cells can be used to grow healthy tissues and treat illness and disease. A constantly evolving science, stem cell therapy aims to prevent or treat the symptoms of medical conditions.

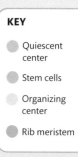

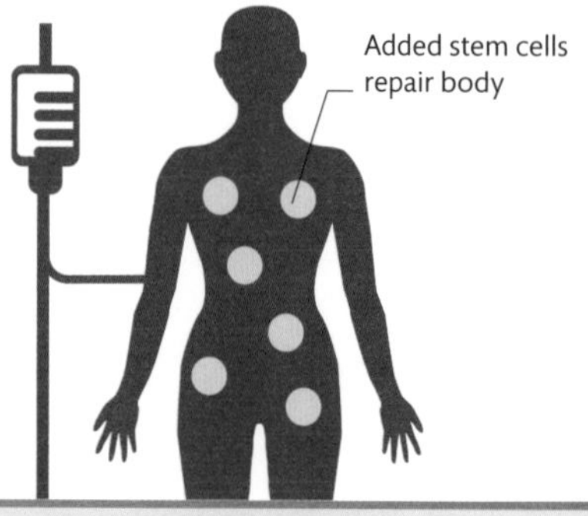

Stem cells in plants

In plants, stem cells are found in meristems—areas of unspecialized cells unique to plants that allow them to grow and change shape constantly. This is possible because plants, unlike animals, can produce an unlimited number of stem cells. Meristems are found in a plant's roots and shoots, but also in the xylem (a tissue essential to transporting water) and the phloem (a tissue that transports nutrients).

A plant's stem cells allow it to survive damage and repair, to grow existing organs and develop new ones, and to propagate new plants from any cuttings that contain meristem tissue.

KEY

- Quiescent center
- Stem cells
- Organizing center
- Rib meristem

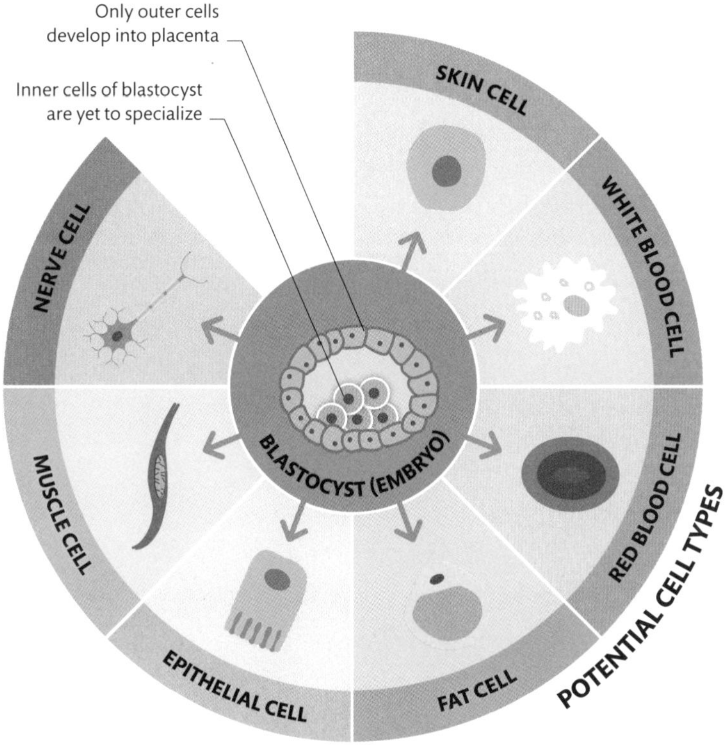
Only outer cells develop into placenta
Inner cells of blastocyst are yet to specialize
SKIN CELL
WHITE BLOOD CELL
RED BLOOD CELL
FAT CELL
EPITHELIAL CELL
MUSCLE CELL
NERVE CELL
BLASTOCYST (EMBRYO)
POTENTIAL CELL TYPES

Early embryo
As the embryo develops, a hollow sphere called a blastocyst forms, and for the first time some of the embryo's cells are specialized. These outer-layer cells help form the placenta in most mammals. The blastocyst's inner cells are pluripotent as they can differentiate into many—but not all—cell types.

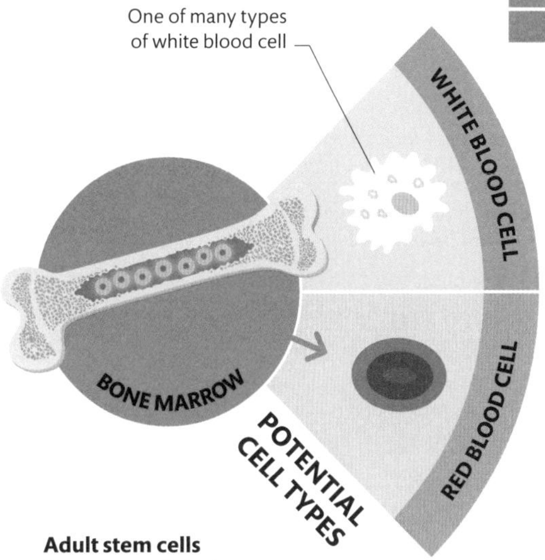
One of many types of white blood cell
WHITE BLOOD CELL
RED BLOOD CELL
BONE MARROW
POTENTIAL CELL TYPES

Adult stem cells
Organisms retain stem cells over the span of their lives, but they become increasingly rare. In adult humans, stem cells are found in bone marrow, skin, eyes, and other parts. These cells are described as multipotent as they can only transform into a limited number of cell types. Bone marrow stem cells can differentiate into red blood cells and white blood cells (among other cells), making them essential in fighting illness or injury.

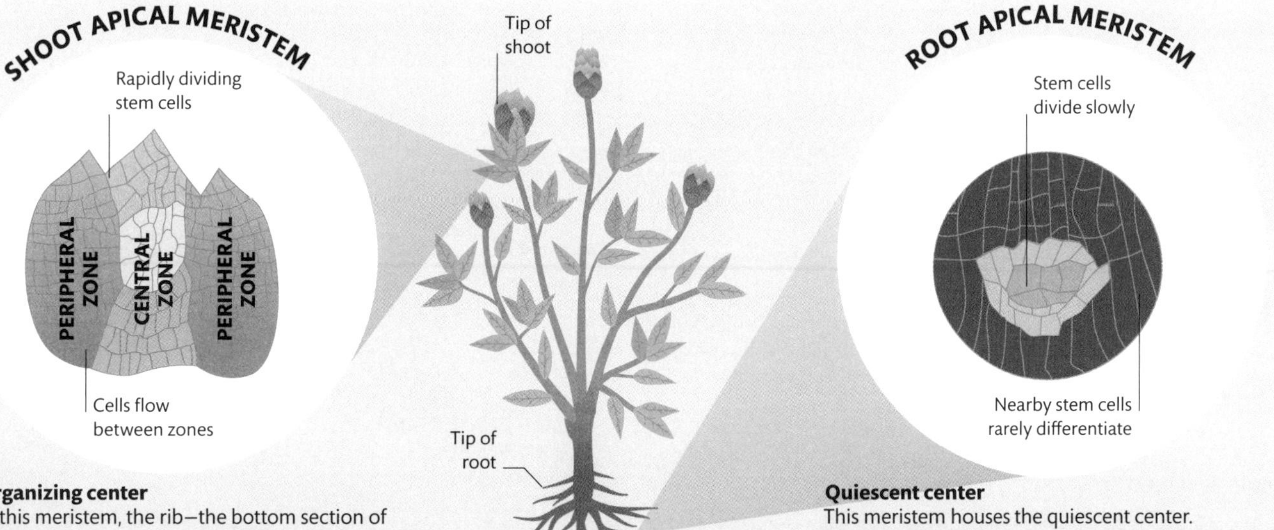
SHOOT APICAL MERISTEM
Rapidly dividing stem cells
PERIPHERAL ZONE
CENTRAL ZONE
PERIPHERAL ZONE
Cells flow between zones
Tip of shoot
Tip of root
FLOWERING PLANT
ROOT APICAL MERISTEM
Stem cells divide slowly
Nearby stem cells rarely differentiate

Organizing center
In this meristem, the rib—the bottom section of the central zone—supports a heavily organized center of cells. Above this, stem cells differentiate to aid processes such as leaf growth.

Quiescent center
This meristem houses the quiescent center. There, unspecialized cells rarely differentiate, and also stop neighboring cells from doing so, protecting the structure of the root.

Reading genes

Each gene in our DNA contains coded instructions to build a single protein or other chemical. The building process involves the code being read by an enzyme and then carried in the form of RNA to the cytoplasm where it is translated into a protein.

WHAT HAPPENS TO mRNA AFTER TRANSLATION?

A strand of mRNA may be translated into a protein many times before it eventually degrades within the cell.

Transcription

Before a gene is translated into a protein, enzymes must first transcribe (copy) a DNA strand into a messenger RNA (mRNA, see p.36) that carries the code to make a protein. This starts when RNA polymerase attaches to the double helix and uses one of the DNA strands as a template for the mRNA. Then the mRNA moves from the nucleus into the cytoplasm.

50 BASES ARE ADDED EVERY SECOND WHEN DNA IS COPIED IN A HUMAN CELL

1 Initiation
The enzyme RNA polymerase binds to a sequence in the gene and breaks the hydrogen bonds between complementary base pairs, unzipping the DNA.

2 Elongation
The RNA polymerase then moves along the gene, making the mRNA as it goes, using a strand of DNA within a gene as a template for the mRNA.

3 Termination
When the mRNA reaches the end of the gene, the polymerase detaches. The mRNA moves into the cytoplasm, where the information encoded in it is used by a ribosome to make a protein.

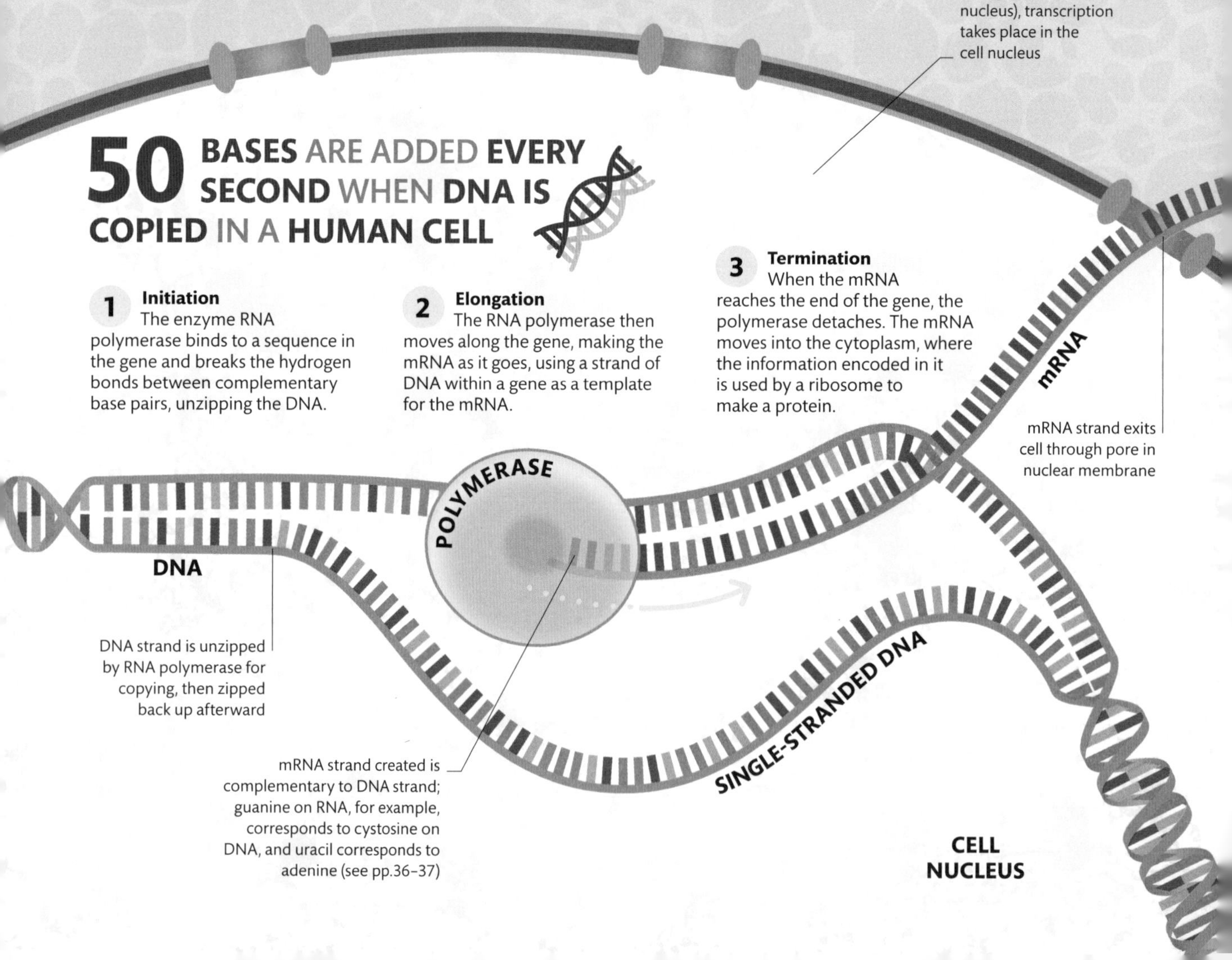
In eukaryotes (cells or organisms with a nucleus), transcription takes place in the cell nucleus
mRNA
mRNA strand exits cell through pore in nuclear membrane
POLYMERASE
DNA
DNA strand is unzipped by RNA polymerase for copying, then zipped back up afterward
SINGLE-STRANDED DNA
mRNA strand created is complementary to DNA strand; guanine on RNA, for example, corresponds to cytosine on DNA, and uracil corresponds to adenine (see pp.36–37)
CELL NUCLEUS

Translation

The mRNA fixes to a ribosome (a protein-building unit). Information stored in the mRNA acts as a code for the protein to be made. Each set of three nucleic acid bases (a codon) on the mRNA matches three bases (an anticodon) on a transfer RNA (tRNA) molecule. As each tRNA brings an amino acid, the base sequence translates into a chain. The genetic code of 64 codons includes three start codons, which begin the process, and one stop codon, which ends it.

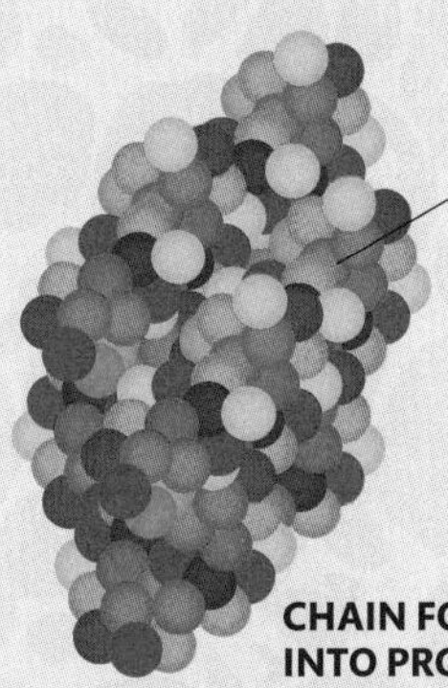
Protein, made up of 20 types of amino acid, folds into different shapes
CHAIN FOLDED INTO PROTEIN

4 Protein is folded
When the stop codon is reached, the peptide chain is released from the tRNA. Organelles within the cell then fold it up and make other modifications to form a protein molecule (see pp.38–39).

1 Initiation
The mRNA travels to a ribosome, attaches to it, and attracts tRNA molecules that correspond to the start codon.

2 Elongation
Another tRNA molecule brings the amino acid that corresponds to the codons. The two amino acids bind together and the first tRNA exits the ribosome.

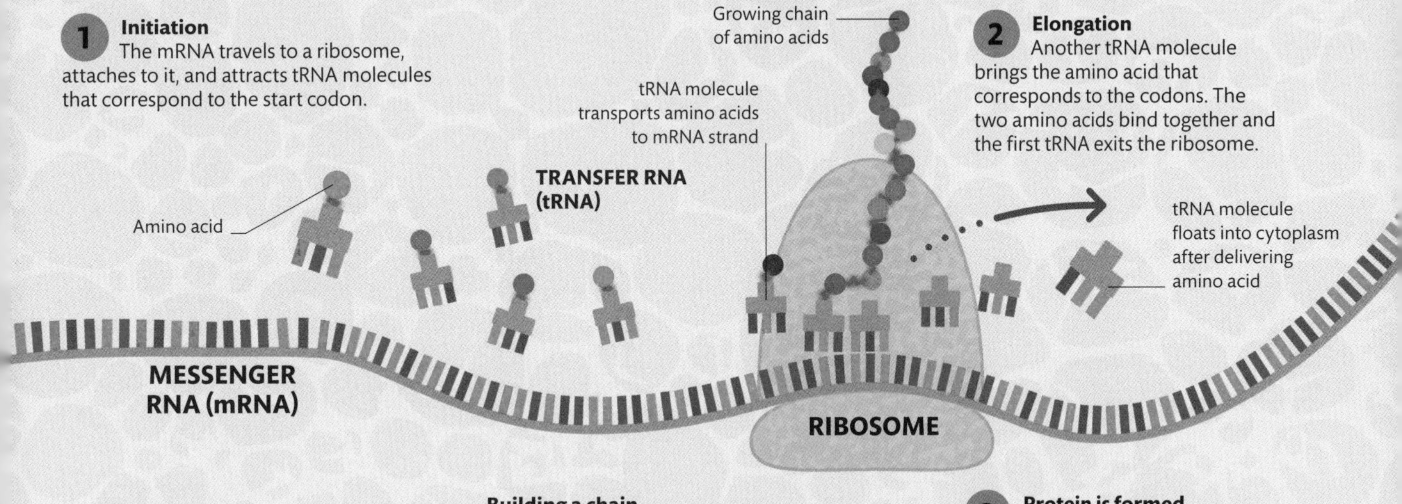
Growing chain of amino acids
tRNA molecule transports amino acids to mRNA strand
TRANSFER RNA (tRNA)
Amino acid
tRNA molecule floats into cytoplasm after delivering amino acid
MESSENGER RNA (mRNA)
RIBOSOME

CYTOPLASM

Building a chain
As the ribosome moves along the mRNA strand, the tRNA molecules attach to the mRNA in a specific order determined by the matching up of codons and anticodons on the tRNA molecule.

3 Protein is formed
As tRNAs enter and exit the ribosome bringing amino acids, a chain of amino acids forms and lengthens. The process continues until the stop codon is reached.

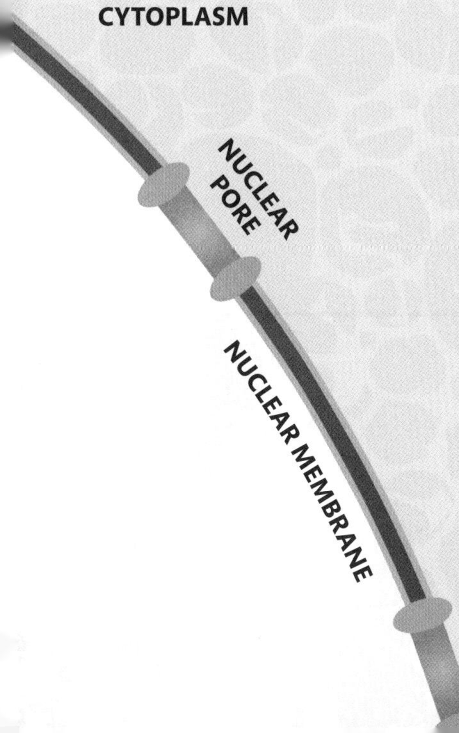
NUCLEAR PORE
NUCLEAR MEMBRANE

GENE SIZES

An average human gene consists of 3,000 DNA base pairs but there is huge variation—from a few hundred to more than 2 million base pairs. Longer genes are linked to brain, heart, and muscle functions, while shorter genes relate to some functions of the immune system or the skin.

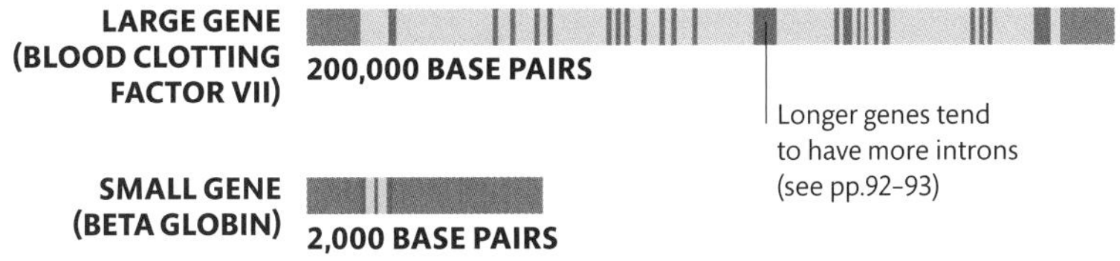
LARGE GENE (BLOOD CLOTTING FACTOR VII)
200,000 BASE PAIRS
Longer genes tend to have more introns (see pp.92–93)
SMALL GENE (BETA GLOBIN)
2,000 BASE PAIRS

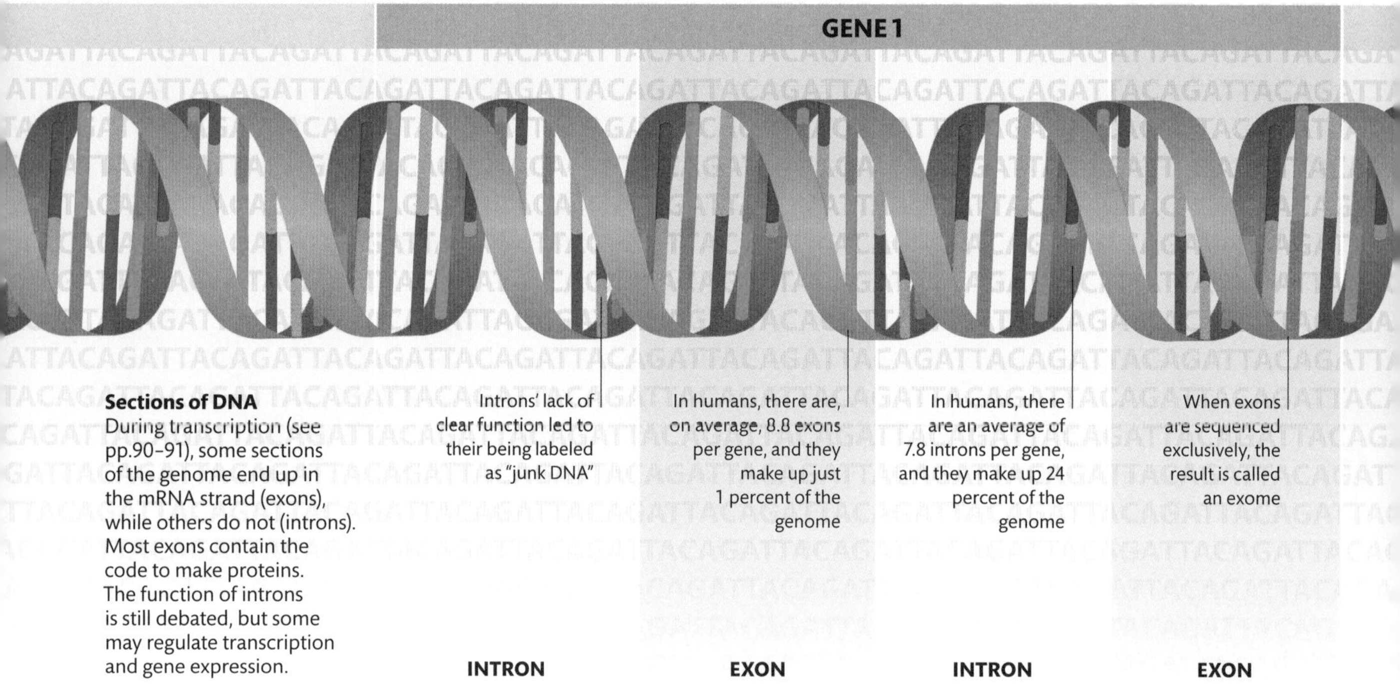
GENE 1
Sections of DNA
During transcription (see pp.90–91), some sections of the genome end up in the mRNA strand (exons), while others do not (introns). Most exons contain the code to make proteins. The function of introns is still debated, but some may regulate transcription and gene expression.
Introns' lack of clear function led to their being labeled as "junk DNA"
In humans, there are, on average, 8.8 exons per gene, and they make up just 1 percent of the genome
In humans, there are an average of 7.8 introns per gene, and they make up 24 percent of the genome
When exons are sequenced exclusively, the result is called an exome
INTRON
EXON
INTRON
EXON

The organization of genes

Genes are sections of DNA that code for specific proteins (see pp.90–91). In bacteria, DNA moves freely within the cytoplasm of a cell. However, in more complex organisms, such as humans, animals, or plants, very long strands of DNA are tightly packed into chromosomes (see pp.58–59) housed in the cell nucleus. Each gene has a particular position on a chromosome, on a coding portion of the genome's DNA. Coding genes are separated by various types of noncoding DNA known as intergenic DNA, introns, and a small number of exons (see above).

70 PERCENT OF HUMAN DNA IS "JUNK DNA"

Genomes

Every organism contains genetic information in its DNA molecules. The full set of an organism's genetic information is called its genome. A genome contains all of the instructions required for an organism to develop and function. Analyzing genomes can allow us to pinpoint certain genes and understand how they work.

WHAT IS THE WORLD'S LARGEST GENOME?

The genome of a Japanese flower, *Paris japonica*, has 149 billion base pairs—about 50 times more than the human genome.

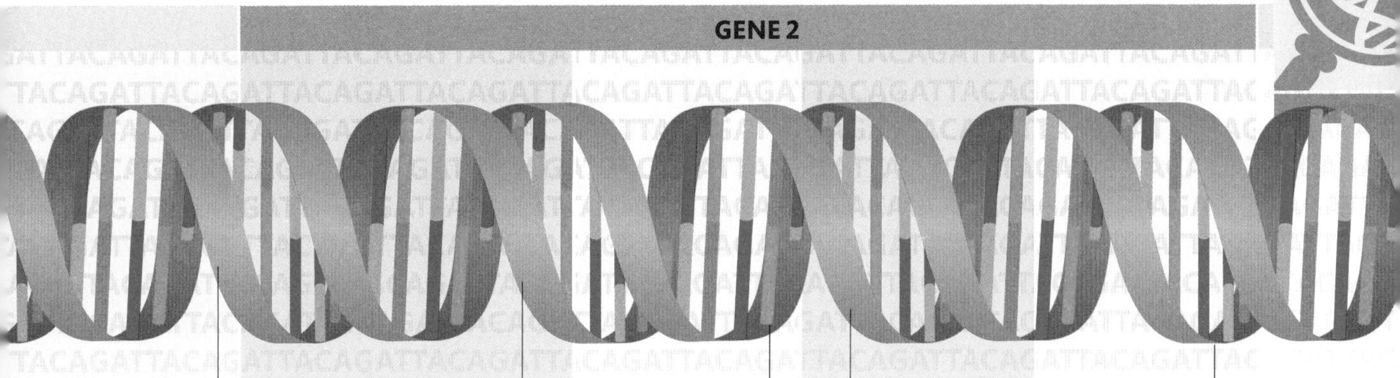
GENE 2
In humans, 75 percent of genome is intergenic DNA—sections between genes that do not code for proteins
Minority of exons do not code for proteins, instead containing regulatory elements that aid in other genetic processes
Introns may be result of evolution "shuffling" sections of genetic code, creating spacers between coding stretches of DNA
Differences in the sequence of bases within an exon give rise to different variants of a gene, adding to variety within species
An intron can either be a section of DNA or its corresponding section in the RNA transcript
INTERGENIC DNA
EXON
INTRON
EXON
INTRON

Codons and anticodons

A codon is a sequence of three bases found on an mRNA strand (see p.36) that pairs up with a complementary set of three bases (an anticodon) on a tRNA strand to code for a specific amino acid. Most of the 64 codons found in human DNA code for one of 20 types of amino acid (each represented by a single-letter abbreviation). This includes the start codon, which begins the process. Only stop codons do not code for amino acids, instead signaling a halt to protein-making.

ATG = M METHIONINE

CCA = P PROLINE

TTT = F PHENYLALANINE

CCC = P PROLINE

TTG = W TRYTOPHAN

TAG = X STOP CODON

Coding for amino acids
Adenine (A), cytosine (C), guanine (G), and thymine (T) are the bases (see p.37) found in codons. These chemicals, combined in different ways, produce specific amino acids. A sample group of codon combinations and the resulting amino acids is shown above.

READING OUR GENOME

Starting in 1990, the Human Genome Project set out to read the entire sequence of the human genome and to map the location and function of all its genes. The full sequence of about 3 billion base pairs was completed in 2003, although only 92 percent of genes had been identified at that time. By 2022, all genes had been defined.

Inheritance

In organisms that reproduce sexually, the traits passed on from parent to offspring are based on the combination of genes produced during fertilization.

Dominant and recessive alleles
One version of an allele is often dominant over the other. Here, each parent has one dominant (D) and one recessive (d) allele. The recessive trait will be visible only when two recessive alleles are inherited.

Alleles

An allele is a variant of a particular gene. Alleles govern a trait in the offspring. They usually come in pairs, with one passed on from each parent. The combination of alleles in an organism is called its genotype, and the organism's observable traits make up its phenotype. The genotype and phenotype of offspring depend on the genotype of the parents.

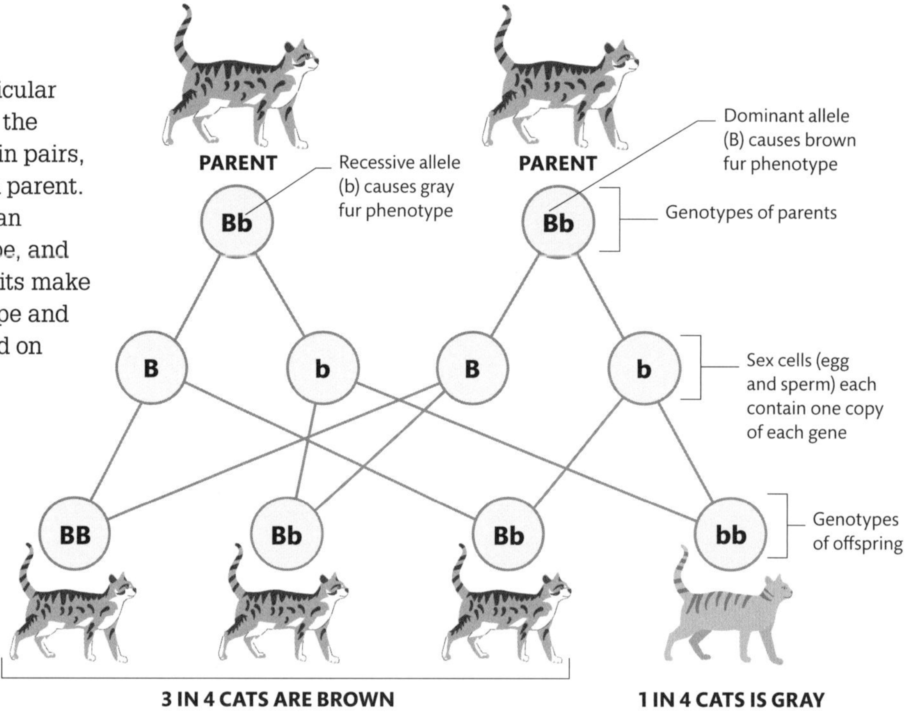
PARENT
PARENT
Recessive allele (b) causes gray fur phenotype
Dominant allele (B) causes brown fur phenotype
Bb
Bb
Genotypes of parents
B
b
B
b
Sex cells (egg and sperm) each contain one copy of each gene
BB
Bb
Bb
bb
Genotypes of offspring
3 IN 4 CATS ARE BROWN
1 IN 4 CATS IS GRAY

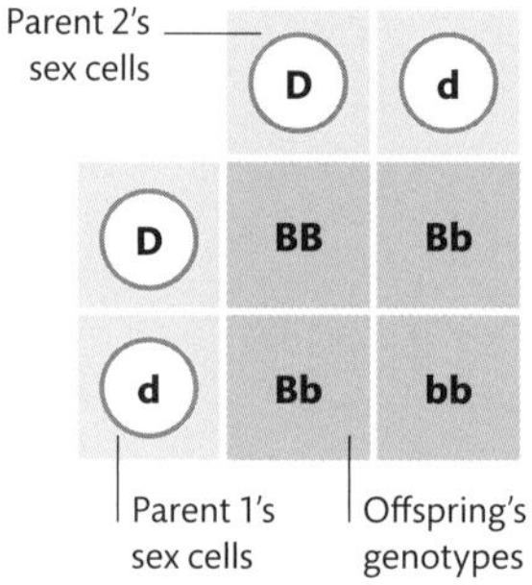
Parent 2's sex cells
D
d
D
BB
Bb
d
Bb
bb
Parent 1's sex cells
Offspring's genotypes

Punnett square
A Punnett square shows the variety of genotypes that each offspring can inherit. In effect, this also gives the odds for each outcome.

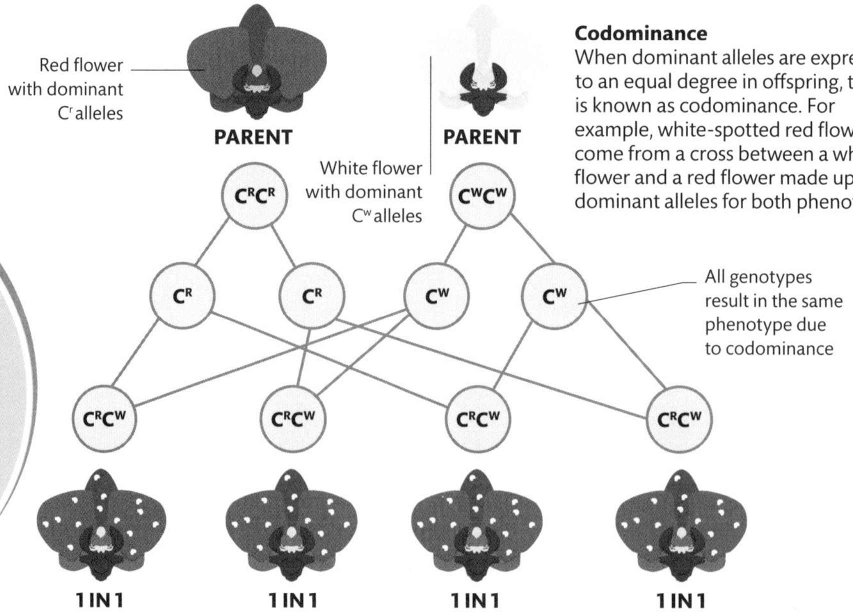
Red flower with dominant C^r alleles
PARENT
PARENT
White flower with dominant C^w alleles
C^RC^R
C^WC^W
C^R
C^R
C^W
C^W
All genotypes result in the same phenotype due to codominance
C^RC^W
C^RC^W
C^RC^W
C^RC^W
1 IN 1
1 IN 1
1 IN 1
1 IN 1

Codominance
When dominant alleles are expressed to an equal degree in offspring, this is known as codominance. For example, white-spotted red flowers come from a cross between a white flower and a red flower made up of dominant alleles for both phenotypes.

HOW DO WE DETERMINE THE RISK FOR DISEASE?

Genome-wide association studies testing hundreds of thousands of genetic variants are used to determine those statistically associated with diseases.

Sex-linked inheritance

The prevalence of some disorders is linked to biological sex. For example, disorders influenced by genes carried on the X chromosome (see pp.98–99) are often less prevalent in women. This is because women have two X chromosomes, and the second, with its dominant allele, overrides the first when it is faulty. However, as men only have one X chromosome, a son carrying the faulty gene will exhibit the disorder.

Inheriting color vision
Color blindness is a recessive trait on the X chromosome. It can be carried (but not exhibited) when the dominant allele for full color vision is present. When this allele is not present, but the recessive allele is, the child is affected.

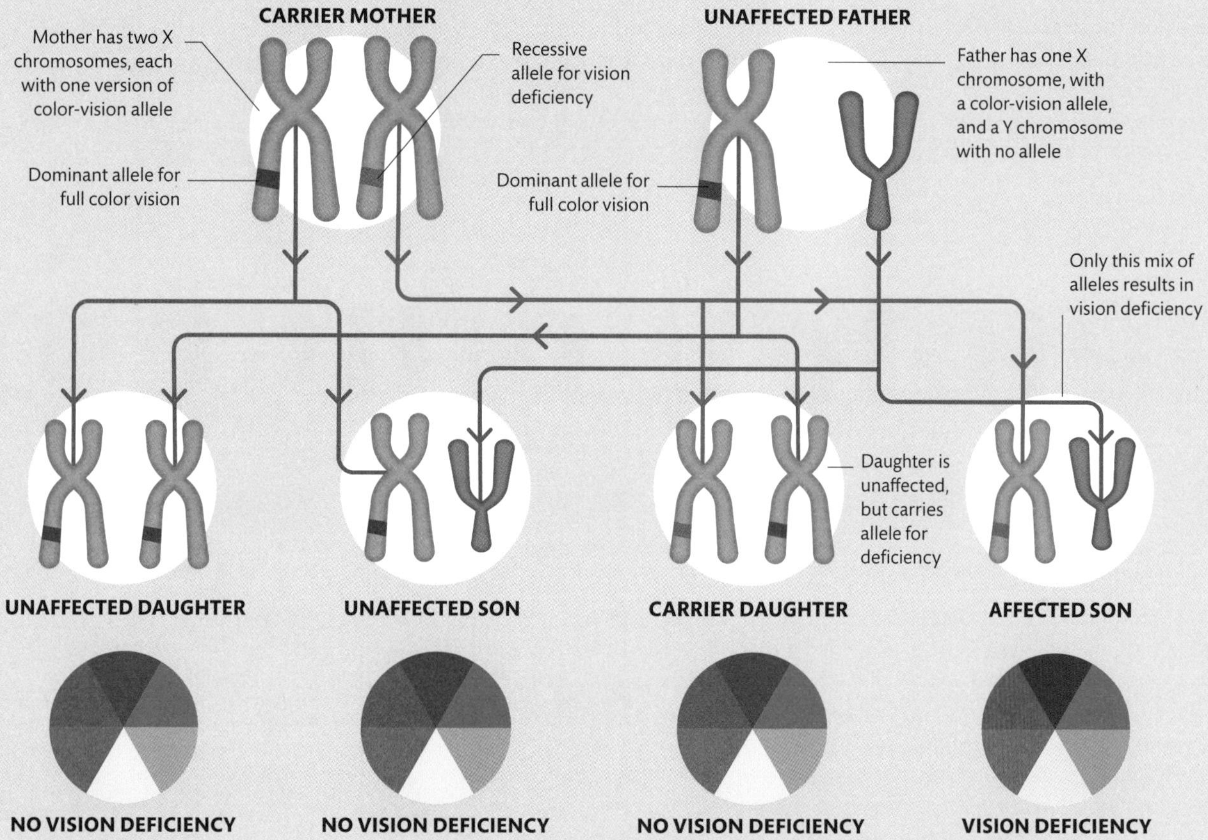
CARRIER MOTHER
UNAFFECTED FATHER
Mother has two X chromosomes, each with one version of color-vision allele
Recessive allele for vision deficiency
Father has one X chromosome, with a color-vision allele, and a Y chromosome with no allele
Dominant allele for full color vision
Dominant allele for full color vision
Only this mix of alleles results in vision deficiency
Daughter is unaffected, but carries allele for deficiency
UNAFFECTED DAUGHTER
UNAFFECTED SON
CARRIER DAUGHTER
AFFECTED SON
NO VISION DEFICIENCY
NO VISION DEFICIENCY
NO VISION DEFICIENCY
VISION DEFICIENCY

COLOR BLINDNESS AFFECTS ABOUT 1 IN 12 MEN BUT ONLY ABOUT 1 IN 200 WOMEN

HYBRID SPECIES

When two species interbreed, the result is a hybrid species. For example, a liger is the offspring of a male lion and a female tiger. Hybrid species are rare due to reproductive barriers. Hybrids' genetic incompatibilities also increase their risk of infertility, injury, and neurological disorders.

LIGER

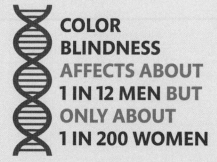

Mutations

Mutations are permanent changes in genes. External causes of mutations, such as radiation, chemicals, and infectious or biological agents, are called mutagens. Mutations also occur during the replication of DNA in cell division and can lead to genetic conditions.

SARS-COV-2, THE VIRUS THAT **CAUSES COVID-19,** WAS FOUND **TO MUTATE** APPROXIMATELY **ONCE A WEEK**

Types of mutation

Mutations fall into two categories. In a frame-shift mutation, a base pair is inserted or deleted, causing the reading frame (see below) to shift, impacting the whole sequence of base pairs. In a point mutation, a single base pair is substituted (and most of the time the mutation is benign).

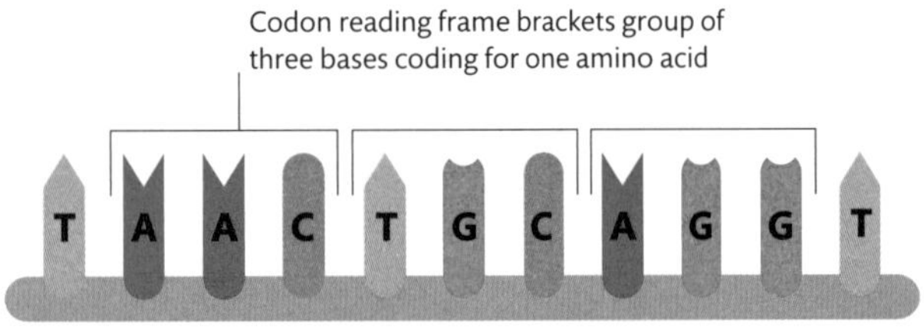
Codon reading frame brackets group of three bases coding for one amino acid
T A A C T G C A G G T

ORIGINAL DNA SEQUENCE

Unaffected DNA
An individual's DNA sequence acts like instructions for coding proteins in a specific order. Because we have two copies of most genes, this reading order is not always affected by a mutation in one gene.

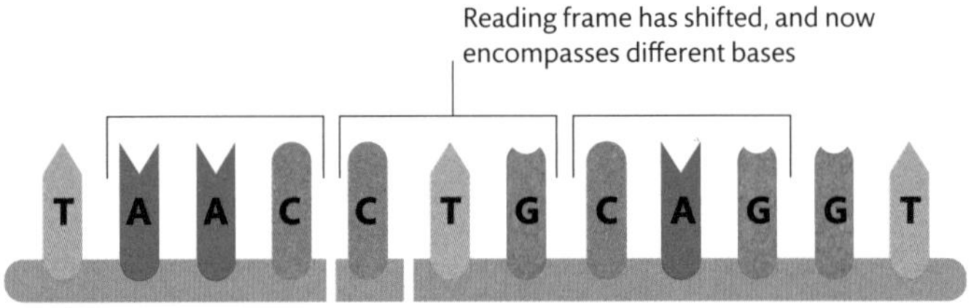
Reading frame has shifted, and now encompasses different bases
T A A C C T G C A G G T

Insertion
Insertion is when one or more bases are added to the DNA sequence. When this occurs during DNA replication or meiosis (see pp.82–83), the effects can be dangerous. Insertion (and deletion) change the way the bases are read and the amino acid sequence.

Causes of mutation

All forms of genetic variation are caused by mutation, meaning they are essential to the evolution of a species. Internal mutations may be caused by mistakes during copying. The body checks for copying errors and normally mends them. However, some go undetected, and may be passed on to offspring. External causes of mutation (mutagens) vary considerably. Cancer is caused by mutations, and some mutagens are also carcinogens (substances that increase cancer risk).

External causes
When environmental agents enter a cell nucleus and interact with the DNA, the DNA can become mutated and damaged. This can happen when humans, other animals, or plants are exposed to a variety of harmful substances.

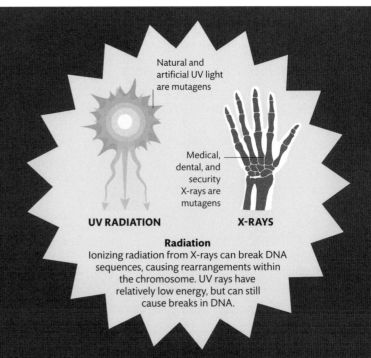

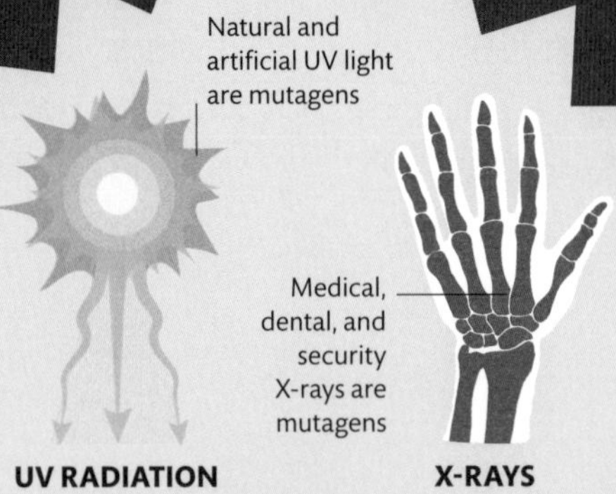
Natural and artificial UV light are mutagens
Medical, dental, and security X-rays are mutagens
UV RADIATION
X-RAYS

Radiation
Ionizing radiation from X-rays can break DNA sequences, causing rearrangements within the chromosome. UV rays have relatively low energy, but can still cause breaks in DNA.

CAN A VIRUS MUTATE INFINITELY?

In effect, yes. A virus can exist for as long as its host does. Over time, a virus mutates and varies—so the best way to protect against a virus is to limit its spread.

MUTATION RATES

Many factors determine the frequency of new mutations in an organism over time. Species with a high metabolic rate are thought to be at greater risk of exposure to mutagens through mitochondrial respiration (see pp.60–61). Additionally, the more generations per unit of time, the greater the odds of DNA replication errors, increasing the mutation rate.

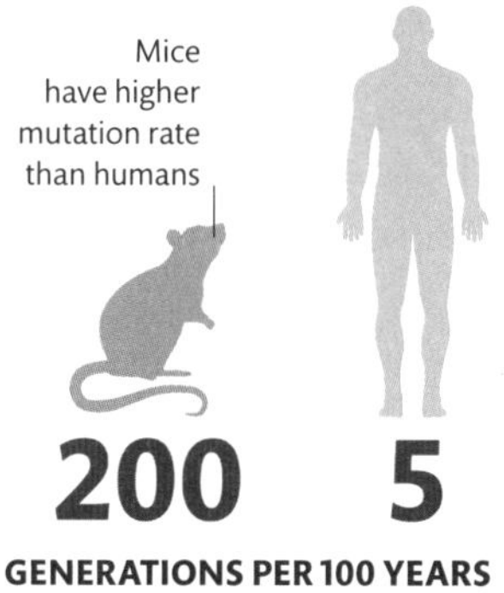
Mice
have higher
mutation rate
than humans
200
5
GENERATIONS PER 100 YEARS

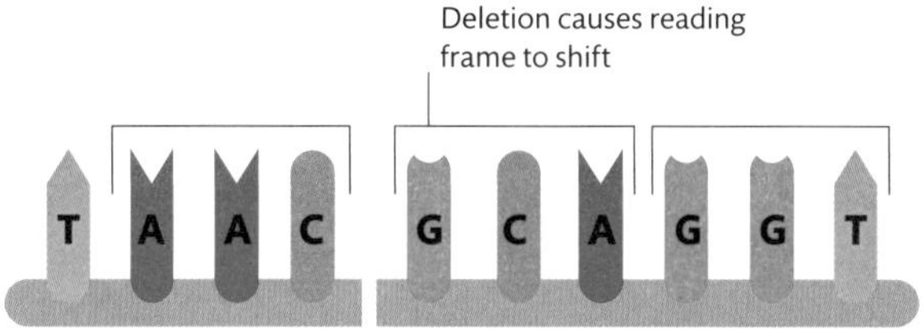
Deletion causes reading
frame to shift
T
A
A
C
G
C
A
G
G
T

Deletion
Sometimes, when genetic material breaks off, the loss of DNA sequences causes a frame-shift. The larger the deletion, the more likely the resulting defect will be. For example, cystic fibrosis is a lung disease caused by a deletional mutation that changes the CFTR gene.

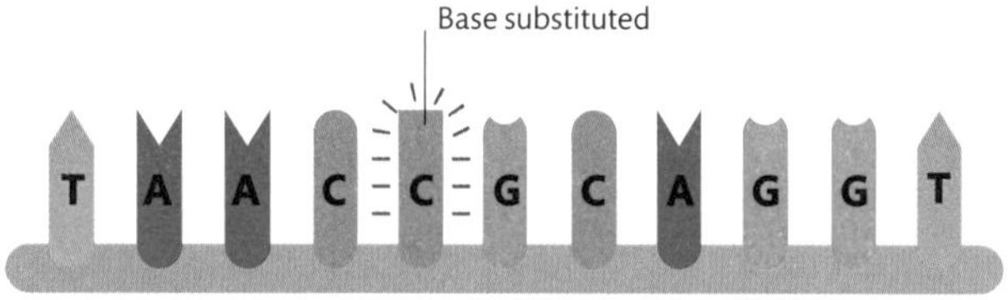
Base substituted
T
A
A
C
C
G
C
A
G
G
T

Substitution
Substitution mutations occur when a base is replaced by another. This type of mutation can have no effect or a drastic effect, depending on which base pair is substituted and with what. For example, sickle cell anemia is caused by a substitution mutation.

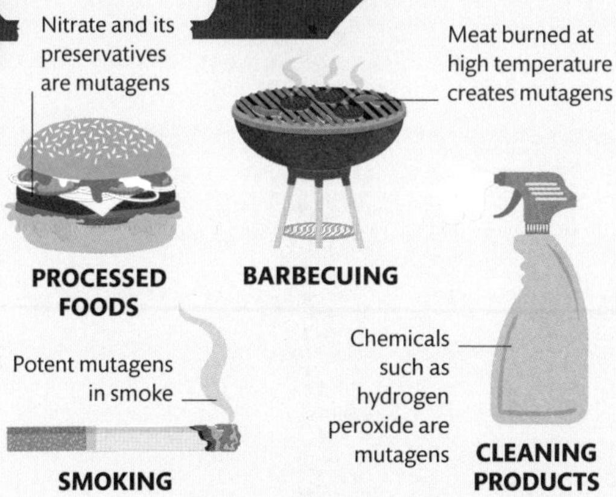
Nitrate and its
preservatives
are mutagens
Meat burned at
high temperature
creates mutagens
PROCESSED
FOODS
BARBECUING
Potent mutagens
in smoke
Chemicals
such as
hydrogen
peroxide are
mutagens
SMOKING
CLEANING
PRODUCTS

Chemicals
Chemical mutagens change the chemistry of DNA. A mutation that takes place in somatic cells (cells other than egg and sperm cells) can lead to cancer. This occurs when the chemical changes the DNA blueprint, causing mutated cells to divide uncontrollably.

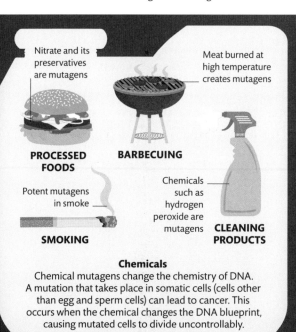

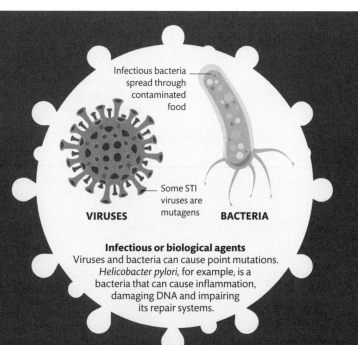

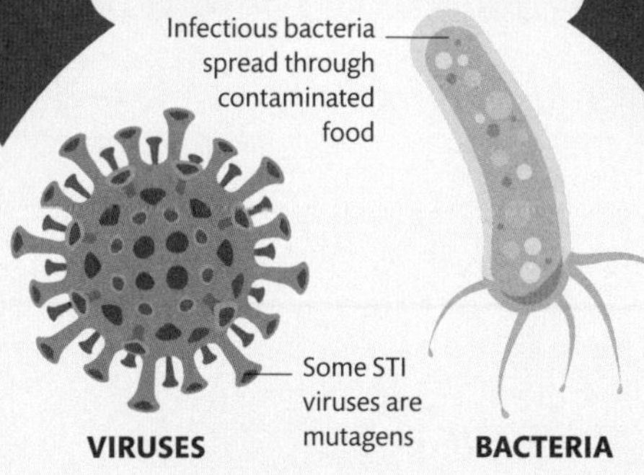
Infectious bacteria
spread through
contaminated
food
Some STI
viruses are
mutagens
VIRUSES
BACTERIA

Infectious or biological agents
Viruses and bacteria can cause point mutations. *Helicobacter pylori*, for example, is a bacteria that can cause inflammation, damaging DNA and impairing its repair systems.

Sex chromosomes

The sex of offspring is determined by sex chromosomes. In humans—and most mammals and plants—females typically have two X chromosomes, while most men have one X and one Y chromosome.

Determining sex

In humans, one of the 23 pairs of chromosomes is called the sex chromosomes because they determine the sex of any offspring. Human offspring inherit one sex chromosome from the mother and one from the father. All female sex cells (eggs) carry an X chromosome but only half of male sex cells (sperm) do, while the other half carry a Y chromosome. However, in many insects, half of male sex cells have an X chromosome, resulting in female offspring, and the other half have no sex chromosome, resulting in male offspring.

Human sex chromosomes
For human offspring, there are only four unique combinations of X and Y chromosomes that can be passed down by the parents at fertilization (see pp.84–85).

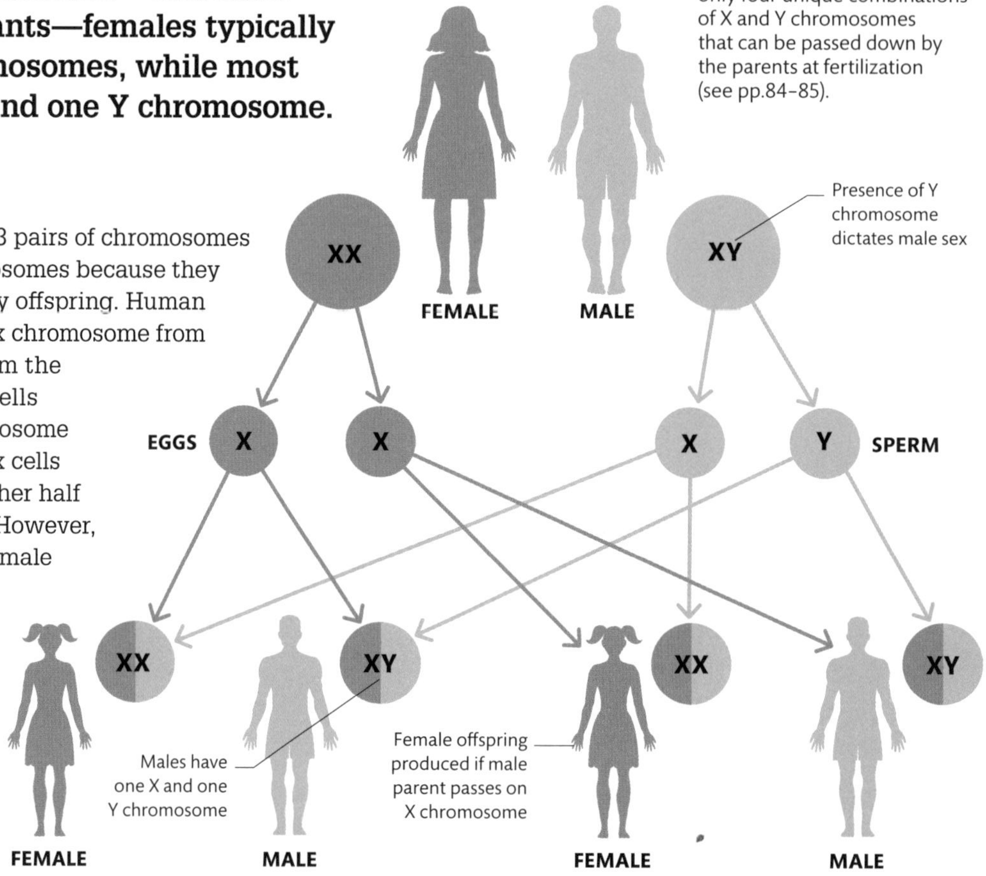

XX
FEMALE
MALE
XY
Presence of Y chromosome dictates male sex
EGGS
X
X
X
Y
SPERM
XX
XY
XX
XY
Males have one X and one Y chromosome
Female offspring produced if male parent passes on X chromosome
FEMALE
MALE
FEMALE
MALE

Bird and insect sex chromosomes
In birds, females carry the mismatched chromosome pair (ZW) and males carry the identical pair (ZZ). In many insects, females carry two X chromosomes (XX) while males only carry one (XO) - the O representing the lack of a second X chromosome.

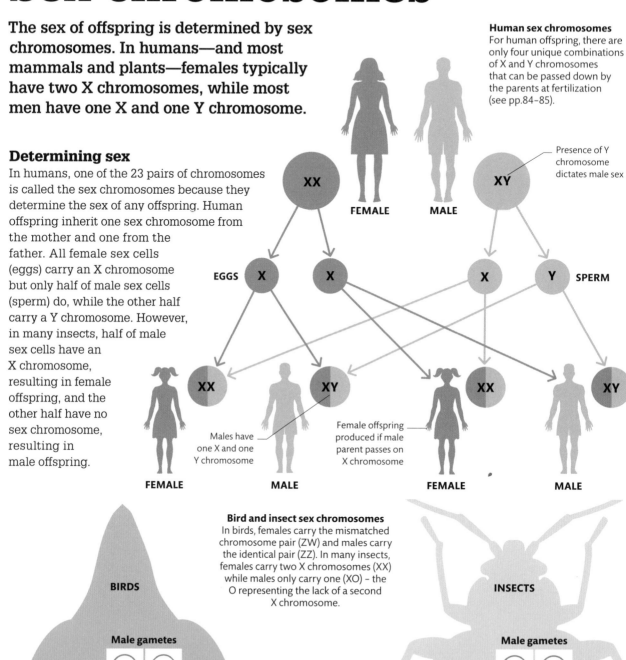

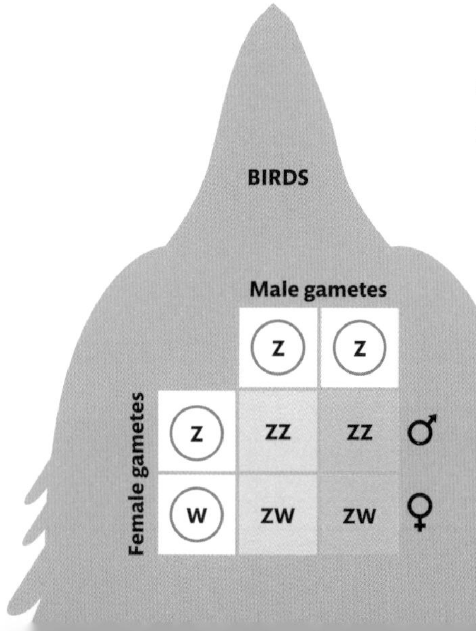

BIRDS
Male gametes
Z
Z
Female gametes
Z
ZZ
ZZ
♂
W
ZW
ZW
♀

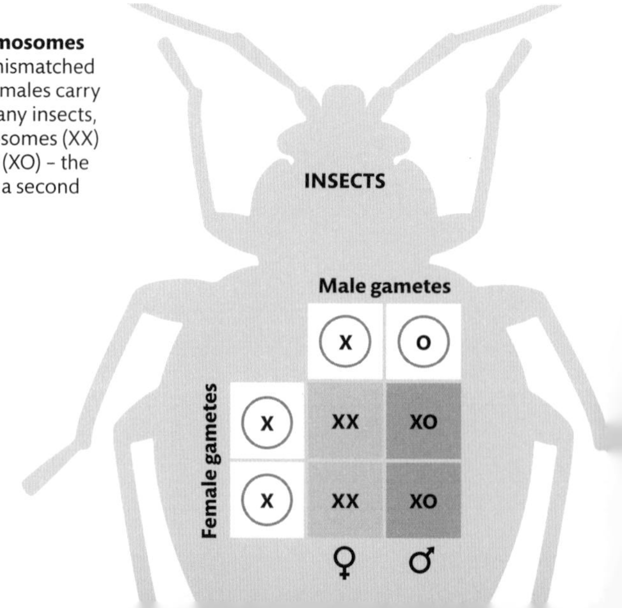

INSECTS
Male gametes
X
O
Female gametes
X
XX
XO
X
XX
XO
♀
♂

The SRY gene

The Y chromosome contains the SRY gene—a sex-determining gene that enables typical sex development for males. This gene, thought to have developed on the Y chromosome 200–150 million years ago (MYA), holds instructions for making a protein that binds to specific regions of DNA and controls the activity of other genes. This triggers the development of male testes and inhibits that of female sex traits such as the uterus.

THE DUCK-BILLED PLATYPUS HAS 10 SEX CHROMOSOMES—THE MOST OF ANY KNOWN ANIMAL

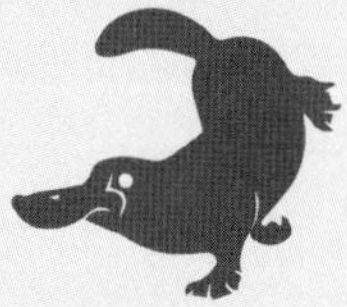

The story of the Y chromosome
The Y chromosome formed when the SRY gene arose as the result of a mutation. Over time, the Y chromosome has become smaller and now contains many fewer genes than it did originally.

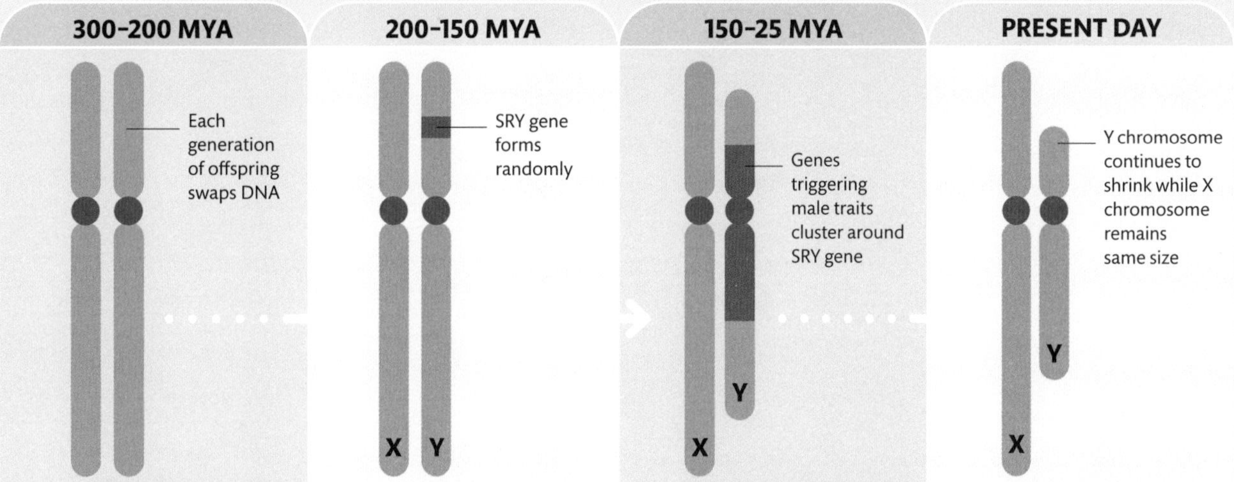
300-200 MYA
Each generation of offspring swaps DNA
200-150 MYA
SRY gene forms randomly
X
Y
150-25 MYA
Genes triggering male traits cluster around SRY gene
X
Y
PRESENT DAY
Y chromosome continues to shrink while X chromosome remains same size
X
Y

1 DNA swaps
Before human sex chromosomes exist, two matched chromosomes exchange bits of DNA.

2 SRY gene forms
A random mutation results in the formation of the SRY gene, and with it, the male sex chromosome.

3 Other genes amass
As the Y chromosome evolves, it no longer matches with the X chromosome and begins to lose parts.

4 Mass of Y reduces
As a result, the Y chromosome has become much smaller and continues to shrink today.

WILL THE Y CHROMOSOME DIE OUT?

Scientists are divided on whether it will continue to shrink at the present rate. If it does, the Y chromosome could die out within 5 million years.

NON-GENETIC SEX DETERMINATION

In most species, sex is determined during fertilization. However, some organisms' sex can be caused by temperature, humidity, or social interactions. In turtles, alligators, and crocodiles, the temperature of the developing eggs triggers genes that determine the offspring's sex. For example, crocodile eggs produce females at around 86°F (30°C) and males at around 93°F (34°C).

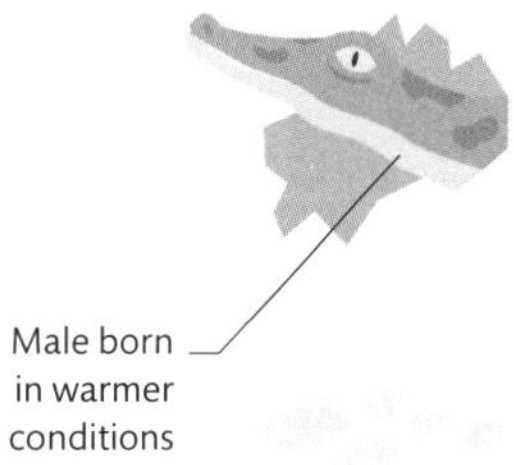
Male born in warmer conditions

CROCODILE BIRTH

Hermaphrodites

A hermaphrodite is an organism that is able to produce both female and male sex cells during its life cycle. Hermaphrodites can be plants or animals. Some hermaphrodites reproduce sexually, others reproduce asexually, and many can do both.

Hermaphroditism in plants

Most plants are hermaphrodites. In flowering plants, this means that they bear flowers containing both male sex organs (stamens, including anthers and filaments) and female sex organs (including ovaries). When these parts both feature in the same flower, the plant is referred to as bisexual. Other plants that have flowers that are either male or female are defined as unisexual. There are two types of unisexual plant: monoecious plants, which are hermaphrodites, feature both male and female flowers (albeit separately); and dioecious plants, which are not hermaphrodites, only grow male or female flowers.

ANTHER

STYLE

OVULE

Pollen from anther falls down onto stigma

If pollen is accepted, it tunnels down style to ovules in ovary

Fertilization occurs within each individual ovule

OVULE

Pollen is fed in through small opening (micropyle)

Egg cells populate embryo sac

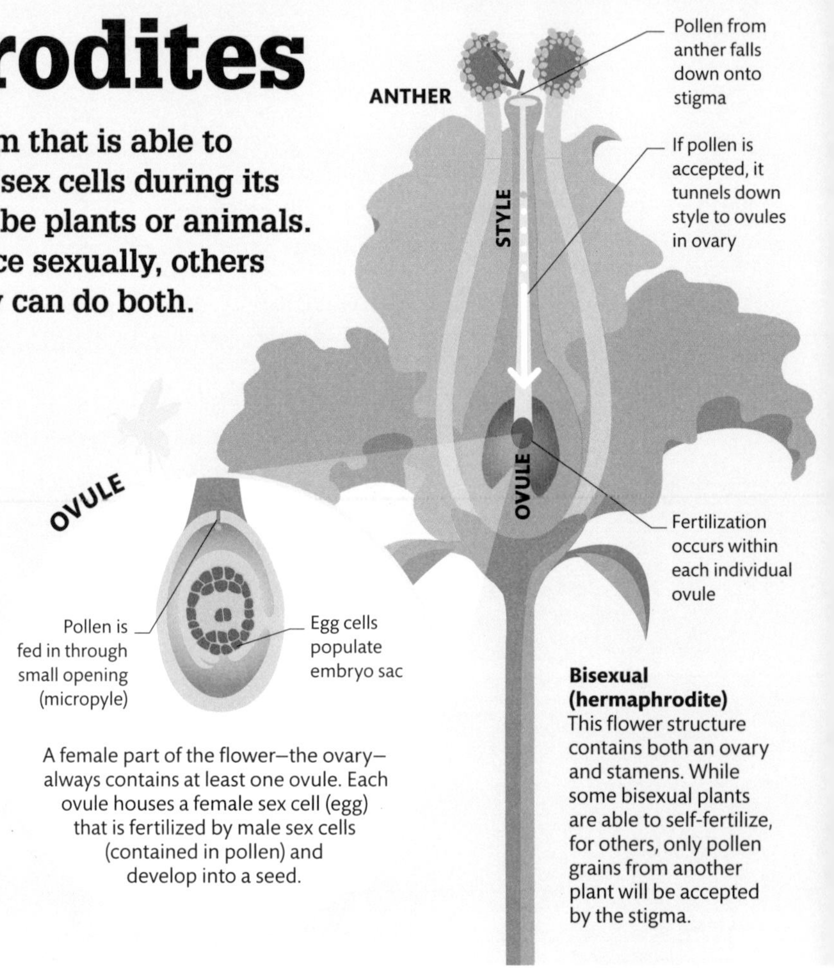

A female part of the flower—the ovary—always contains at least one ovule. Each ovule houses a female sex cell (egg) that is fertilized by male sex cells (contained in pollen) and develop into a seed.

Bisexual (hermaphrodite)
This flower structure contains both an ovary and stamens. While some bisexual plants are able to self-fertilize, for others, only pollen grains from another plant will be accepted by the stigma.

Hermaphroditism in animals

All animal groups (except birds and mammals) contain hermaphrodites. Some types of invertebrates, such as worms, mollusks, and jellyfish, are especially likely to be simultaneous hermaphrodites, meaning they can develop female and male sexual organs at the same time. Animals like this tend to be sedentary, slow moving, or widely dispersed, and so interact less frequently. As a result, they need to maximize their breeding potential by being able to provide both sperm and eggs. Hermaphroditic vertebrates, such as certain fish, amphibians, and reptiles, tend to be sequential hermaphrodites (see panel).

Sexual anatomy of a snail
Most snails that live on land are simultaneous hermaphrodites. During mating, a genital protrusion sticks out from a small pore beside the snail's head. This protrusion contains both a penis, which delivers a sac of sperm to the mate, and a vagina, which receives one.

VAGINA

PENIS

Penis delivers sperm to other snail

Vagina receives sperm from other snail

Genital pore opens during mating

SNAIL

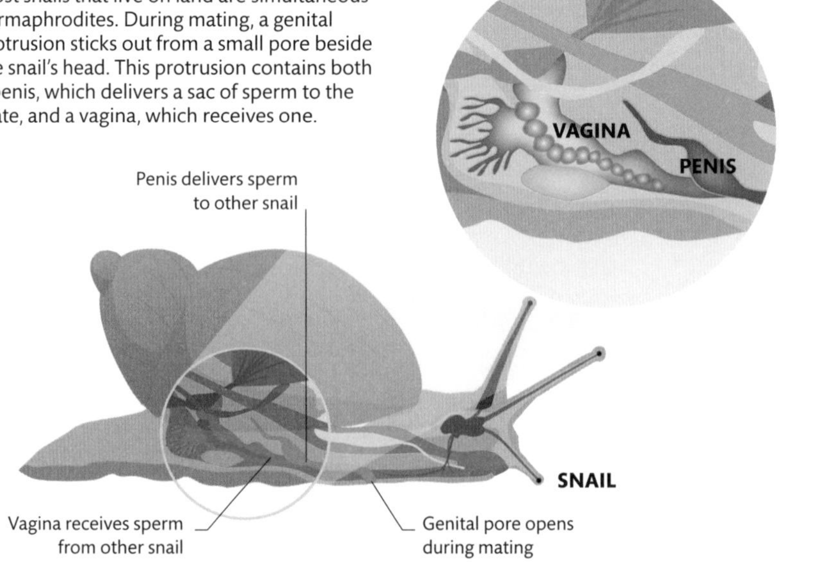

Male flower produces pollen that can fertilize female flowers

Female flower can collect pollen sent by male flower on same plant or another plant

POLLEN

Pollen is transferred from male to female plants

Male plant only produces pollen

ANTHER

OVULE

Female flower only produces ovules

ARE MOST SPECIES HERMAPHRODITIC?

About 95 percent of plants are hermaphrodites. Although only 5 percent of animals are hermaphrodites, this figure jumps to 30 percent when insects are excluded.

MONOECIOUS

DIOECIOUS (FEMALE)

DIOECIOUS (MALE)

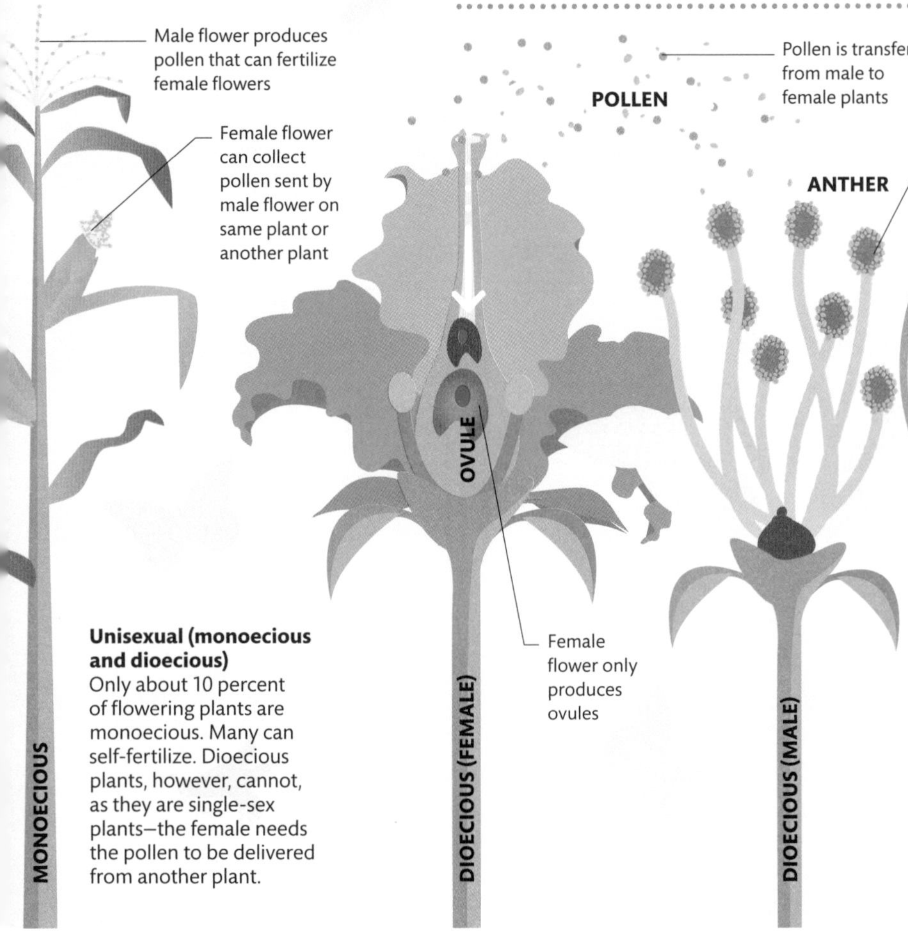

Unisexual (monoecious and dioecious)
Only about 10 percent of flowering plants are monoecious. Many can self-fertilize. Dioecious plants, however, cannot, as they are single-sex plants—the female needs the pollen to be delivered from another plant.

THE **CHALK BASS FISH** CAN CHANGE **ITS SEX** UP TO **20 TIMES** A DAY

EARTHWORM

Earthworms
The eggs of an earthworm are stored in the thicker middle section, while its male sex organs are found at one tip of its body.

SEQUENTIAL HERMAPHRODITES

Organisms that change sex are sequential hermaphrodites. Protogynous species start female and become male. Most protogynous fish have a breeding system where a large male controls a harem of females. Protandrous species start male and become female. In protandrous fish (such as clownfish) the larger female form can produce large numbers of eggs.

SEA SPONGE

Sea sponges
These seabed animals reproduce by releasing eggs and sperm into the water at different times to minimize self-fertilization.

Dominant female
If the dominant female clownfish dies, the largest male becomes female.

Nonrdominant males
Many males (and one dominant female) develop.

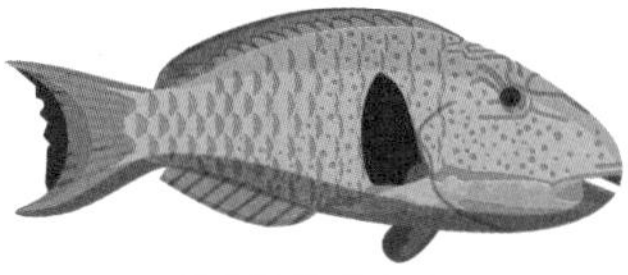

PARROTFISH

Parrotfish
The offspring of parrotfish hatch female, and most are female for life. However, the largest will eventually become males.

Undifferentiated
At birth, clownfish have both sets of sex organs.

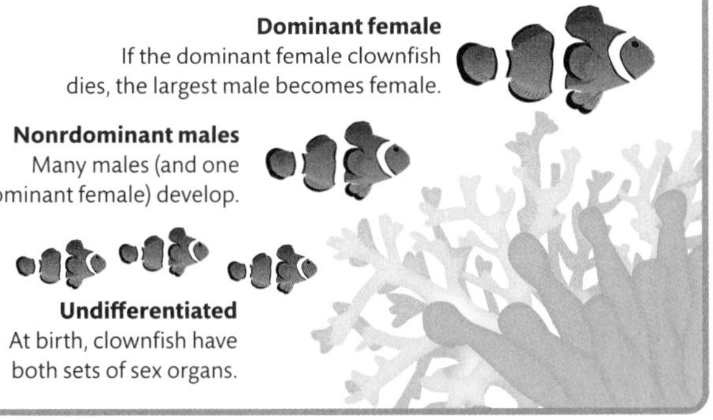

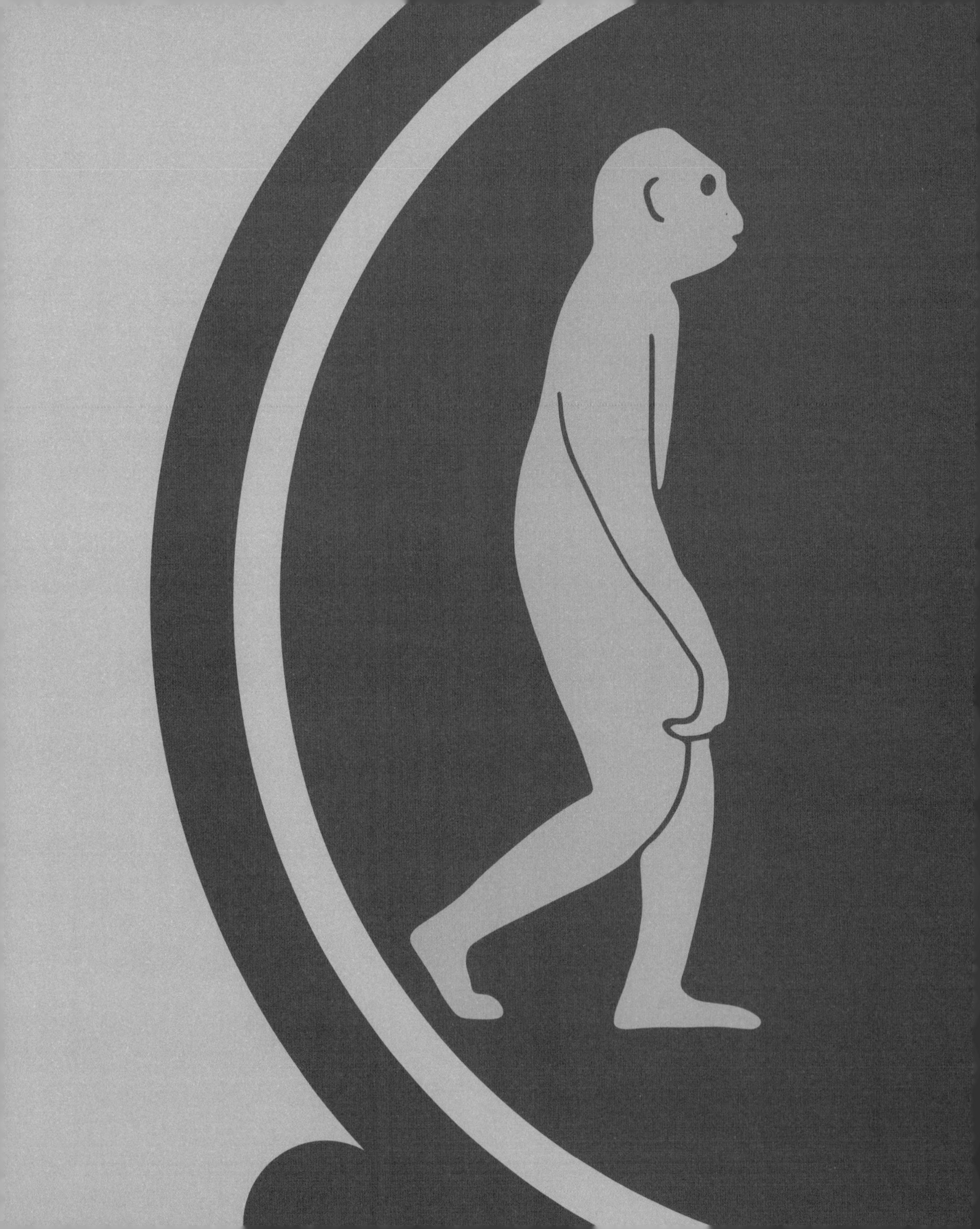

EVOLUTION

Adaptation and natural selection

Members of a population of any organism often vary in the traits that they inherit from their parents. When these variations mean some individuals are better able to survive and reproduce in a particular set of environmental conditions, the species adapts.

> **CAN INDIVIDUALS EVOLVE?**
>
> No, only populations of organisms, rather than individual organisms, can evolve, and it happens over the course of many generations.

Natural selection

Random genetic mutations (see pp.96–99) can lead to differences in the offspring of an organism. Some differences have a negative impact on an organism's ability to survive, others are neutral, and some are beneficial. Individuals that inherit beneficial traits will thrive and be able to produce more offspring, so over time they make up a larger proportion of the species' total population and become dominant. This is evolution by means of natural selection. An inherited characteristic that makes survival and reproductive success more likely is called an adaptation.

1 Mutation and variation
When organisms like this cricket reproduce, mutations sometimes arise in the genes of their offspring, leading to variations in features such as color. Such changes can increase genetic diversity and help the species survive in a constantly changing world.

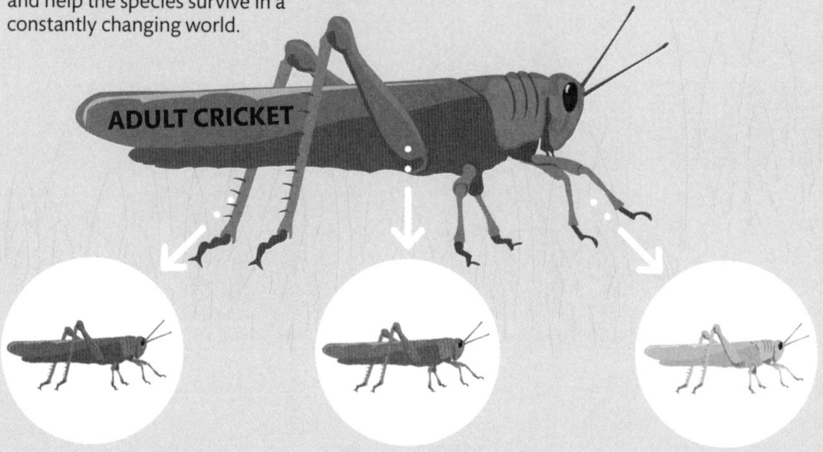

BROWN OFFSPRING

GREEN OFFSPRING

YELLOW OFFSPRING

THE **FIRST LAND VERTEBRATES EVOLVED** FROM LOBE-FINNED **FISH** 375 MILLION YEARS AGO

ANTIBIOTIC RESISTANCE

While antibiotic drugs have saved millions of lives since they were first prescribed in the 1940s, overuse has led to infectious bacteria evolving resistance to the drugs in less than 50 years. As a result, scientists must constantly develop new antibiotics in order to beat bacterial evolution.

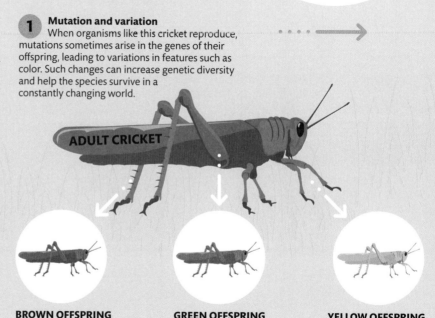

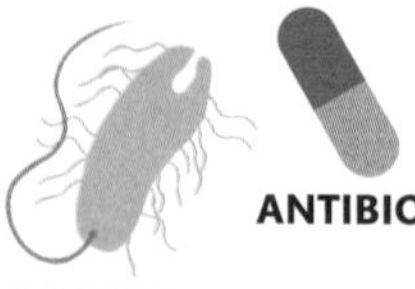

ANTIBIOTIC

BACTERIUM

Types of adaptation

The way a species fits into its environment is known as its niche. A niche comprises many factors, from the availability of food, water, and sites for shelter to the prevalence of predators and toxins. All organisms display a range of adaptations that make them better suited to survive and reproduce within their niche. Adaptations can involve behavior (including modifying their environment), body processes, and physical features.

2 Survival of the fittest

In a wet climate, vegetation is green all year. Birds predate mostly yellow and brown crickets because, unlike the green individuals, they are not camouflaged. Over time, green crickets come to dominate the population because they are more likely to survive and pass on their genes. This process is known as survival of the fittest or natural selection.

3 A changing environment

Environments change over time. If rainfall decreases and the environment becomes more arid, green crickets are no longer camouflaged and, along with the yellow crickets, they become prey to the birds. The camouflaged brown crickets now escape the attention of predators and dominate the population—the whole population adapts to the environment's change.

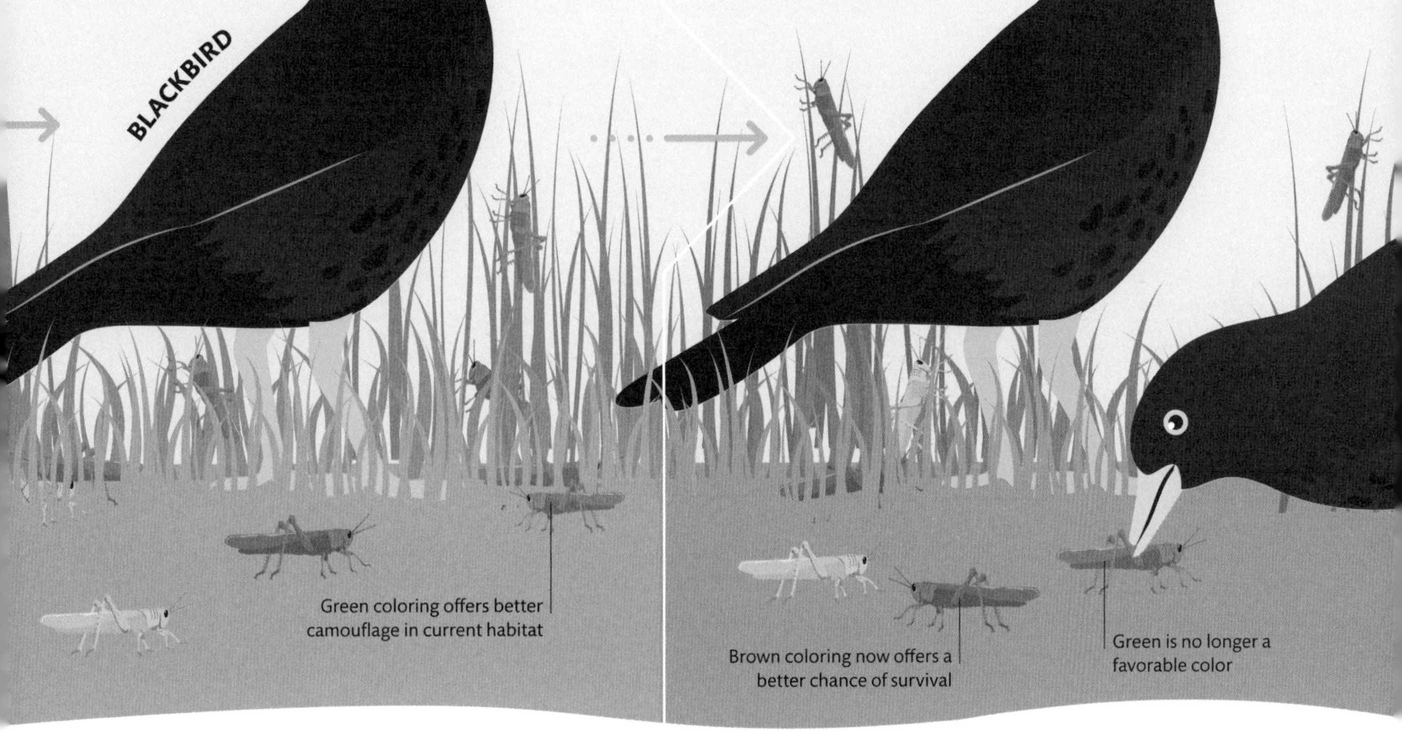
Bird spots yellow and brown crickets more easily
BLACKBIRD
Green coloring offers better camouflage in current habitat
Bird sees yellow and green crickets more easily
Brown coloring now offers a better chance of survival
Green is no longer a favorable color

BEHAVIORAL ADAPTATIONS

The evolution of behavior to offer an animal a better chance of survival is a behavioral adaptation. Many dolphins hunt in pods to improve their feeding success, for example.

PHYSIOLOGICAL ADAPTATIONS

Changes to internal body processes to better suit the environment are physiological adaptations. For example, camels produce more concentrated urine and sweat less than most mammals to save water.

STRUCTURAL ADAPTATIONS

The evolution of physical features that improve an organism's chance of survival are structural adaptations. Cacti leaves evolved into spines, which lose less water and offer protection from herbivores, for example.

Speciation

The process that results in the evolution of new species is called speciation. It arises when genetic changes in part of a species' population lead to reproductive isolation—an inability to breed successfully with the rest of the population.

Allopatric speciation

When a population is divided into geographically isolated subpopulations, for example by the emergence of a river or mountain range, allopatric speciation may occur. The separation allows the gene pools of the two different populations to grow apart over the course of many generations. Eventually the differences are so great that the two populations can no longer interbreed, which means they can be classified as different species. The movement of continents can also create the geographical isolation needed for this type of speciation and can be seen in the frog species of Madagascar and India, which were once joined.

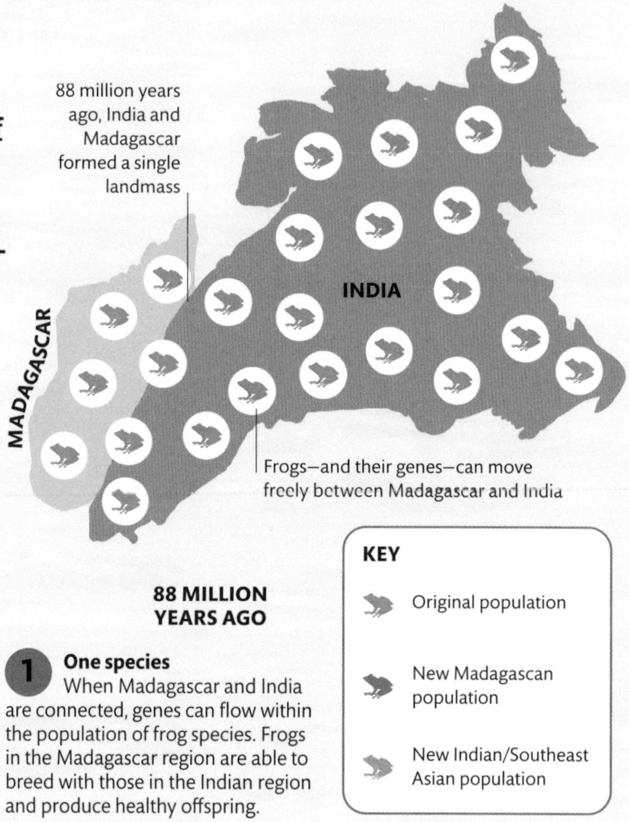

88 million years ago, India and Madagascar formed a single landmass
INDIA
MADAGASCAR
Frogs—and their genes—can move freely between Madagascar and India
88 MILLION YEARS AGO
KEY
Original population
New Madagascan population
New Indian/Southeast Asian population

1 One species
When Madagascar and India are connected, genes can flow within the population of frog species. Frogs in the Madagascar region are able to breed with those in the Indian region and produce healthy offspring.

Sympatric speciation

Unlike allopatric speciation, sympatric speciation occurs when there are no physical barriers preventing members of a species from mating with each other—instead the movement of genes (called gene flow) between subpopulations is restricted by factors such as habitat differentiation or polyploidy. The former may arise when a subpopulation starts to exploit a habitat (or its food resources) not used by the parent population. Polyploidy—where some individuals have an extra set of chromosomes—can result in self-fertilization or the opportunity to mate with other individuals with extra chromosomes. In just one generation, polyploidy can result in reproductive isolation without geographic separation.

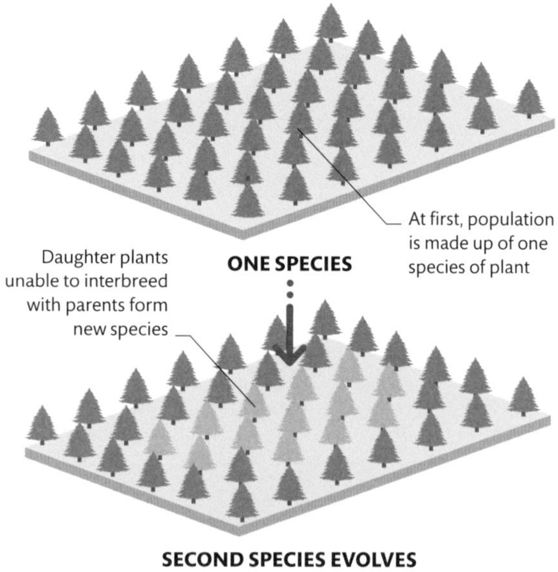

At first, population is made up of one species of plant
ONE SPECIES
Daughter plants unable to interbreed with parents form new species
SECOND SPECIES EVOLVES

Polyploidy in plants
Polyploidy is common in plants, as well as some amphibian and fish species. Cell division "mistakes" lead to the multiplication of chromosomes in daughter plants, so they are unable to interbreed with the parent plant and a new species arises.

AT LEAST **ONE-THIRD OF ALL PLANT SPECIES** PROBABLY EVOLVED THROUGH **POLYPLOIDY**

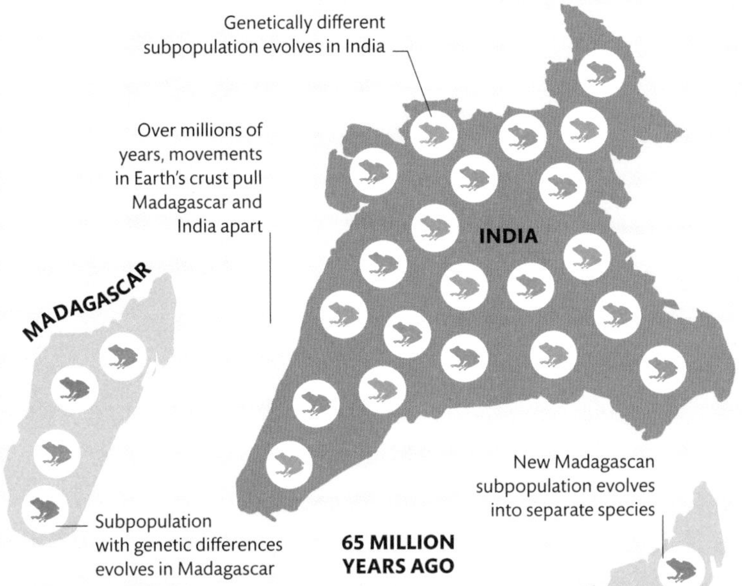
Genetically different subpopulation evolves in India
Over millions of years, movements in Earth's crust pull Madagascar and India apart
MADAGASCAR
INDIA
Subpopulation with genetic differences evolves in Madagascar
65 MILLION YEARS AGO

2 Population divides
The separation of the landmass divides the population in two, so genes can no longer flow between Indian and Madagascan frogs. Mutation, natural selection, and adaptation to the changing environment occur, so the populations begin to diverge genetically.

New Madagascan subpopulation evolves into separate species
MADAGASCAR

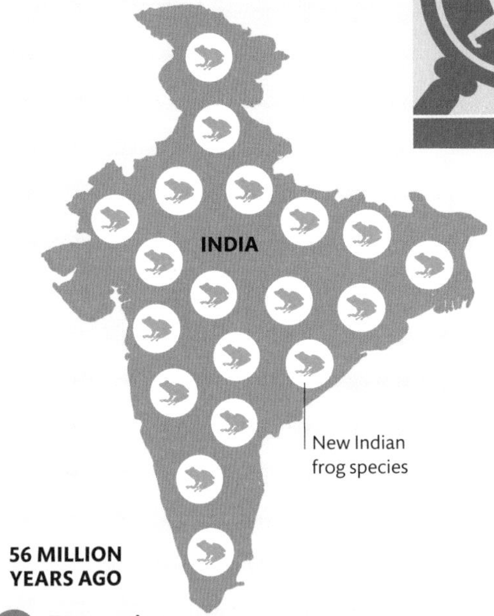
INDIA
New Indian frog species
56 MILLION YEARS AGO

3 Two species
Allopatric speciation occurs within the now-separated populations of the ancestral frog, as the two populations can no longer interbreed. Over millions of years, more than 200 other animal and plant species evolve in Madagascar, and many more in India.

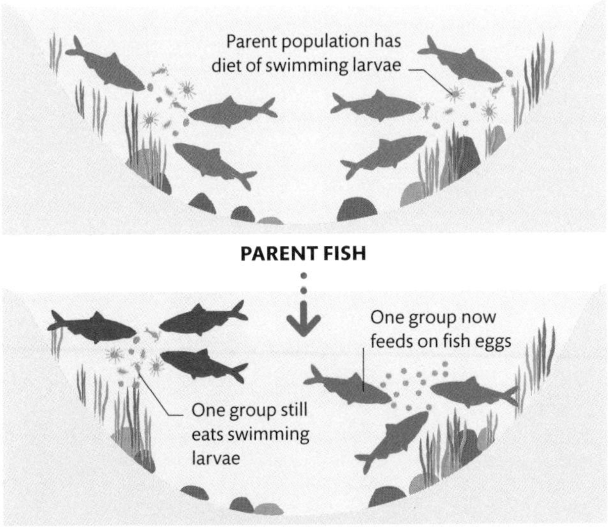
Parent population has diet of swimming larvae
PARENT FISH
One group now feeds on fish eggs
One group still eats swimming larvae
DESCENDANT POPULATIONS

Habitat differentiation in animals
Mutation and inheritance of new characteristics in a subpopulation of a species may result in it being able to exploit a source of food different from the main population. Over time, this results in further genetic divergence and, eventually, sympatric speciation.

WHAT IS A SPECIES?

The most common definition of a species is the biological species concept (BSC). The definition states that a species is a group of organisms whose members can interbreed and produce fertile offspring, but they are unable to do this with members of other groups. According to the BSC, mules cannot be considered to be a species because they are sterile.

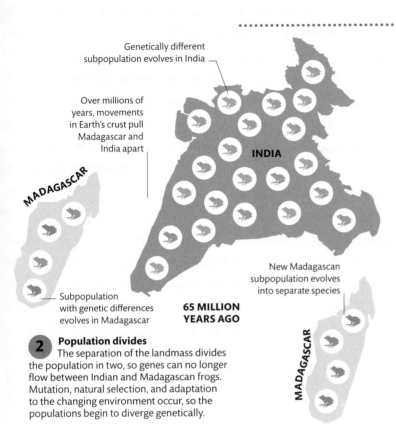

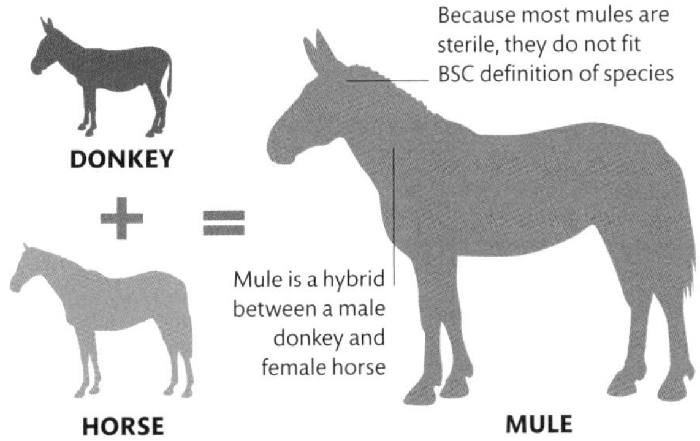
Because most mules are sterile, they do not fit BSC definition of species
DONKEY
+
=
HORSE
Mule is a hybrid between a male donkey and female horse
MULE

Sexual selection

Since mutual attraction between the sexes is an important factor in reproduction, sometimes natural selection is driven by physical characteristics that increase an organism's chance of finding a mate.

DOES SEXUAL SELECTION OCCUR IN PLANTS?

Yes. Pollinating animals have demonstrated a preference for symmetrical over asymetrical flowers, for example, leading to sexual selection.

What is sexual selection?

Individuals with certain inherited characteristics are known to be more likely than others to find mates. This drives a certain kind of natural selection in which these characteristics—including larger size, brighter colors, or more extravagant displays—may be enhanced over generations. This can lead to sexual dimorphism in which males and females of the same species look very different. Sexual selection operates through intrasexual selection or intersexual selection.

Males use teeth and weight to fight for control of colony and dozens of females in it
If larger seal wins, its genes will be passed on to next generation through mating
Smaller, weaker males are rarely successful
LARGER BULL ELEPHANT SEAL
SMALLER BULL ELEPHANT SEAL

Intrasexual selection
Direct competition between members of one sex—usually males—to establish the opportunity to mate with members of the opposite sex is called intrasexual selection. Competition may involve ritualized displays, shows of aggression, or actual fighting—as with bull elephant seals, for example.

Intersexual selection
When members of one sex are choosy in selecting a mate of the opposite sex, sexual selection occurs. This kind of selection is based on factors such as the clarity and loudness of songs—as with birds and frogs—or the showiness of their displays, for example those of peacock spiders.

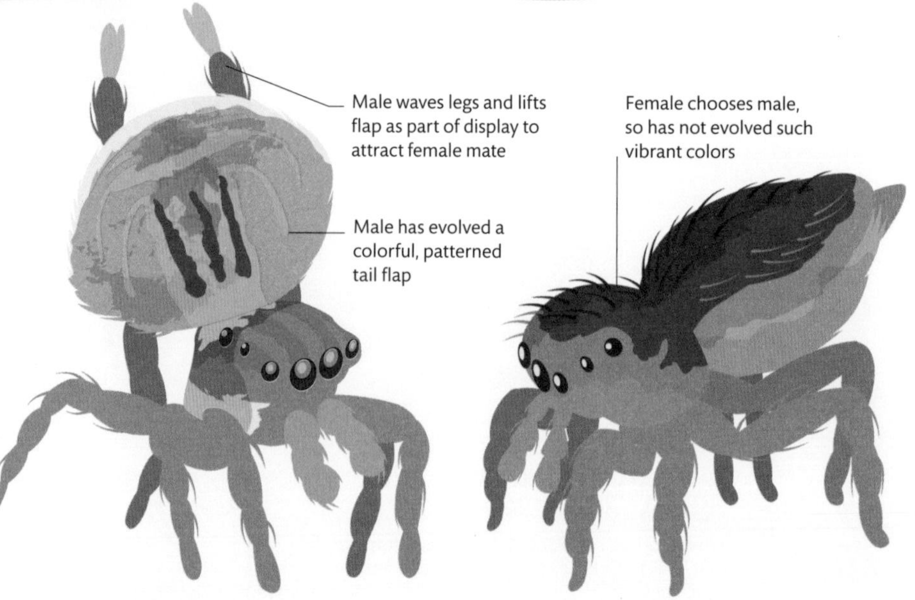

Male waves legs and lifts flap as part of display to attract female mate
Male has evolved a colorful, patterned tail flap
Female chooses male, so has not evolved such vibrant colors
MALE PEACOCK SPIDER
FEMALE PEACOCK SPIDER

MALE DEER HAVE **LARGE ANTLERS** AS A RESULT OF **INTRASEXUAL SELECTION**

Reproductive success

The males and females of most species are similar in size, shape, and coloration. However, in some species conspicuous physical traits have evolved that give some individuals greater success in finding a mate. These characteristics have usually evolved in males and range from the enormous antlers of some deer to the colorful crests and tails seen in many birds. Male peafowl (peacocks) have evolved their magnificent long tails with brightly colored "eyes," or ocelli, as a result of sexual selection, as female peafowl (peahens) have demonstrated a preference for males with more ocelli in their tail feathers.

SEXUAL DIMORPHISM

The differences in appearance between males and females of the same species are called sexual dimorphism. Small differences are widespread in nature, but sexual selection has driven them to extremes in animals such as birds of paradise and peafowl.

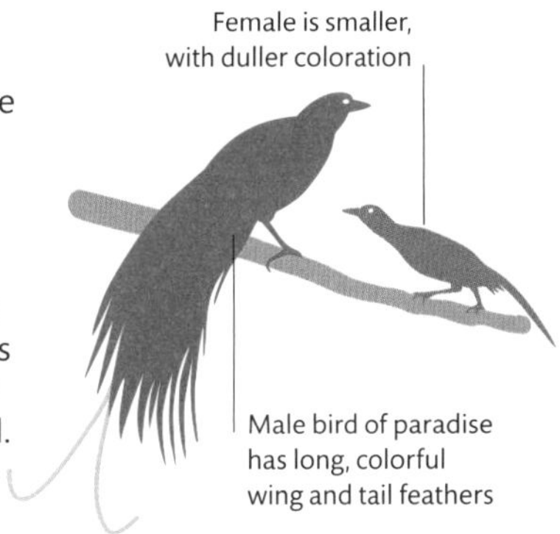
Female is smaller, with duller coloration
Male bird of paradise has long, colorful wing and tail feathers

1st generation
Female peafowl prefer peacocks whose tails bear more ocelli, so peacocks with more numerous ocelli mate more frequently and produce more offspring. Male offspring inherit the genes for a many-eyed tail from their father.

Peahen selects mate from several available males
Chosen peacock is one with most tail ocelli

2nd generation
More males of the next generation will have an increased number of ocelli. Second-generation peahens will again choose mating partners with the most ocelli, and yet more male offspring will inherit this gene.

Next generation has more numerous ocelli

3rd generation
The process is repeated in the next generation. On average, peacocks now have a greater number of ocelli in their tail feathers, but females continue to choose the males who possess the most.

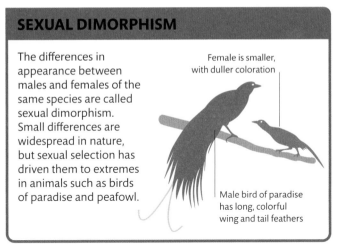

Number of ocelli increases with each generation

Coevolution

HAVE HUMANS COEVOLVED WITH ANOTHER SPECIES?

Dogs have coevolved with humans over thousands of years to become their friends, shepherds, guards, and hunting partners.

A species, or group of species, evolves to occupy a particular ecological niche, or role, within a habitat. Sometimes that niche sees the species living alongside another unrelated one, and the two coevolve into an unwitting partnership as they adapt to survive.

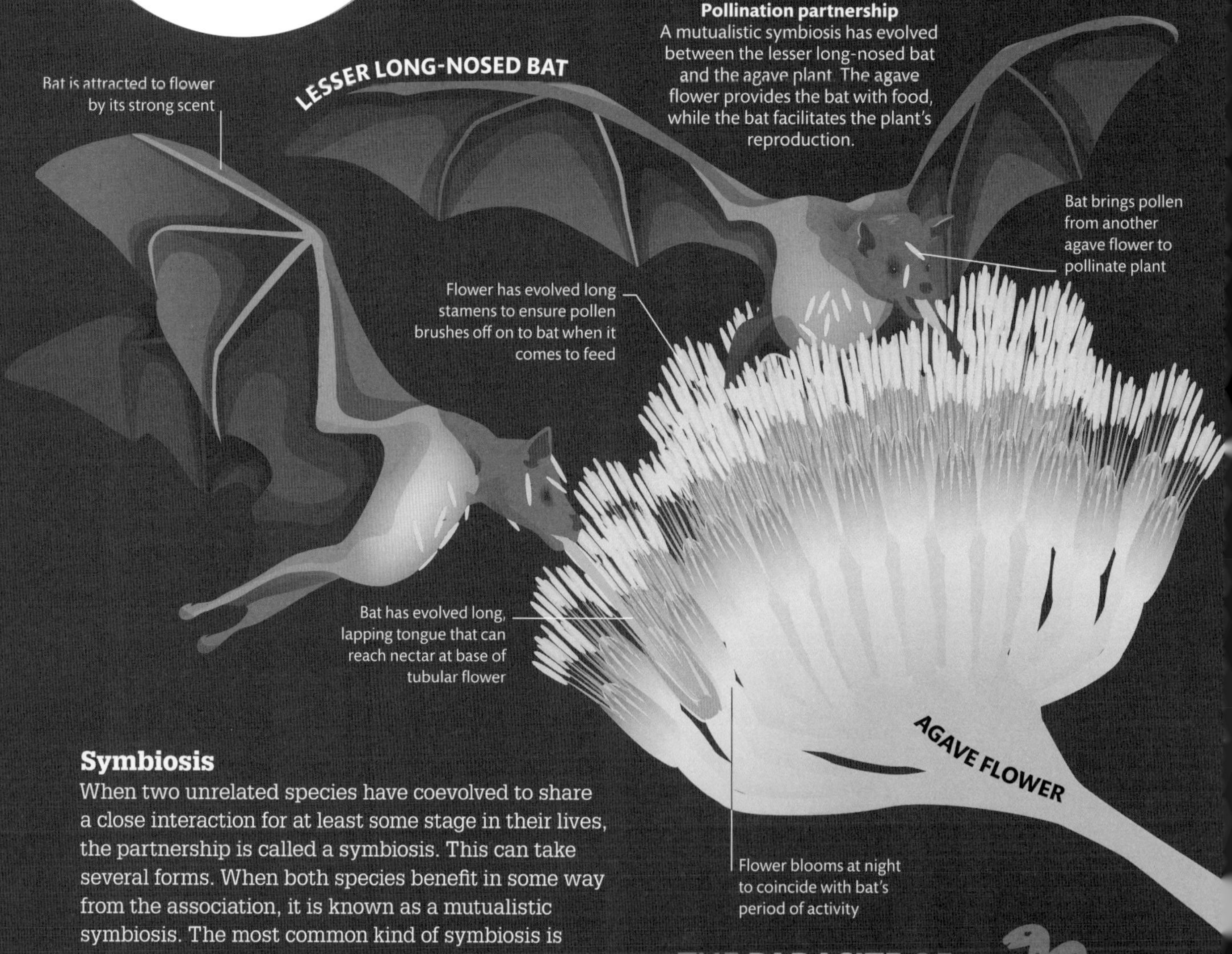

LESSER LONG-NOSED BAT
Bat is attracted to flower by its strong scent
Pollination partnership
A mutualistic symbiosis has evolved between the lesser long-nosed bat and the agave plant. The agave flower provides the bat with food, while the bat facilitates the plant's reproduction.
Bat brings pollen from another agave flower to pollinate plant
Flower has evolved long stamens to ensure pollen brushes off on to bat when it comes to feed
Bat has evolved long, lapping tongue that can reach nectar at base of tubular flower
AGAVE FLOWER
Flower blooms at night to coincide with bat's period of activity

Symbiosis

When two unrelated species have coevolved to share a close interaction for at least some stage in their lives, the partnership is called a symbiosis. This can take several forms. When both species benefit in some way from the association, it is known as a mutualistic symbiosis. The most common kind of symbiosis is parasitism (see opposite), where the host species is actually harmed by the presence of its partner. A rarer combination where only one species benefits, but the other remains unharmed, is known as commensalism.

THE PARASITE OF ANOTHER PARASITE IS CALLED A HYPERPARASITE

Zombie snail
The tiny amber snail hosts a highly evolved parasitic flat worm, the green-banded broodsac, which takes over the host's eye stalks and turns the snail into a zombie.

SNAIL EATS PARASITE

1 The snail forages on leaves for algae and bacteria and becomes infected with the worm parasite by eating bird feces containing eggs of the parasitic worm.

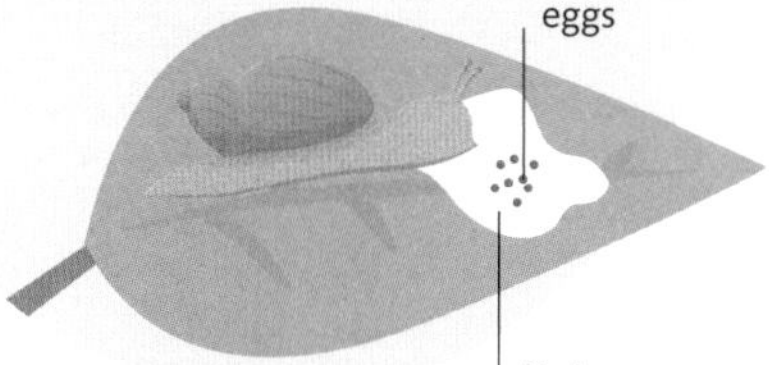
Worm eggs
Bird droppings

PARASITE'S EGGS HATCH

2 The worm hatches and grows in the snail, stealing nutrients from its digestive system. The parasite then takes over the snail's body and the tentacles move to its head.

Parasite sends tentacles into snail's eye stalks

PARASITE TAKES OVER

3 The parasite's tentacles contain young forms of the parasite, which create slowly pulsating colors as they move around inside the tentacles.

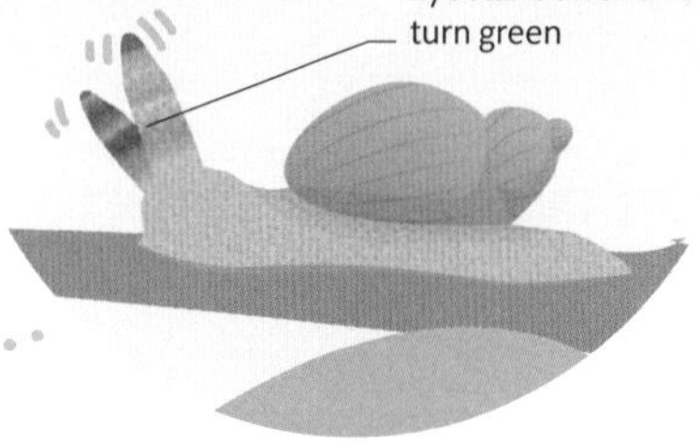
Eye stalks swell and turn green

BIRD EATS SNAIL

4 Amber snails normally stay in dark places. When infected with parasites the snails look for bright sunny areas, where songbirds mistake the eye stalks for tasty caterpillars.

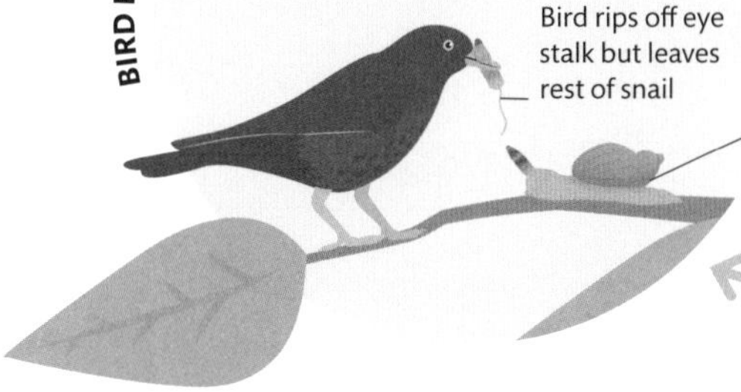
Bird rips off eye stalk but leaves rest of snail
Snail survives attack and may even grow a new eye stalk

PARASITE REPRODUCES

5 The parasite breeds in the songbird's stomach, then moves to its rectum, where it adds eggs to the host's droppings.

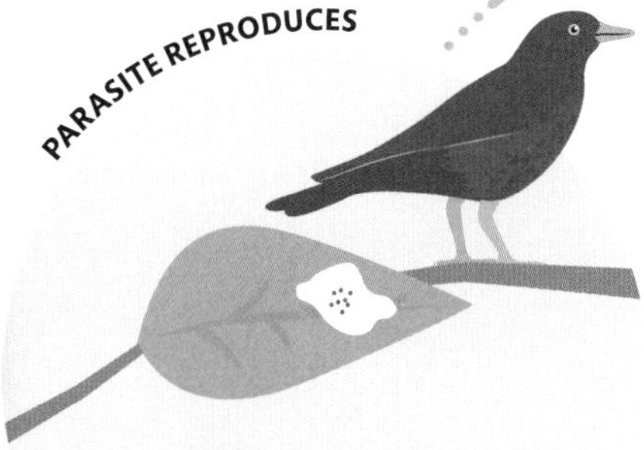

Parasitism

Parasitism is a form of symbiosis in which one organism lives on, or inside, another species. It is estimated that about 40 percent of species are parasitic on one or more hosts. The parasite will aim to steal the resources of the host in amounts that will only weaken it but not kill it.

Mimicry

In another type of coevolution one species evolves to mimic another. Mimicry is most often seen when an animal copies a species that uses a visual signal to warn of potential defense mechanisms, such as a sting or venom. For example, after being stung once by a striped wasp, an animal learns to avoid anything that resembles that wasp.

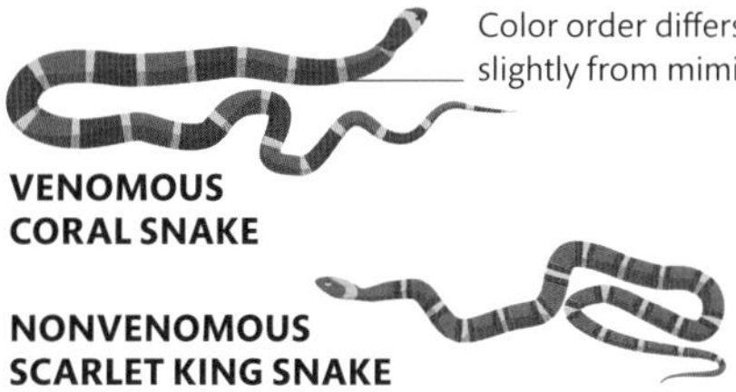
Color order differs slightly from mimic
VENOMOUS CORAL SNAKE
NONVENOMOUS SCARLET KING SNAKE

Batesian mimicry
This form of mimicry sees a harmless species adopt the look and coloring of a highly dangerous one, such as a harmless king snake copying the toxic coral snake.

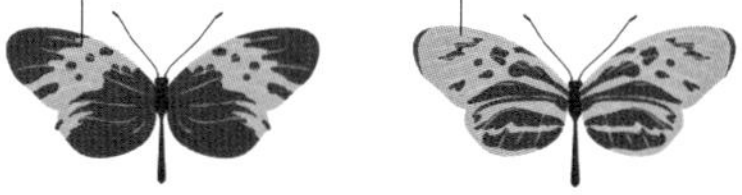
Similar shape
Similar warning coloring
HELICONID BUTTERFLIES

Mullerian mimicry
Both the patterns and shapes of these two forms of foul-tasting butterfly species are similar. Each benefits from the warning colorings used by the other one.

Microevolution

Every species is evolving, or changing its pool of genes, all the time. Microevolution describes the changes that do not create any obvious visible differences or new behaviors. This process is driven as much by random fluctuations as natural selection.

The Red Queen effect

A kind of microevolution in which predators and prey are in constant competition to survive, by either being more successful at hunting or better at escaping from hunters, is known as the Red Queen effect. Despite the constant adaptations to each other's evolution, the two species stay in the same place relative to each other, rather like the Red Queen in the Alice in Wonderland stories, who runs all the time but never moves. Although the microevolution of the Red Queen effect is largely hidden from view, some animals will signal that they are making a success by growing eyecatching features, such as vibrant body coloring or, in birds, an especially long tail. These are "honest" signals that their genes are making the animals fit and healthy.

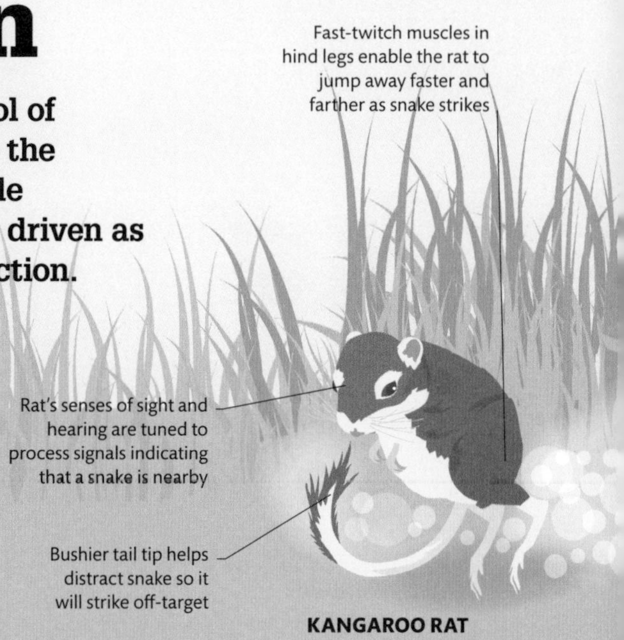
Fast-twitch muscles in hind legs enable the rat to jump away faster and farther as snake strikes
Rat's senses of sight and hearing are tuned to process signals indicating that a snake is nearby
Bushier tail tip helps distract snake so it will strike off-target
KANGAROO RAT

Prey survival
The aim of the kangaroo rat is to avoid detection by its predator, the rattlesnake. If that fails, it will first try to escape and, lastly, defend itself. Natural selection ensures that the rats that can do these things best will survive.

Gene flow

Another mechanism for microevolution is gene flow, in which the gene pool of a group of organisms is altered by the transfer of individuals and their genes to and from different populations. This can have markedly different effects. First, a gene common in the first population could represent a new variant in the second. This variant could out-compete the original so the second population becomes more genetically similar to the first. By contrast, the migrant could carry a rare gene that is lost from the first population but which could be added to the second. This can increase the genetic differences between the two populations, making it more likely that they diverge into separate species.

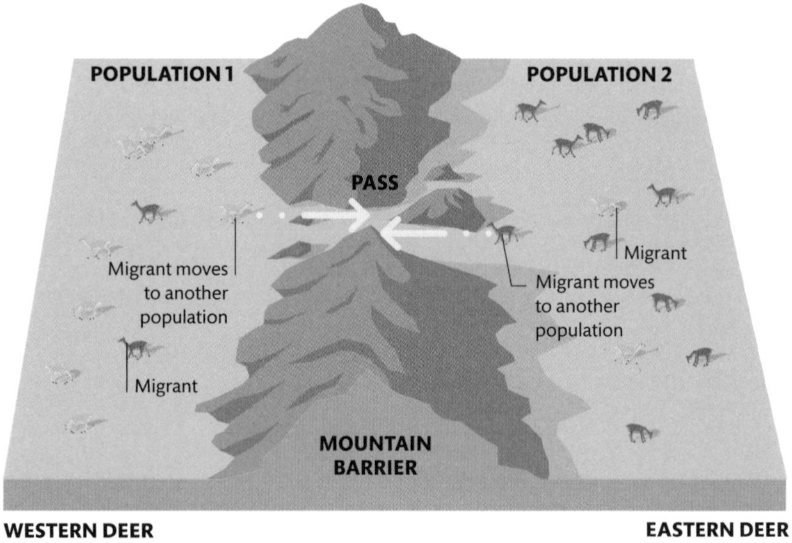
POPULATION 1
POPULATION 2
PASS
Migrant moves to another population
Migrant
Migrant moves to another population
Migrant
MOUNTAIN BARRIER
WESTERN DEER
EASTERN DEER

Movement of genes
Gene flow occurs simply by a member of one population of an animal finding a way to move to another. The genes this migrant carries can flow to the new population. The impact of gene flow is greater when the populations involved are small and not genetically diverse.

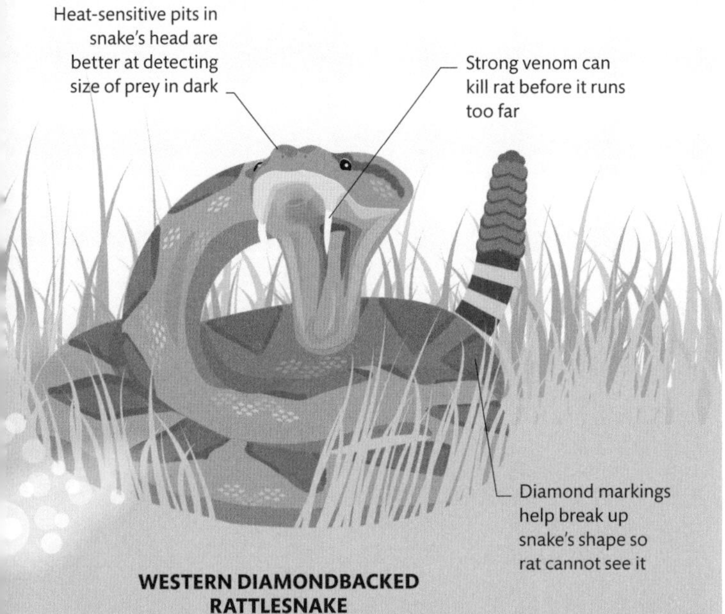
Heat-sensitive pits in snake's head are better at detecting size of prey in dark
Strong venom can kill rat before it runs too far
Diamond markings help break up snake's shape so rat cannot see it

WESTERN DIAMONDBACKED RATTLESNAKE

Predator survival
The rattlesnake must kill rats or starve. Over time, it develops adaptations that enable it to overcome its prey's constantly evolving defence abilities. Tiny differences between outwardly similar snakes can mean the difference between success and failure.

GENETIC DRIFT

Random changes to the gene pool, known as genetic drift, can have a significant effect on the evolution of a species, especially when populations are small and fragmented. When a single population is split, the individual members—and their genes—will not be divided up equally. Some genes will be completely absent from the resulting populations, purely by chance.

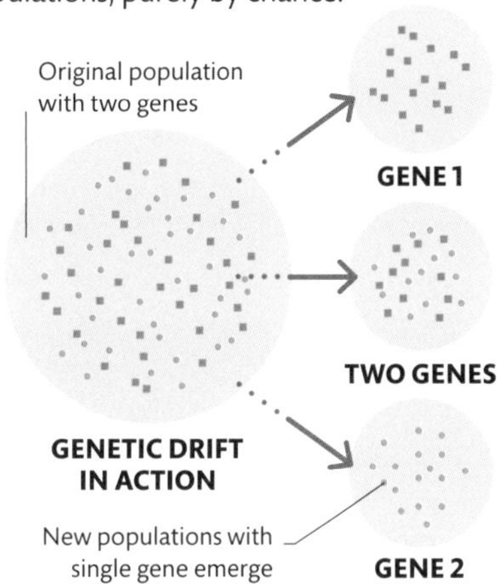
Original population with two genes
GENE 1
TWO GENES
GENE 2
New populations with single gene emerge

GENETIC DRIFT IN ACTION

1 Separate populations Although the two populations of deer belong to the same species, there are some differences in their genes. As they mingle, random chance will act to turn them into a genetically homogenous group.

Colored dots represent genes

TWO GENE POPULATIONS

2 Gene flow begins Genes are exchanged between the populations and move in both directions. The most common genes in a population are more likely to flow to the neighboring population than rare genes.

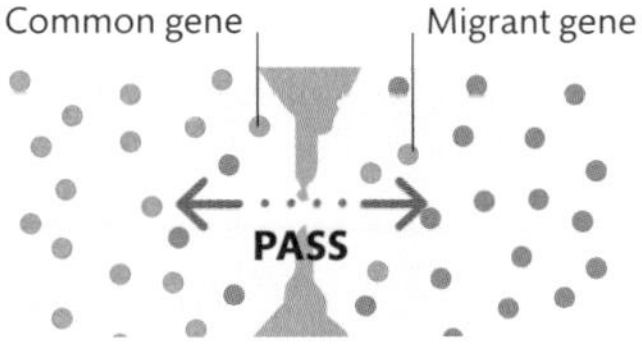
Common gene
Migrant gene
PASS

GENE MIGRATION BETWEEN POPULATIONS

3 Mixed populations If the gene flow rate is high, genes will mix and the two populations will be genetically similar. If the flow is slow, evolution has time to work, and a single migrant gene could create a significant difference.

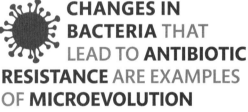

SIMILAR GENE POPULATIONS

CHANGES IN BACTERIA THAT LEAD TO **ANTIBIOTIC RESISTANCE** ARE EXAMPLES OF **MICROEVOLUTION**

IS THE EVOLUTION OF A NEW SPECIES ALWAYS OBVIOUS?

In a process called cryptic speciation, two populations can evolve so they no longer interbreed and so become two species but still look the same.

Extinction

When the last members of a species die out, the species has become extinct—it will never be seen alive again. Extinction is a natural process, and most of the species that have ever lived are extinct.

Types of extinction

Extinction is an important part of evolution by natural selection. Just as new species evolve to adapt to changes in the environment, others fail to adapt and die out—this is known as background extinction. Rapid changes to the environment that leave animals no time for adaptation can cause extinction. In these situations, many species often die out in what are termed mass extinctions, which can affect large taxonomic groups, such as, famously, the dinosaurs or trilobites (see pp.116–17). In addition, single genes can become extinct once they are no longer available to be passed to the next generation (see pp.112–13).

OVER 99 PERCENT OF SPECIES THAT HAVE EVER LIVED ON EARTH HAVE GONE EXTINCT

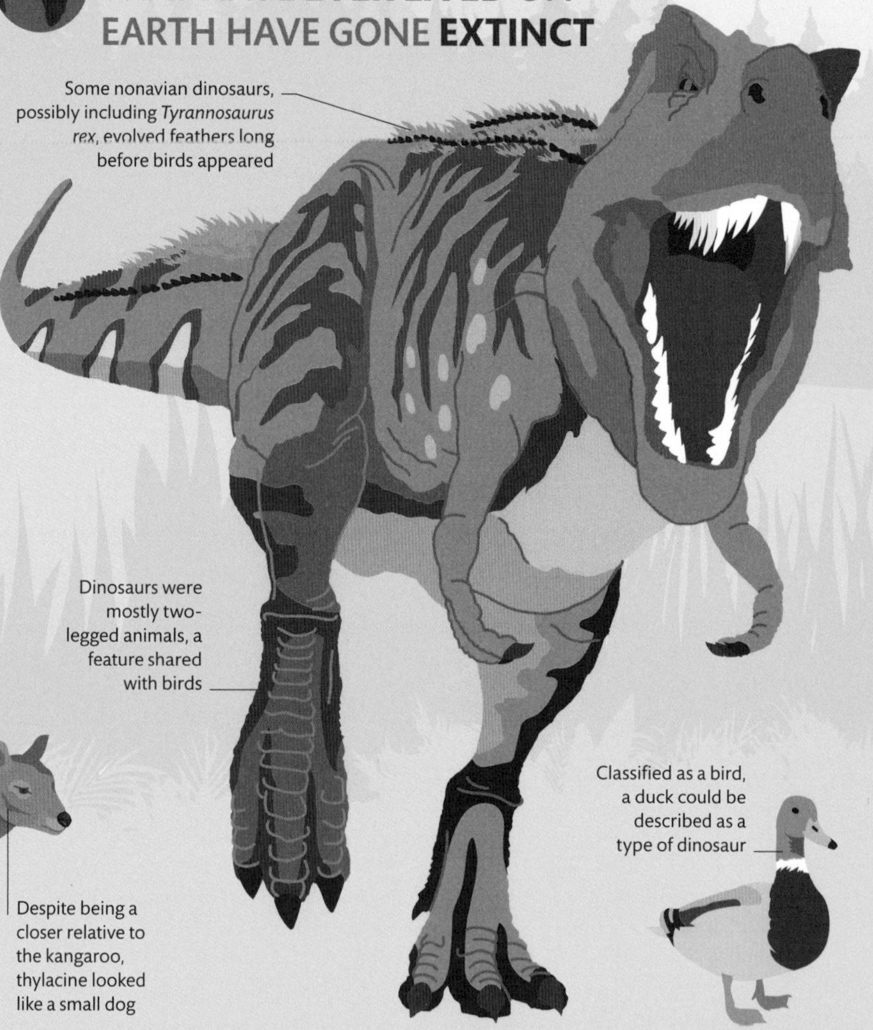

Some nonavian dinosaurs, possibly including *Tyrannosaurus rex*, evolved feathers long before birds appeared
Dinosaurs were mostly two-legged animals, a feature shared with birds
Classified as a bird, a duck could be described as a type of dinosaur

TYRANNOSAURUS REX

Stripes helped with camouflage as they do on tigers
Despite being a closer relative to the kangaroo, thylacine looked like a small dog

THYLACINE

DUCK

True extinction
This type of extinction follows the standard definition. The thylacine, or Tasmanian wolf, a marsupial predator, is truly extinct. Eradicated through human activity (excessive hunting, habitat destruction, and disease) by the 1930s, it is an evolutionary dead end as no species diverged from the thylacine.

Pseudoextinction
All the dinosaurs became extinct as a result of the Cretaceous mass extinction, 66 million years ago (see p.116). But birds, which are direct descendants of dinosaurs, survived. Nonavian dinosaurs (all dinosaurs except for birds) could therefore be described as pseudoextinct, because their descendants live on.

CAN A SPECIES BE BROUGHT BACK FROM EXTINCTION?

Scientists are trying to edit the genes of species genetically similar to the thylacine in order to resurrect the species.

Bird had wingspan of about 8 ft 6 in (2.6 m)
At 33 lb (15 kg), Haast's eagle was largest ever bird of prey
HAAST'S EAGLE
Moa could grow to 12 ft (3.6 m) tall and weigh around 440 lb (200 kg)
Tall, flightless bird could run fast on its powerful legs
MOA

Coextinction

Animals that have coevolved—for example, certain predator-prey pairings (see pp.110–11)—can become coextinct because one species cannot survive without the other. For example, when the moa, a group of large flightless birds, were wiped out in New Zealand, their predator, Haast's Eagle, was also lost.

Why do species die out?

As a result of natural selection, species adapt to changes in their environment. However, some changes are so dramatic that a species cannot adapt, all the organisms die, and the species goes extinct. Factors that increase the risk of extinction include human activity and natural disasters.

THE MAIN CAUSES OF EXTINCTION

Habitat loss
Natural climate changes, such as ice ages, shift habitat distribution. Species that cannot move will die out.

Climate change
Fluctuations in temperature, humidity, and salinity have a marked effect on the biosphere, especially in the oceans.

Slow adaptation
Species that breed slowly may be unable to produce new generations to adapt to change, so the population dwindles.

Sudden appearance of new species
New species in the ecosystem, including humans, upset the ecological balance so others do not have a niche to occupy.

Spread of disease
It is rare for one disease to make a species extinct, but in combination with other factors, it can be a major cause.

Changing into new species
A species becomes extinct if some of its members evolved into species better adapted for survival. But these organisms cannot breed with the original members due to geographical, behavioral, physiological, or genetic barriers or differences.

LIVING FOSSILS

Occasionally species are found that belong to groups thought to be extinct. While all the other members have died out, the "living fossil" has survived unseen. For example, the coelacanth is the only living relative of fish that, 400 million years ago, evolved paired fins capable of the motion seen in four-limbed terrestrial animals.

Sturdy fins made of thick bones are used for "walking" on the seabed
Fish can be 6 ft (2 m) long and lives in deep-sea caves
COELACANTH

Mass extinctions

CAN WE STOP THE CURRENT MASS EXTINCTION?

Biodiversity loss can be stopped by protecting wilderness areas across the globe and rewilding as much land as possible.

A mass extinction is a global event in which a major proportion of the species living all over the world become extinct in a short amount of geological time. The actual causes of these events are often unclear.

MORE THAN **40 PERCENT** OF **AMPHIBIAN** SPECIES ARE CURRENTLY AT **RISK OF EXTINCTION**

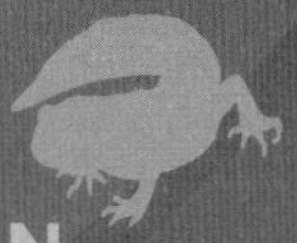

Timeline of mass extinctions

The diversity of life on Earth has fluctuated a great deal over time, and its natural history has been punctuated by mass extinctions, which mark the boundaries between the periods in the geological timescale. Evidence of these events can be clearly seen in the fossil record, where the remains of some organisms are suddenly no longer evident. Since complex life first evolved on Earth there have been five major mass extinction events (see below).

END ORDOVICIAN–444 MILLION YEARS AGO (MYA)

A period of global cooling reduced plant cover on land and lowered sea levels, and species of marine animals such as trilobites reduced in number. Then a sudden warming of the planet wiped out species that had adapted to cold.

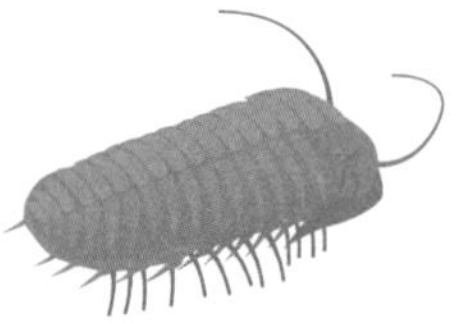

TRILOBITE

END DEVONIAN–359 MYA

A theory suggests that land plants evolved deeper root systems that reached more minerals in soil. The chemicals ended up in the oceans, creating algal blooms that caused many fish species, including the giant Dunkleosteus, to die out.

DUNKLEOSTEUS

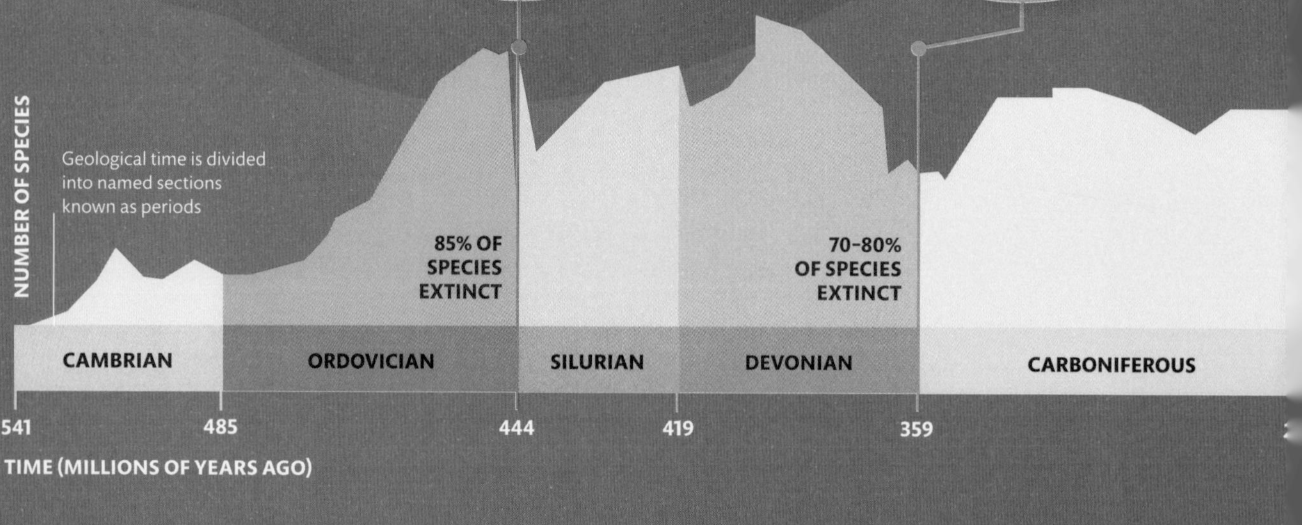
NUMBER OF SPECIES
Geological time is divided into named sections known as periods
85% OF SPECIES EXTINCT
70–80% OF SPECIES EXTINCT
CAMBRIAN
ORDOVICIAN
SILURIAN
DEVONIAN
CARBONIFEROUS
541
485
444
419
359
TIME (MILLIONS OF YEARS AGO)

A SIXTH MASS EXTINCTION?

Scientists have proposed that Earth is in the midst of a sixth mass extinction due to large-scale destruction of habitats by humans. An estimated 2,000 species are becoming extinct each year—this is 10,000 times faster than a natural extinction rate from 200 years ago.

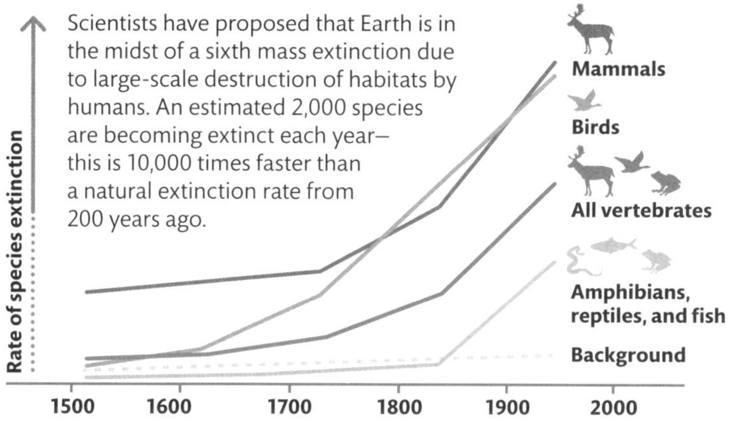
Rate of species extinction
Mammals
Birds
All vertebrates
Amphibians, reptiles, and fish
Background
1500
1600
1700
1800
1900
2000

END CRETACEOUS—66 MYA

This is the event that wiped out most dinosaurs, such as triceratops, although ancestors of modern birds survived. A large asteroid impact created global devastation and perhaps triggered immense volcanic eruptions.

TRICERATOPS

END PERMIAN—252 MYA

Known as the Great Dying, the cause of this extinction is thought to be volcanic activity that led to significant climate changes. Oceans also became too acidic for most species, including shellfish such as ammonites, to survive.

AMMONITE

END TRIASSIC—201 MYA

Possibly caused by climate changes or an asteroid strike, the Triassic extinction wiped out many of the early relatives of today's mammals, leaving the way clear for dinosaurs to dominate Earth.

PLATEOSAURUS

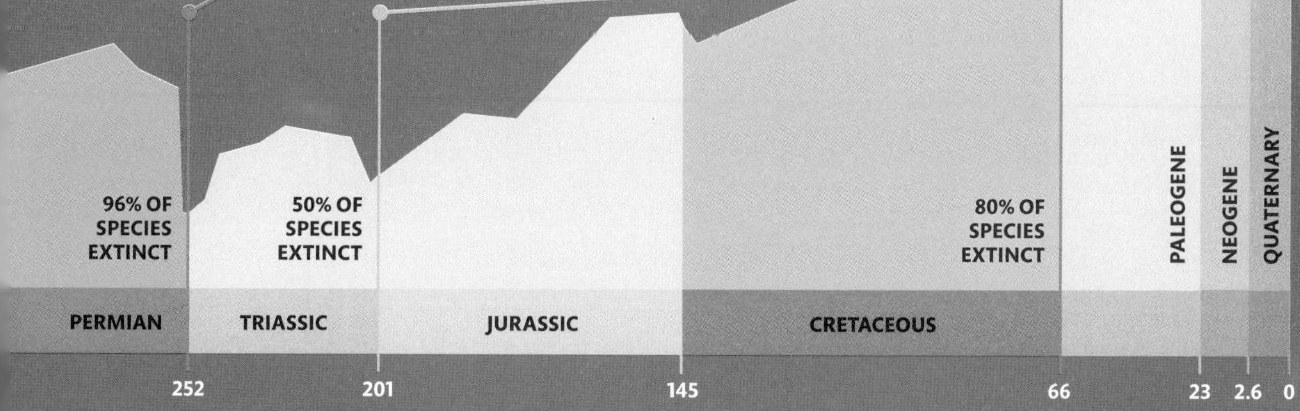
96% OF SPECIES EXTINCT
50% OF SPECIES EXTINCT
80% OF SPECIES EXTINCT
PALEOGENE
NEOGENE
QUATERNARY
PERMIAN
TRIASSIC
JURASSIC
CRETACEOUS
252
201
145
66
23
2.6
0

THE TREE OF LIFE

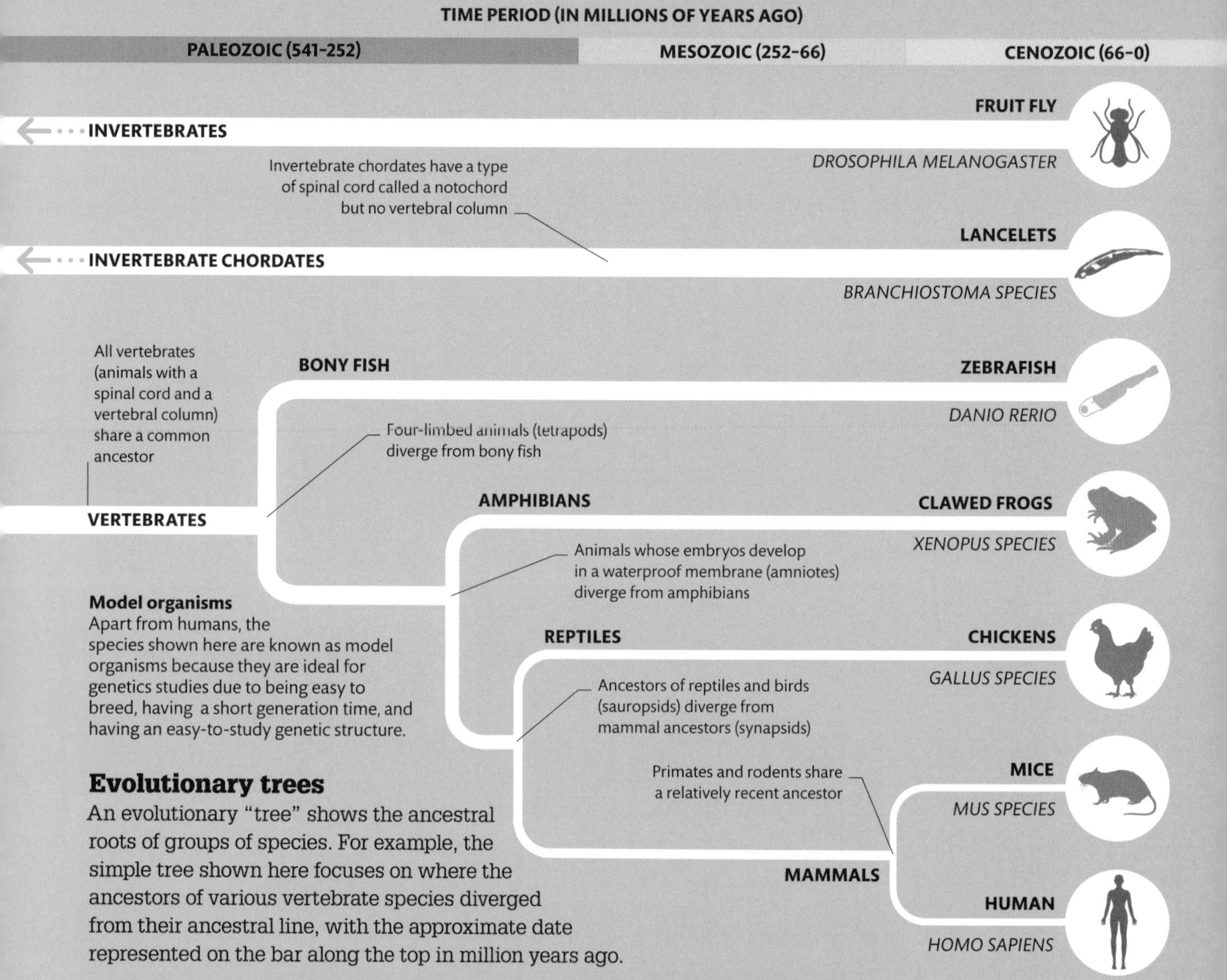

TIME PERIOD (IN MILLIONS OF YEARS AGO)
PALEOZOIC (541–252)
MESOZOIC (252–66)
CENOZOIC (66–0)
INVERTEBRATES
FRUIT FLY
DROSOPHILA MELANOGASTER
Invertebrate chordates have a type of spinal cord called a notochord but no vertebral column
INVERTEBRATE CHORDATES
LANCELETS
BRANCHIOSTOMA SPECIES
All vertebrates (animals with a spinal cord and a vertebral column) share a common ancestor
BONY FISH
ZEBRAFISH
DANIO RERIO
Four-limbed animals (tetrapods) diverge from bony fish
VERTEBRATES
AMPHIBIANS
CLAWED FROGS
XENOPUS SPECIES
Animals whose embryos develop in a waterproof membrane (amniotes) diverge from amphibians
REPTILES
CHICKENS
GALLUS SPECIES
Ancestors of reptiles and birds (sauropsids) diverge from mammal ancestors (synapsids)
Primates and rodents share a relatively recent ancestor
MICE
MUS SPECIES
MAMMALS
HUMAN
HOMO SAPIENS

Model organisms
Apart from humans, the species shown here are known as model organisms because they are ideal for genetics studies due to being easy to breed, having a short generation time, and having an easy-to-study genetic structure.

Evolutionary trees

An evolutionary "tree" shows the ancestral roots of groups of species. For example, the simple tree shown here focuses on where the ancestors of various vertebrate species diverged from their ancestral line, with the approximate date represented on the bar along the top in million years ago.

How living things are classified

Relationships between species were once based on what they looked like or how they behaved. Scientists now use a combination of genetic material and analysis of the fossil record. The evolution of shared physical characteristics can also be used to group together species. This information helps build a tree of life, which can be represented in various ways.

WHAT IS CONVERGENT EVOLUTION?

This occurs when similar environmental pressures and natural selection produce similar adaptations in species with different ancestries.

ALL BIRDS EVOLVED FROM A GROUP OF MEAT-EATING DINOSAURS THAT WERE ALSO ANCESTORS OF *TYRANNOSAURUS REX*

LAST UNIVERSAL COMMON ANCESTOR

The last universal common ancestor (LUCA) is the most recent ancestor from which all species alive today are descended. Although the identity of LUCA is not known, it probably lived about 4 billion years ago.

LUCA

Cladograms
These branching diagrams depict points of species divergence from common ancestral lines but without information about the degree of evolutionary deviation.

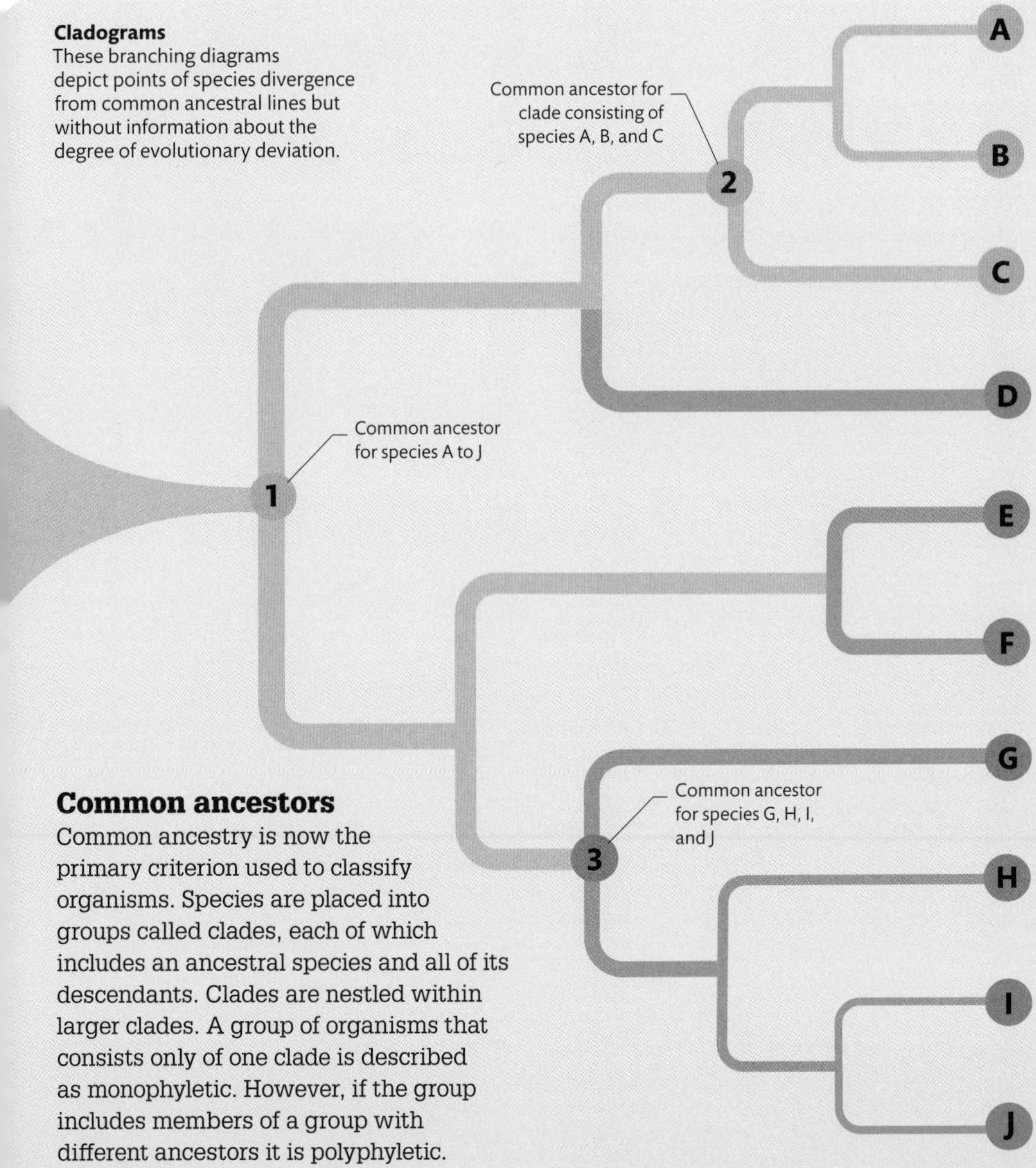
Common ancestor for clade consisting of species A, B, and C
Common ancestor for species A to J
Common ancestor for species G, H, I, and J
1
2
3
A
B
C
D
E
F
G
H
I
J

Clade
A clade is a group of species that includes a single common ancestor and all its descendants. Species A, B, and C all share the common ancestor 2, and they are all contained within the group. A clade is also known as a monophyletic group.

Polyphyletic group
This is a group of species in which the most recent common ancestor of every member of the group is not included. The species E and F have a relatively recent common ancestor, but this is not shared with species D.

Paraphyletic group
When a group of species contains a common ancestor and some (but not all) of its descendants, it is called paraphyletic. The species G, H, I, and J all share common ancestor 3, but J is not included in the group. Reptiles are an example of such a group when birds are not included.

Common ancestors

Common ancestry is now the primary criterion used to classify organisms. Species are placed into groups called clades, each of which includes an ancestral species and all of its descendants. Clades are nestled within larger clades. A group of organisms that consists only of one clade is described as monophyletic. However, if the group includes members of a group with different ancestors it is polyphyletic.

DO PROKARYOTES CAUSE DISEASE?

Yes, although only a few hundred of the 30,000 or so described bacteria are known to cause disease in humans.

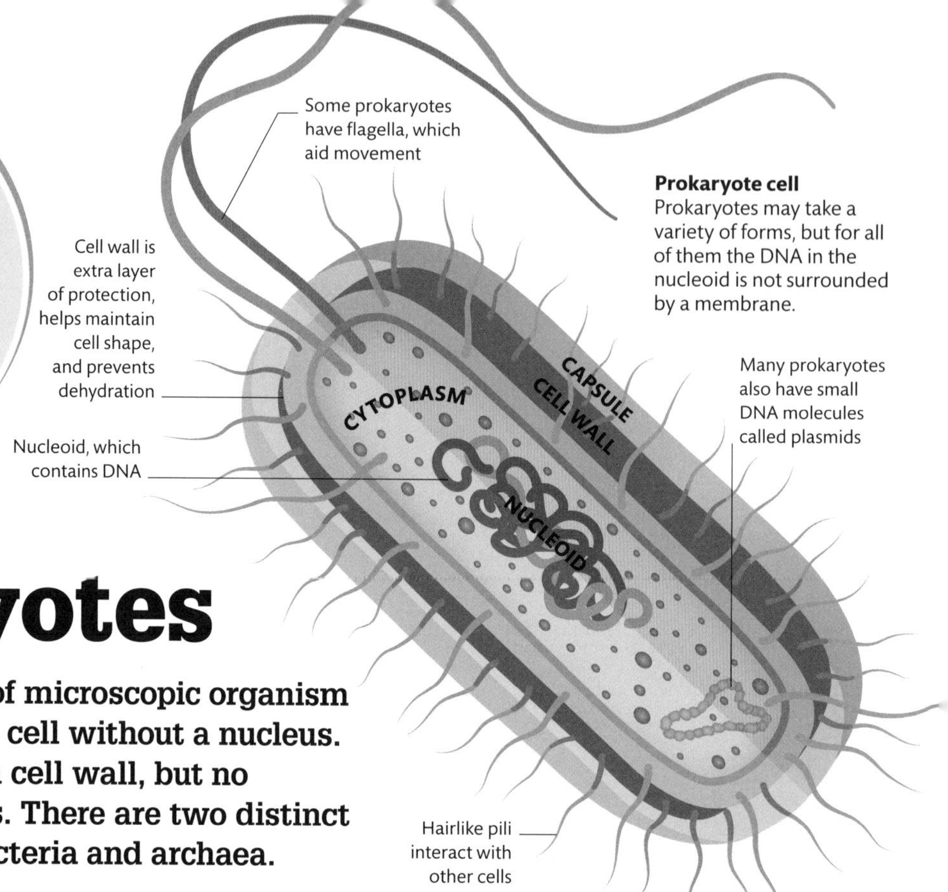
Some prokaryotes have flagella, which aid movement
Cell wall is extra layer of protection, helps maintain cell shape, and prevents dehydration
Nucleoid, which contains DNA
CYTOPLASM
CAPSULE
CELL WALL
NUCLEOID
Many prokaryotes also have small DNA molecules called plasmids
Hairlike pili interact with other cells

Prokaryote cell
Prokaryotes may take a variety of forms, but for all of them the DNA in the nucleoid is not surrounded by a membrane.

Prokaryotes

A prokaryote is a type of microscopic organism that consists of a single cell without a nucleus. Prokaryotic cells have a cell wall, but no interior cell membranes. There are two distinct types of prokaryote: bacteria and archaea.

Bacteria

Bacteria are microscopic, mostly free-living, single-celled organisms. While some are pathogens, most play a positive role, such as enabling animals to digest food and driving the global circulation of carbon, nitrogen, sulfur, and phosphorus. The ways in which they differ from archaea (see opposite) include the composition of their cell walls, the range of energy sources they exploit, and their metabolism. The first known bacteria lived about 2.4 billion years ago, but they probably evolved long before then.

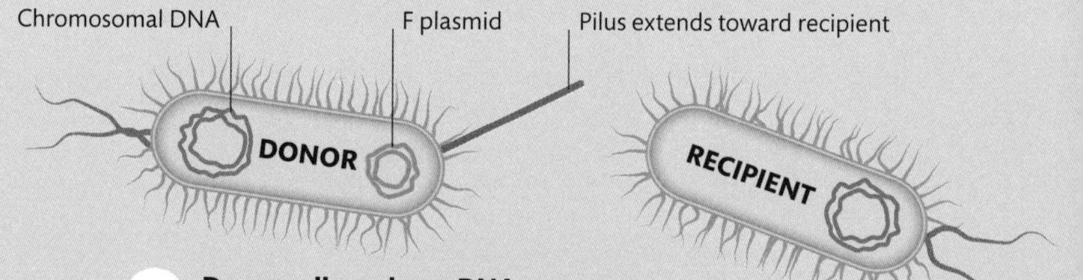
Chromosomal DNA
F plasmid
Pilus extends toward recipient
DONOR
RECIPIENT

1 Donor cell produces DNA
In addition to its chromosomal DNA, the donor bacterium carries a DNA sequence called a fertility factor, or F plasmid.

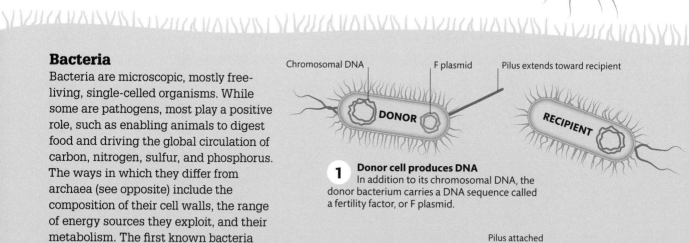

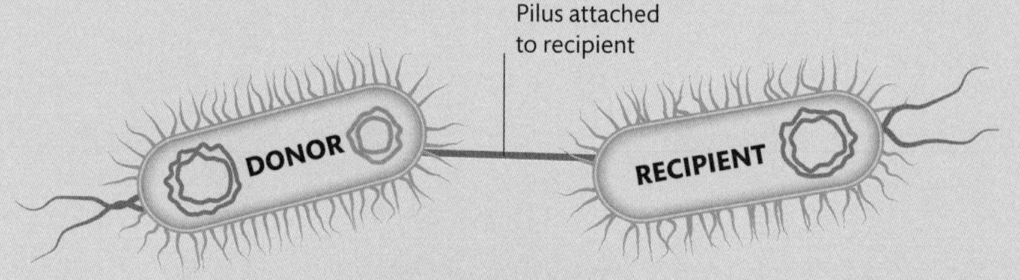
Pilus attached to recipient
DONOR
RECIPIENT

2 Donor attaches to recipient
The F plasmid enables the donor to produce a thin tubelike structure called a pilus, or mating bridge, which it uses to contact a recipient bacterium lacking an F plasmid.

How bacteria transfer DNA
Bacteria transfer DNA between each other in a process of direct contact called conjugation. This gene-transfer mechanism, alongside rapid reproduction rates, allows bacteria to adapt and evolve quickly to environmental changes.

Archaea

Probably the first forms of life to evolve, archaea are known to have lived more than 3.5 billion years ago. Unlike bacteria, they do not have the ability to form spores. Some are able to tolerate extreme environments. For example, thermophiles may live in temperatures greater than 212°F (100°C)–environments that would cause other organisms' DNA to unravel.

SOME SPECIES OF ARCHAEA DO NOT NEED OXYGEN TO SURVIVE

EXAMPLES OF ARCHAEA	SHAPE	DISTRIBUTION
Methanobrevibacter	Very short rods, or coccobacilli	These archaea are anaerobic (live without oxygen), and are found in the digestive systems of animals, including humans.
Methanospirillum	Rods, or bacilli	These widespread organisms occur in marine and terrestrial animals, plants, and soil. Some are aerobic, others anaerobic.
Pyrodictium	Disk-shaped cells held together by hollow tubules called cannulae	*Pyrodictium* are thermophiles and found in deep-ocean hydrothermal vents.
Methanococcus	Spherical, or coccoid	These organisms are mesophiles (preferring moderate temperatures) and are found close to deep-ocean hydrothermal vents.

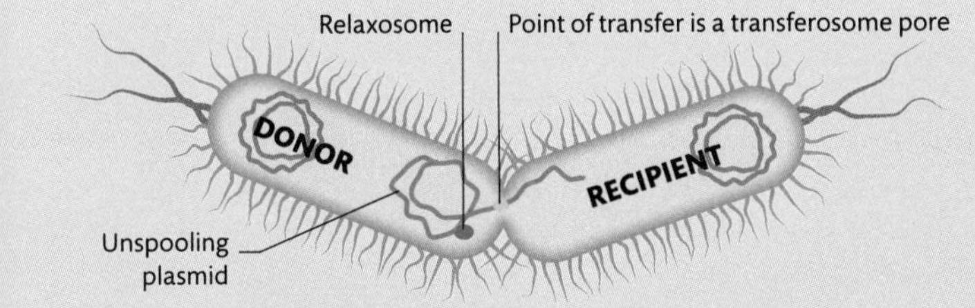
Relaxosome
Point of transfer is a transferosome pore
DONOR
RECIPIENT
Unspooling plasmid

3 DNA transferred
Enabled by a protein called relaxosome, a strand of the plasmid moves into the recipient. As the transfer continues, the donor plasmid unravels so more passes to the recipient.

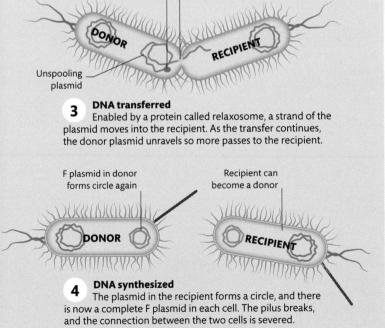

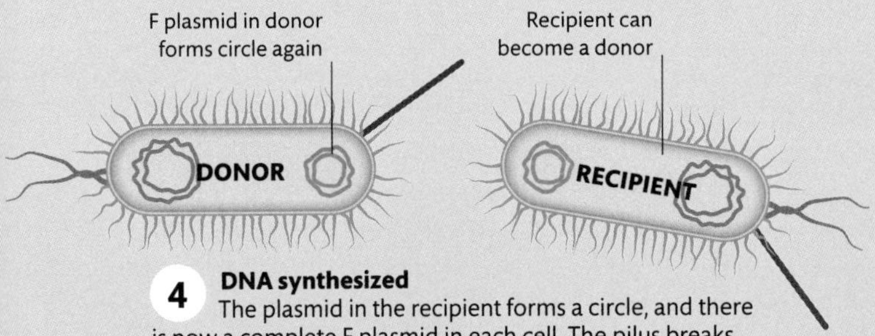
F plasmid in donor forms circle again
Recipient can become a donor
DONOR
RECIPIENT

4 DNA synthesized
The plasmid in the recipient forms a circle, and there is now a complete F plasmid in each cell. The pilus breaks, and the connection between the two cells is severed.

BACTERIAL DISEASE

Pathogenic bacteria cause about half of all human diseases. They usually cause illness by producing toxins. Some bacteria are transmitted by other species (vectors) such as animal ticks.

DISEASE	IMPACT
Lyme disease	Ticks spread *Borrelia* bacteria, which cause fever, headaches, and fatigue. If untreated, it may cause facial paralysis and chronic pain.
Tuberculosis	Inhaled *Myobacterium tuberculosis* bacteria usually infect the lungs. It is fatal if untreated.
Anthrax	Toxins from *Bacillus anthracis* may cause nausea, diarrhea, and breathlessness. It is commonly fatal, even with treatment.
Granville wilt	*Ralstonia solanacearum* attacks the xylem cells of crop plants, such as potatoes. It causes plants to wilt, collapse, and die.

Eukaryotes

The members of the domain Eukarya, also known as eukaryotes, form one of the three major divisions of living things. Their cells have a nucleus enclosed within a nuclear envelope and typically contain specialized, membrane-bound structures called organelles.

The origin of eukaryotes

Eukaryotes probably evolved through a process called endosymbiosis. Before 1.6–2.1 billion years ago, all organisms are thought to have been prokaryotic (see pp.122–23). By means not fully understood, archaean cells engulfed bacterial cells, which remained within their hosts. Over time, a mutually beneficial relationship developed between the engulfed bacterium (the endosymbiont) and its host. Over many generations, a symbiotic relationship developed so that neither the host nor the endosymbiont could survive without the other.

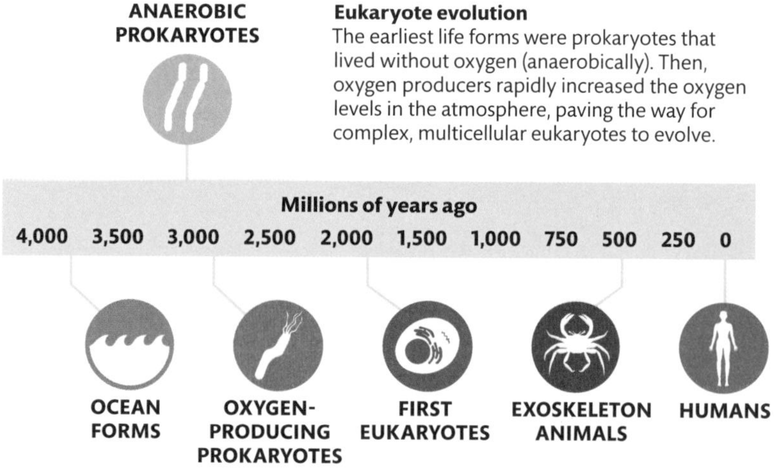

ANAEROBIC PROKARYOTES
Millions of years ago
4,000 3,500 3,000 2,500 2,000 1,500 1,000 750 500 250 0
OCEAN FORMS
OXYGEN-PRODUCING PROKARYOTES
FIRST EUKARYOTES
EXOSKELETON ANIMALS
HUMANS

Eukaryote evolution
The earliest life forms were prokaryotes that lived without oxygen (anaerobically). Then, oxygen producers rapidly increased the oxygen levels in the atmosphere, paving the way for complex, multicellular eukaryotes to evolve.

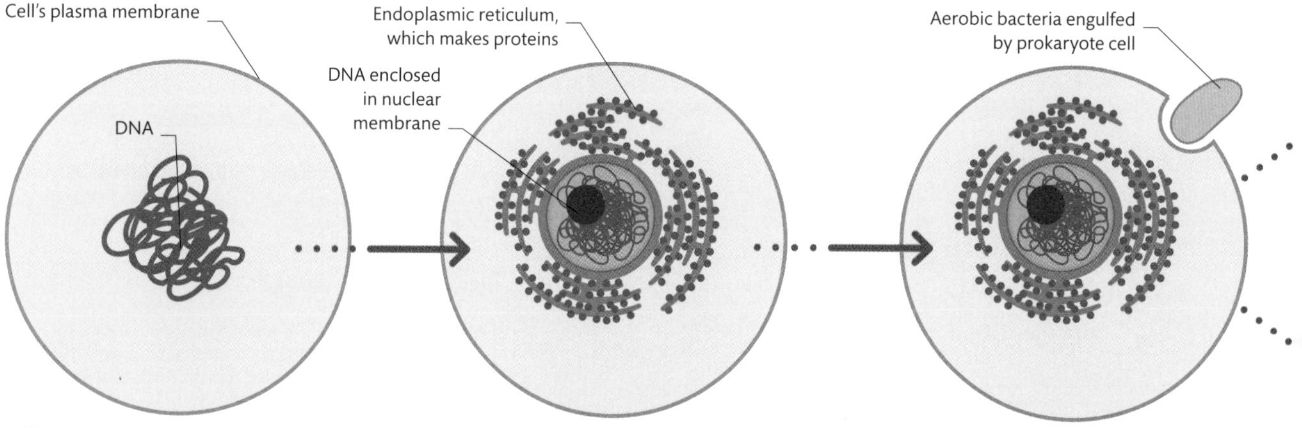

Cell's plasma membrane
DNA
Endoplasmic reticulum, which makes proteins
DNA enclosed in nuclear membrane
Aerobic bacteria engulfed by prokaryote cell

1 Ancestral prokaryote
An ancestral anaerobic (living without oxygen) prokaryotic cell has its own DNA and cytoplasm, which are surrounded by a plasma membrane.

2 Endoplasmic reticulum forms
The cell's plasma membrane folds in on itself, giving rise to new components within the cell, including a nucleus and an endoplasmic reticulum.

3 Consuming bacteria
Some anaerobic prokaryotic cells engulf smaller aerobic bacteria, which rather than being digested, live within their host and become mitochondria.

Endosymbiosis
Eukaryotes evolved when early prokaryote cells engulfed bacteria. The engulfed bacteria allowed their hosts to use oxygen to release energy stored in nutrients for the first time, and the hosts protected the bacteria.

THE **SINGLE CELL** OF THE AQUATIC ALGA ***CAULERPA TAXIFOLIA*** CAN **GROW TO 12 IN (30 CM)** IN LENGTH

Types of eukaryote

The Eukarya is an extraordinarily diverse group, ranging from single-celled protists (see pp.126–27) to giant blue whales. It is traditionally considered to comprise four kingdoms: animals, plants, fungi, and protists. The protists are a diverse group and some are more closely related to members of other kingdoms than they are to other protists. Their diversity is such that they could be classified into several kingdom-level groups.

TYPE	KEY CHARACTERISTICS
Protists	Most—but not all—protists are unicellular. Their cells have a nucleus and other membrane-bound organelles. Protists obtain food by ingesting or engulfing bacteria and other small particles.
Fungi	Most fungi are multicellular, and the cells have a wall. They reproduce by means of spores, either sexually or asexually. Fungi lack chlorophyll, so cannot perform photosynthesis.
Plants	Plants are multicellular and the cells have a wall. Almost all produce their own food by photosynthesis. Most reproduce sexually, with male and female reproductive organs either on the same or different plants.
Animals	Animals are multicellular, but a cell wall is absent. They usually obtain energy from digesting other organisms or their products. Animals reproduce sexually or asexually.

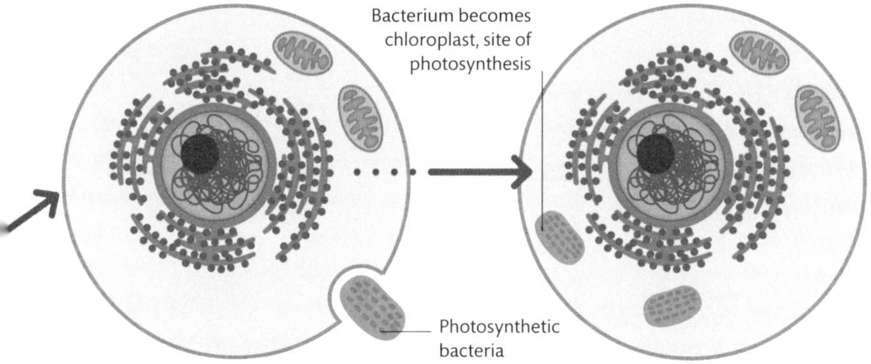
Bacterium becomes chloroplast, site of photosynthesis
Photosynthetic bacteria

4 Adding photosynthesis
Some cells also engulf photosynthetic bacteria, which live within their host, over time becoming chloroplasts.

5 Modern plant cell
A plant cell has mitochondria and chloroplasts, enabling it to perform both cellular respiration and photosynthesis.

DO EUKARYOTES HAVE AN ADVANTAGE OVER PROKARYOTES?

Yes, eukaryotic cells can organize themselves into complex, multicellular organisms—something that prokaryotic cells cannot do.

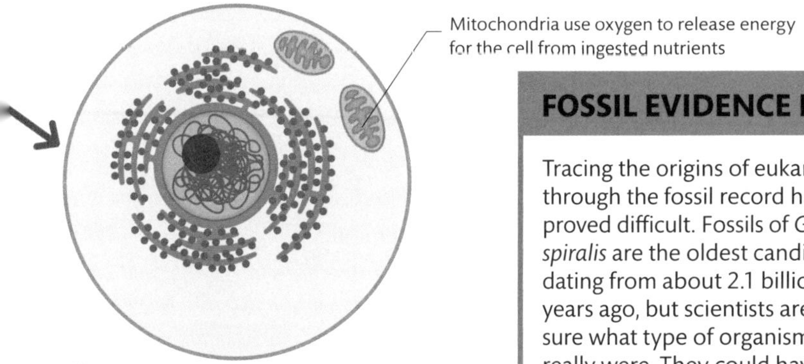
Mitochondria use oxygen to release energy for the cell from ingested nutrients

4 Heterotrophic cell
The cells of animals and fungi are heterotrophic, which means they cannot make their own food and have to ingest nutrients.

FOSSIL EVIDENCE FOR EUKARYOTES

Tracing the origins of eukaryotes through the fossil record has proved difficult. Fossils of *Grypania spiralis* are the oldest candidates, dating from about 2.1 billion years ago, but scientists are not sure what type of organisms they really were. They could have been eukaryotic algae, giant bacteria, or bacterial colonies.

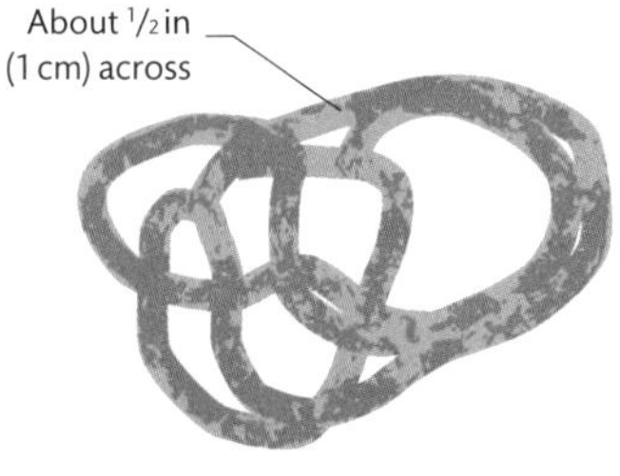
About 1/2 in (1 cm) across

GRYPANIA SPIRALIS **FOSSIL**

Radiolarians

Mostly marine, and usually nonmotile (unable to propel themselves), these single-celled organisms have delicate internal skeletons of silica and often exhibit radial symmetry. The soft anatomy is divided into a central capsule and an extracapsulum, separated by a central capsular wall. Radiolaria obtain energy by catching small plankton with pseudopods (see opposite), although some also have symbiotic relationships with photosynthetic algae.

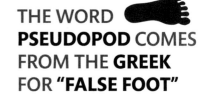

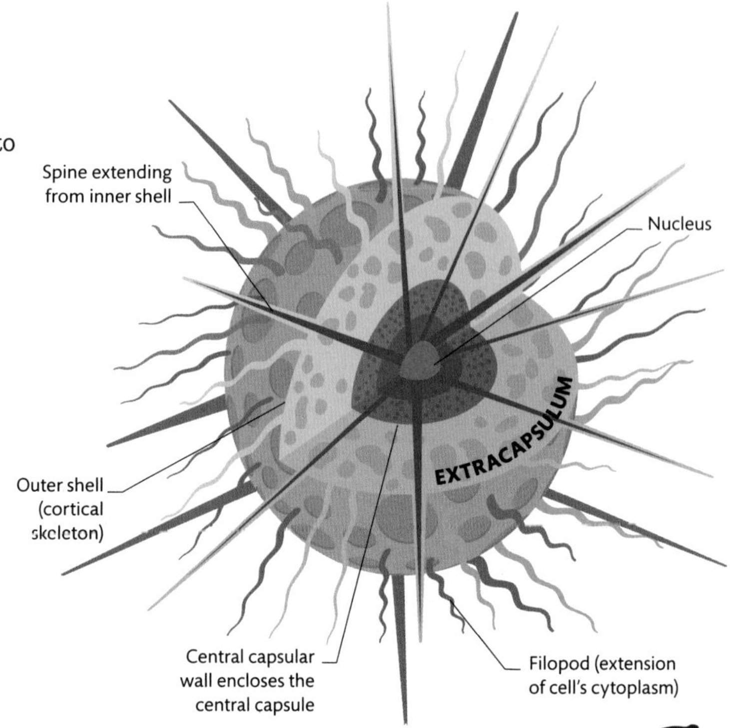
Spine extending from inner shell
Nucleus
EXTRACAPSULUM
Outer shell (cortical skeleton)
Central capsular wall encloses the central capsule
Filopod (extension of cell's cytoplasm)

Radial symmetry
Many radiolarians resemble miniature jewelry. Named for the radial symmetry of many species, some have radial spines of silica that increase drag in the water. The largest radiolarians are $^{3}/_{32}$ in (2 mm) in diameter.

Protists

The term protist is used to group together eukaryotes that do not belong to the animal, plant, or fungi kingdoms. Mostly single-celled organisms, protists have a wide range of structures and life cycles.

THE WORD **PSEUDOPOD** COMES FROM THE **GREEK** FOR **"FALSE FOOT"**

KEY PROTIST GROUPS

Although almost all protists are unicellular, many have the most elaborate of all cells, with organelles in their cells—rather than multicellular organs—carrying out biological functions. Protists exhibit a wide variety of feeding, locomotive, and reproductive behaviors. Fission is a form of asexual reproduction whereby a body divides into two (binary) or more (multiple) copies of itself.

Group	Nutrition	Reproduction	Locomotion
Radiolarians	Consume zooplankton, phytoplankton, and bacteria	Binary fission, multiple fission, or budding (see p.81)	Usually nonmotile, they drift in water currents
Diatoms	Mostly make their own food through photosynthesis	Primarily binary fission; sexual reproduction in a few species	Nonmotile species drift in currents; motile species use flagella to move
Ciliates	Consume bacteria and algae	Binary fission, budding, or sexual	Move by undulating cilia
Dinoflagellates	Eat diatoms and zooplankton	Binary fission or sexual	Move by undulating flagella
Euglenozoans	Photosynthesis, parasitism, or consume other organisms	Binary fission	Move by undulating flagella
Amoebozoans	Feed by process of phagocytosis (see opposite)	Binary fission or sexual	Move with pseudopods

WHICH DISEASES ARE CAUSED BY PROTISTS?

There are several, including malaria, sleeping sickness, Chagas disease, giardiasis, and amoebic dysentery.

Amoebozoans

Members of this large, diverse group are able to change shape, usually by extending or retracting pseudopods (fluid-filled projections of the cell). Some are predators, and others consume detritus. They ingest food by a process called phagocytosis, their pseudopods engulfing live prey. While most are single-celled, slime molds have a multicellular life stage.

Phagocytosis
Amoebozoans engulf their food with pseudopods. Organelles called lysosomes then move into action, releasing enzymes to break down the food.

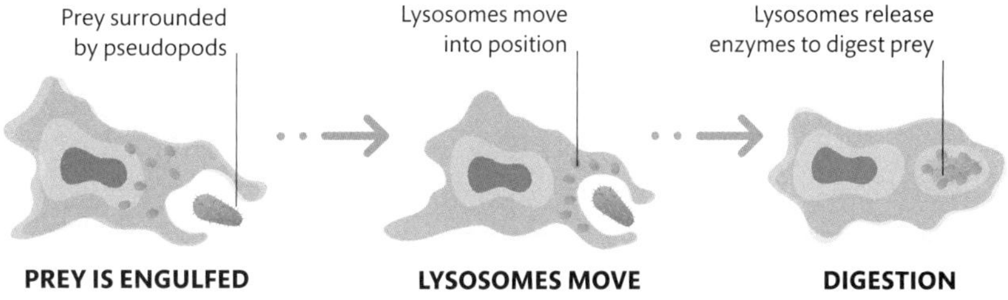
Prey surrounded by pseudopods
Lysosomes move into position
Lysosomes release enzymes to digest prey
PREY IS ENGULFED
LYSOSOMES MOVE
DIGESTION

Flagellates and ciliates

Two large groups of protists use specialized organelles—flagella or cilia—for locomotion or feeding. Cilia are short, hairlike structures, while flagella are long and whiplike. At some stage in their life cycle, flagellate protists possess one or more flagella. Ciliates are thought to have evolved from flagellates. Their surface may be completely covered with cilia, or these may be clustered in a few rows.

Double nuclei
Ciliates are single-celled organisms that use short organelles called cilia to swim. Every ciliate has one or more of two types of nuclei—macronuclei and micronuclei.

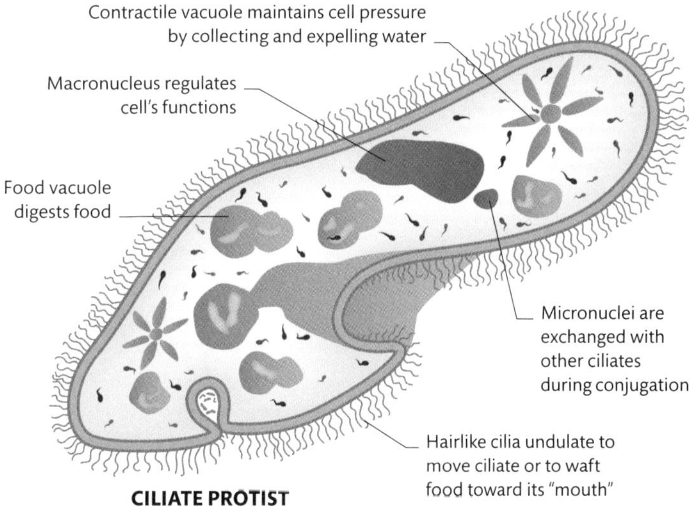
Contractile vacuole maintains cell pressure by collecting and expelling water
Macronucleus regulates cell's functions
Food vacuole digests food
Micronuclei are exchanged with other ciliates during conjugation
Hairlike cilia undulate to move ciliate or to waft food toward its "mouth"
CILIATE PROTIST

MALARIA'S LIFE CYCLE

When an infected *Anopheles* mosquito bites a person, it passes on the protist *Plasmodium* in the sporozoite phase of its life cycle. These protists migrate to liver cells, where they become merozoites. These are released to the bloodstream, where they destroy red blood cells.

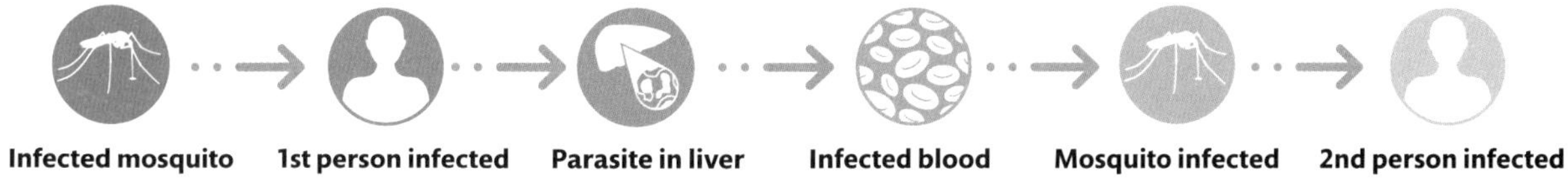
Infected mosquito
1st person infected
Parasite in liver
Infected blood
Mosquito infected
2nd person infected

Fungi

Ranging from mushrooms to microscopic molds and yeast, fungi are mostly multicellular organisms that absorb nutrients by growing in or through their food. This makes them different from plants, which make their own food by photosynthesis, and animals, which ingest food.

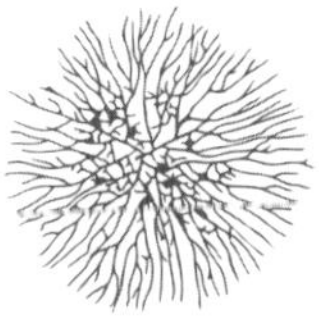

A SINGLE MYCELIUM DISCOVERED IN OREGON, US, **WEIGHS ABOUT 900,000 LB (400,000 KG)**

THE MAIN TYPES OF FUNGI

Until the present century, fungi were classified according to their form and structure. More recently, the advent of DNA analysis has challenged traditional divisions, and the kingdom Fungi is now divided into nine subdivisions (phyla). Four of the main phyla are listed here, along with the grouping Deuteromycota, which is still being studied and classified.

GROUP	KEY CHARACTERISTICS
Chytridiomycota	This group includes more than 750 species. They are decomposers, parasites, or live in symbiotic relationships in animals' digestive systems. One species is the cause of a deadly amphibian disease.
Glomeromycota	Many of the approximately 230 species of Glomeromycota have a symbiotic relationship with bryophytes (see p.132) and land plants, forming mycorrhizas within their roots.
Ascomycota	This group contains more than 64,000 species. These "sac fungi" possess an ascus, a sexual structure in which spores are produced. Present in 98 percent of lichens.
Basidiomycota	This group is made up of about 32,000 species. These "club fungi" include the familiar fruiting bodies of mushrooms, stinkhorns, bracket fungi, and puffballs.
Deuteromycota	This is a group of 25,000 "imperfect" fungi, so named because their sexual form of reproduction has never been observed. The best known is *Penicillium*.

1 Mature fruiting bodies
A mature fruiting body of a fungus produces spores. These are released and carried away by the wind, water, or animals. Once it has released its spores, the fruiting body begins to decompose.

4 Young fruiting body
Tiny hyphal knots form on the mycelium, and these grow into miniature fruiting bodies, or pinheads. Some of these develop into young fruiting bodies, while others grow no further.

3 Mycelium
The fine threads of hyphae in the mycelium consume the organic material around it. In the right conditions, the fungi will produce fruiting bodies.

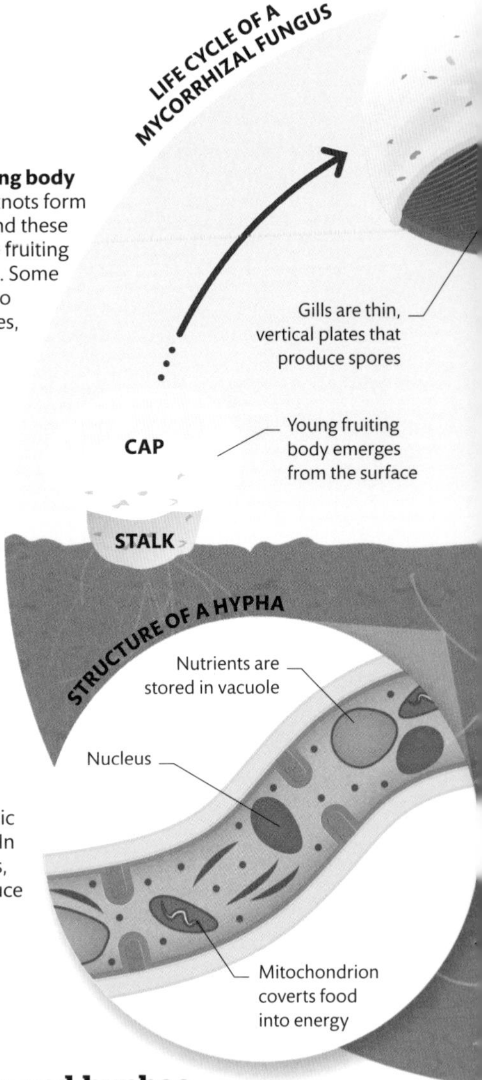
LIFE CYCLE OF A MYCORRHIZAL FUNGUS
Gills are thin, vertical plates that produce spores
CAP
Young fruiting body emerges from the surface
STALK
STRUCTURE OF A HYPHA
Nutrients are stored in vacuole
Nucleus
Mitochondrion coverts food into energy

Fungal cells and hyphae

With the exception of yeasts, fungi are multicellular organisms. Most grow as multicellular filaments called hyphae. Fungal hyphae form an interwoven mass called the mycelium, which grows through the material on which the fungus feeds. Some fungi have specialized hyphae called haustoria, which allow them to exchange nutrients with their hosts. Such mycorrhizal fungi are enormously important for the health of ecosystems.

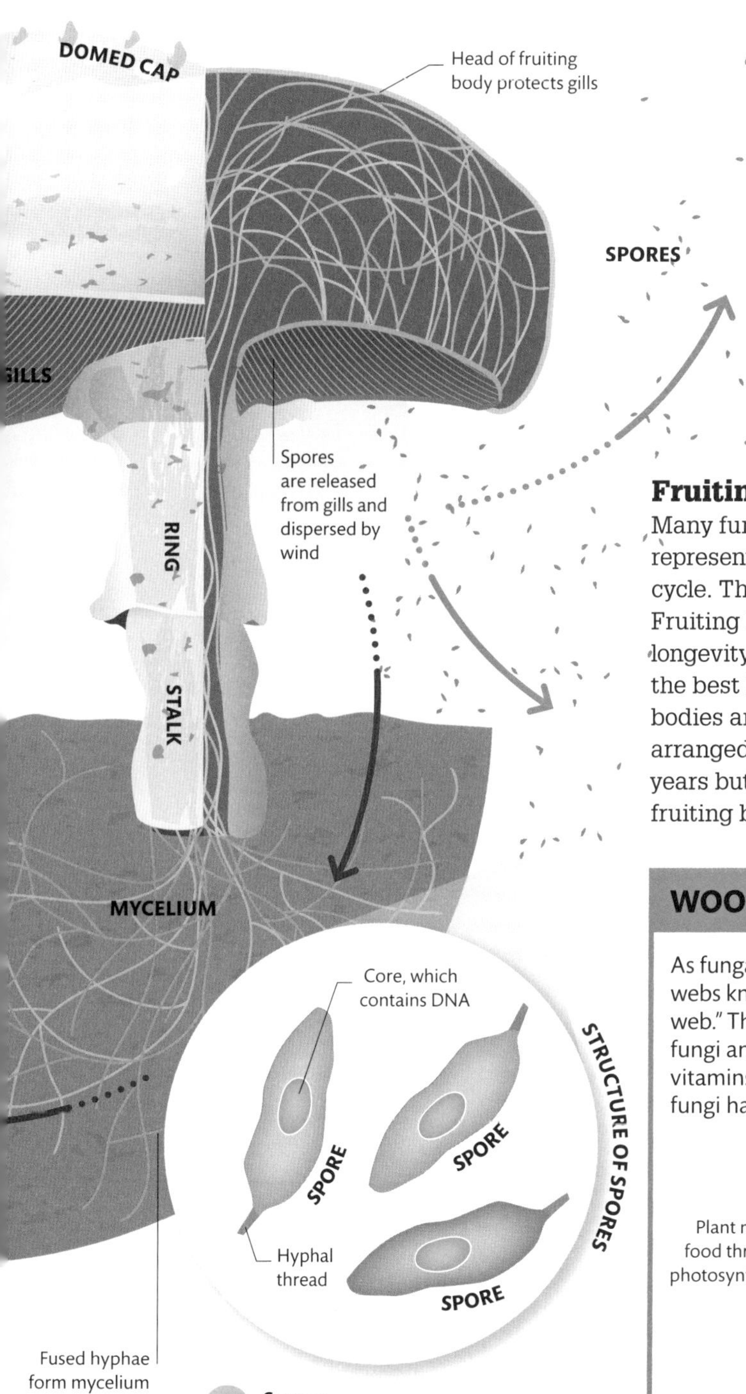
DOMED CAP
Head of fruiting body protects gills
SPORES
GILLS
Spores are released from gills and dispersed by wind
RING
STALK
MYCELIUM
Core, which contains DNA
STRUCTURE OF SPORES
SPORE
SPORE
SPORE
Hyphal thread
Fused hyphae form mycelium

2 Spores
Spores that fall where there is sufficient moisture and food germinate and form hyphae. When the hyphae of one spore meet those of another, they combine and form larger groups of hyphae, called the mycelium.

WHAT ARE LICHENS?

They are symbiotic associations between tiny photosynthetic organisms (usually algae) and a fungus, in which the former are held in a mass of fungal hyphae.

Fruiting bodies

Many fungi produce fruiting bodies, or sporocarps, which represent the sexual reproduction stage of a species' life cycle. These bear spores, which allow fungi to reproduce. Fruiting bodies vary greatly in shape, size, color, longevity, odor, and spore-dispersal mechanism. Some of the best known have a cap and stalk. Many fruiting bodies are solitary, but others are grouped in clumps or arranged in rings. Those of bracket fungi may persist for years but others last only a few days. Highly nutritious, fruiting bodies are often eaten by animals.

WOOD-WIDE WEBS

As fungal hyphae spread, they link up to plant roots, creating webs known as mycorrhizal networks, or a "wood-wide web." This enables an exchange of chemicals between fungi and plants. The plants provide carbohydrates and vitamins. In return, they receive water and minerals that fungi have absorbed from the soil and organic matter.

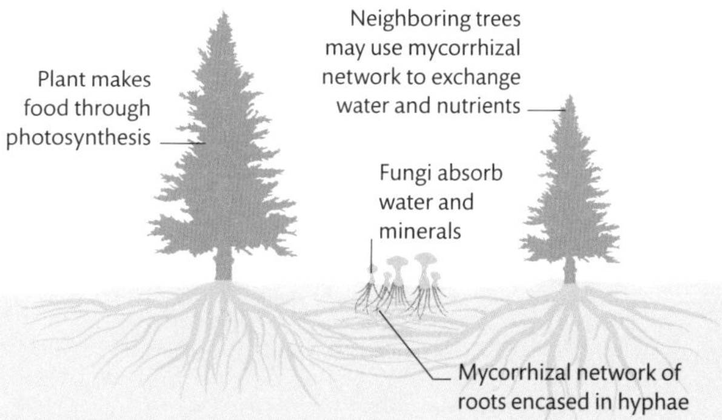
Plant makes food through photosynthesis
Neighboring trees may use mycorrhizal network to exchange water and nutrients
Fungi absorb water and minerals
Mycorrhizal network of roots encased in hyphae

Algae

Algae is a collective name for a diverse range of photosynthesizing organisms that live in the sea or in fresh water. All algae are eukaryotes (their cells have nuclei, unlike bacteria). Many are single-celled and microscopic; others, such as seaweeds, are larger and more complex.

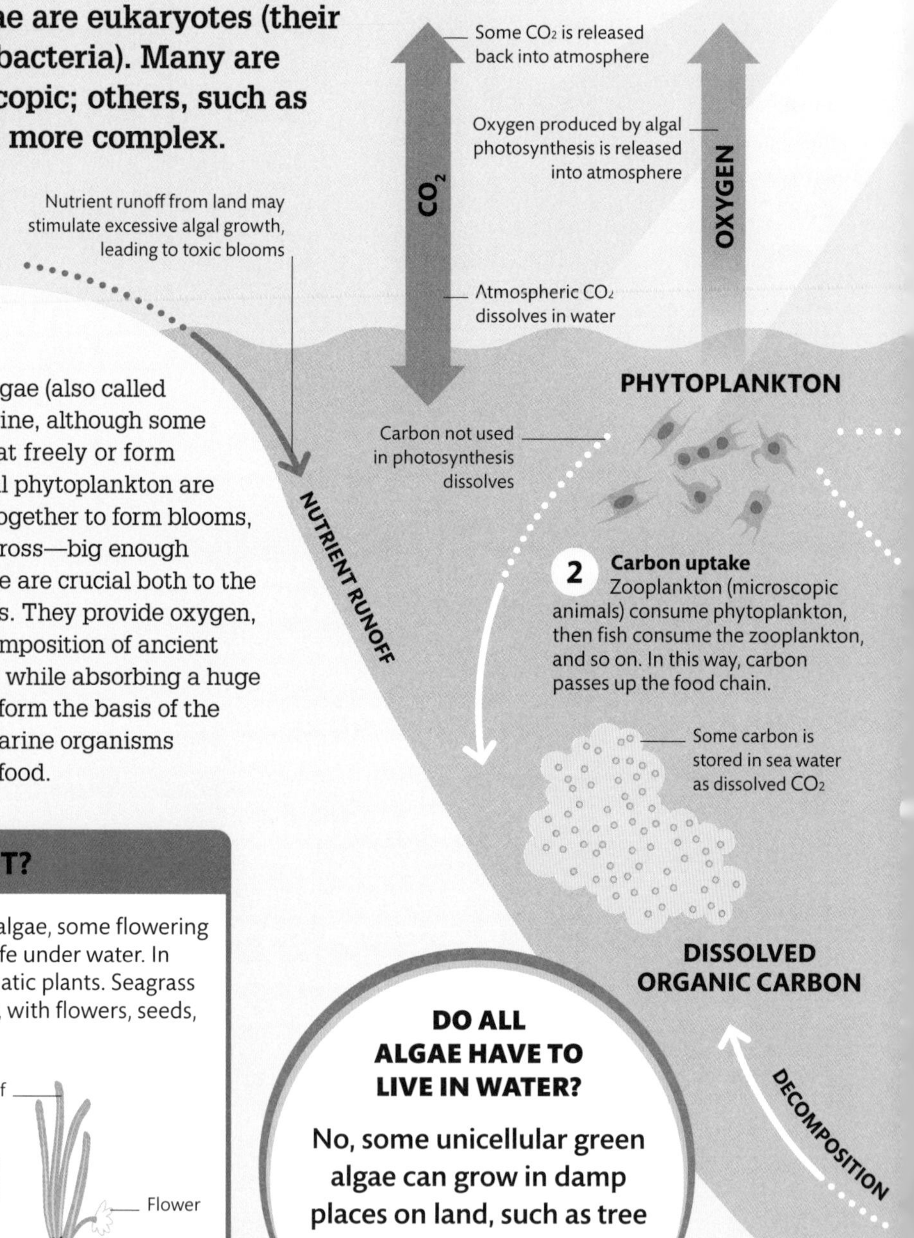

Nutrient runoff from land may stimulate excessive algal growth, leading to toxic blooms
Some CO_2 is released back into atmosphere
Oxygen produced by algal photosynthesis is released into atmosphere
CO_2
OXYGEN
Atmospheric CO_2 dissolves in water
PHYTOPLANKTON
Carbon not used in photosynthesis dissolves
NUTRIENT RUNOFF
2 Carbon uptake
Zooplankton (microscopic animals) consume phytoplankton, then fish consume the zooplankton, and so on. In this way, carbon passes up the food chain.
Some carbon is stored in sea water as dissolved CO_2
DISSOLVED ORGANIC CARBON
DECOMPOSITION

Phytoplankton

Phytoplankton are single-celled algae (also called microalgae). Most species are marine, although some live in fresh water. They may float freely or form large colonies. Although individual phytoplankton are microscopic, when billions come together to form blooms, these can be hundreds of miles across—big enough to be easily seen from space. Algae are crucial both to the world's ecosystems and to humans. They provide oxygen, food, fuel (oil comes from the decomposition of ancient algae), and medicinal compounds, while absorbing a huge amount of carbon. Phytoplankton form the basis of the marine food web, with all other marine organisms ultimately depending on them for food.

ALGA OR AQUATIC PLANT?

Although land plants evolved from algae, some flowering plants (see p.132) have reverted to life under water. In ponds, macroalgae coexist with aquatic plants. Seagrass is not a seaweed but a marine plant, with flowers, seeds, phloem, and xylem.

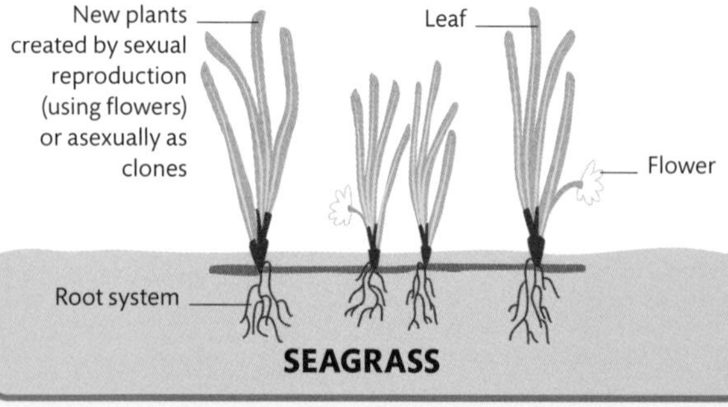

New plants created by sexual reproduction (using flowers) or asexually as clones
Leaf
Flower
Root system
SEAGRASS

DO ALL ALGAE HAVE TO LIVE IN WATER?

No, some unicellular green algae can grow in damp places on land, such as tree trunks, soil, wet rocks, damp bricks, or even animal fur.

SUN

The carbon pump
The oceans play a big part in the world's carbon cycle and are the biggest global store of carbon. In a phenomenon called the carbon pump, phytoplankton extract carbon dioxide from the atmosphere and use it for metabolism or to make organic compounds. Even small changes to this process can affect atmospheric CO_2 levels and therefore climate.

1 Photosynthesis
Phytoplankton photosynthesize in the ocean's sunlit surface waters, using dissolved atmospheric carbon dioxide (CO_2). They are the primary producers in the food chain.

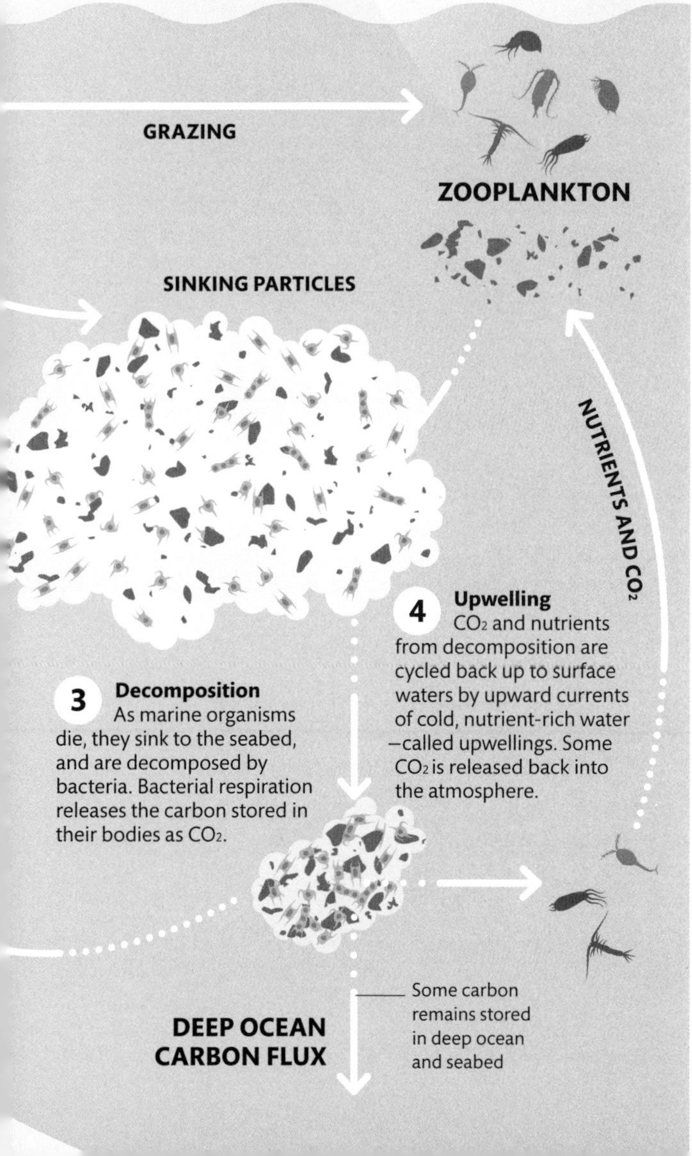
GRAZING
ZOOPLANKTON
SINKING PARTICLES
NUTRIENTS AND CO2
DEEP OCEAN CARBON FLUX
Some carbon remains stored in deep ocean and seabed

4 Upwelling
CO_2 and nutrients from decomposition are cycled back up to surface waters by upward currents of cold, nutrient-rich water—called upwellings. Some CO_2 is released back into the atmosphere.

3 Decomposition
As marine organisms die, they sink to the seabed, and are decomposed by bacteria. Bacterial respiration releases the carbon stored in their bodies as CO_2.

Seaweeds

Seaweeds are large, multicellular, oceanic algae. They fall into three groups—green, red, and brown algae. These use different photosynthetic pigments, causing the different colors. Like plants, seaweeds are autotrophic (make their own food), but lack xylem, phloem, and stomata. Instead of roots, stems, and leaves, they have holdfasts, stipes, and blades. Some also have floats to help them reach the light. Lacking flowers and seeds, seaweeds reproduce using spores.

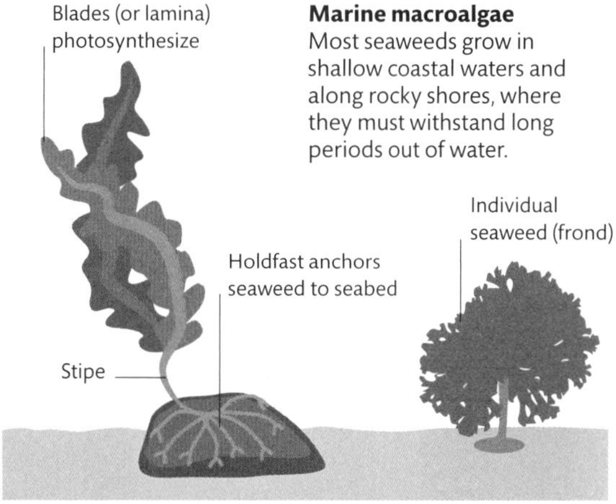
Blades (or lamina) photosynthesize
Holdfast anchors seaweed to seabed
Individual seaweed (frond)
Stipe

Marine macroalgae
Most seaweeds grow in shallow coastal waters and along rocky shores, where they must withstand long periods out of water.

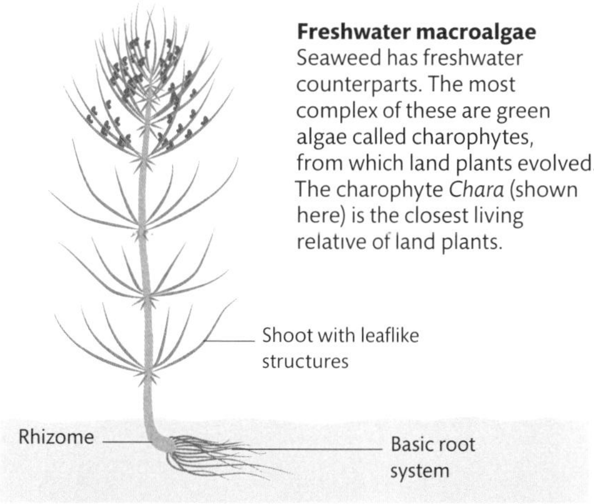
Shoot with leaflike structures
Rhizome
Basic root system

Freshwater macroalgae
Seaweed has freshwater counterparts. The most complex of these are green algae called charophytes, from which land plants evolved. The charophyte *Chara* (shown here) is the closest living relative of land plants.

UP TO 80% OF EARTH'S OXYGEN PRODUCTION OCCURS IN THE OCEAN

Plants

Found mostly on land, plants are multicellular organisms that usually contain the pigment chlorophyll. Typically growing in a permanent position, they absorb water and minerals through their roots and make nutrients by photosynthesis.

Land plants evolve
Plants adapted to life on land by controlling water loss and, later, developing roots to scavenge nutrients from rocks to create soils.

Lycophytes and pteridophytes (seedless vascular plants)
These were the first plants with true roots and a xylem and phloem transport system (see pp.148–49). Still reproducing using spores, they needed damp conditions, but vascular systems meant they could grow tall and develop large leaves—especially ferns.

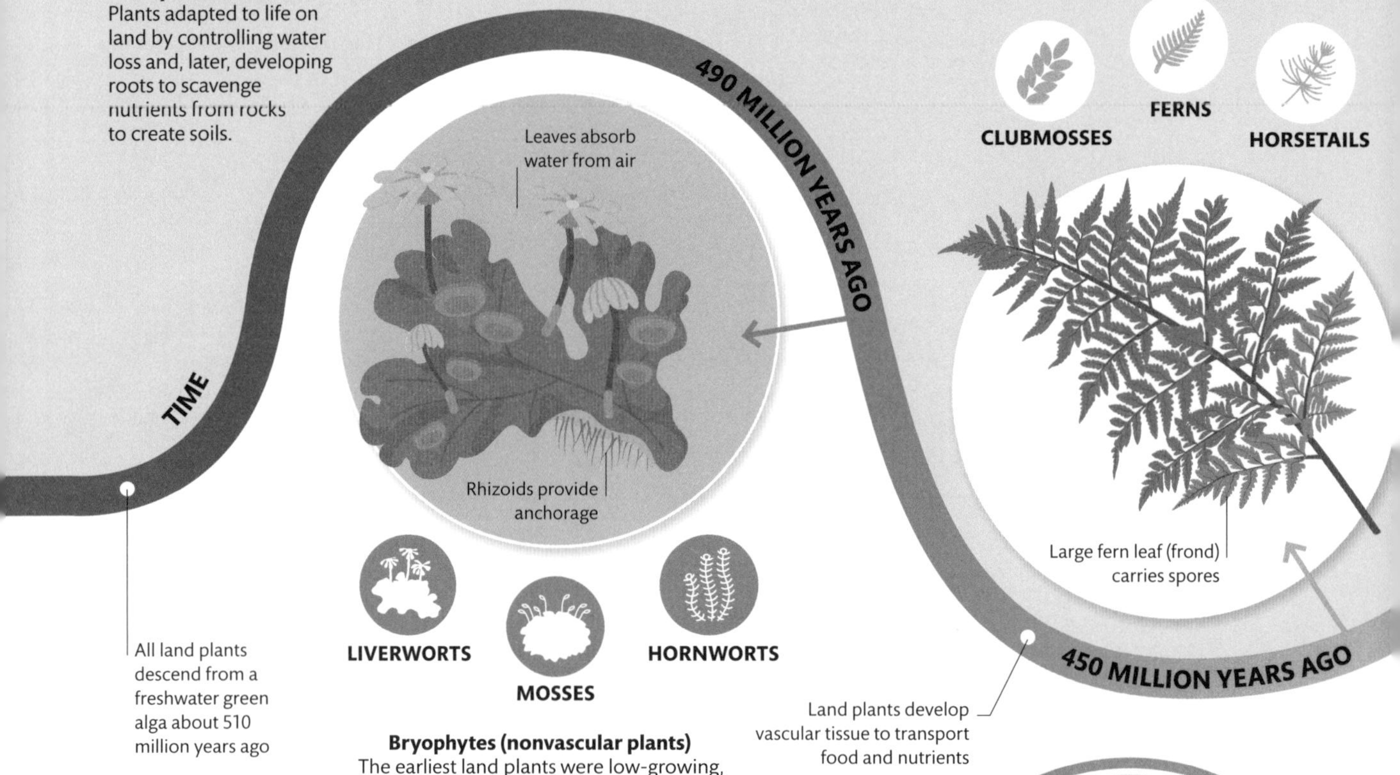
TIME
490 MILLION YEARS AGO
450 MILLION YEARS AGO
Leaves absorb water from air
Rhizoids provide anchorage
CLUBMOSSES
FERNS
HORSETAILS
Large fern leaf (frond) carries spores
LIVERWORTS
MOSSES
HORNWORTS
All land plants descend from a freshwater green alga about 510 million years ago
Land plants develop vascular tissue to transport food and nutrients

Bryophytes (nonvascular plants)
The earliest land plants were low-growing, nonvascular plants, meaning they had no transport system for water and nutrients. They reproduced using dustlike spores and were confined to damp areas.

Nonflowering plants

The colonization of land by plants nearly half a billion years ago was one of the most significant events in Earth's history. All the earliest land plants were nonflowering. The most primitive groups, such as mosses and ferns, did not have seeds, instead reproducing using spores that were spread by water or wind. Gymnosperms were the first plants to have pollen grains and seeds. No longer dependent on water to reproduce, they could colonize new areas.

HOW MANY SPECIES OF PLANT CURRENTLY EXIST?

It is estimated that there are currently around 450,000 plant species. Around 382,000 have been discovered and named. About 40 percent are at risk of extinction.

Vascular plants develop seeds for reproduction
320 MILLION YEARS AGO
Seeds carried in woody cone
Narrow leaves suited to cold weather

CONIFERS
CYCADS
GINKGOS

Gymnosperms (vascular seed plants)
These were the first seed plants, producing pollen and having naked seeds (not contained in ovaries). Modern gymnosperms are dominated by conifers (woody trees such as pine and fir), with about 600 species.

Flowering plants

The development of flowers and fruits made angiosperms hugely successful, leading to massive diversification into around 370,000 species that have colonized all but the most extreme habitats on Earth. Angiosperms make up the vast majority of plant species today, dominating most ecosystems on land. They take on a huge variety of forms, from towering trees to tiny herbaceous plants, and have formed complex symbiotic relationships with animals through pollination.

Angiosperms (flowering plants)
After the evolution of gymnosperms came the huge evolutionary innovation of flowers and fruits (mature ovaries). Flowering plants are known as angiosperms. Angiosperm seeds are contained within fruits.

MONOCOTS
DICOTS

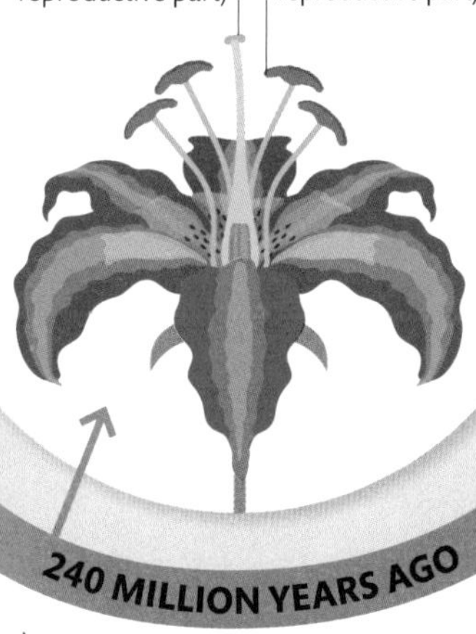
Carpel (female reproductive part)
Stamen (male reproductive part)
240 MILLION YEARS AGO

MONOCOTS AND DICOTS

Flowering plants fall into two groups: dicotyledons (about 70 percent of species) and monocotyledons (30 percent). Although dicots are more numerous and diverse, monocots include some of the largest plant families and most of the staple crops.

MONOCOTS	DICOTS
 The seed has one cotyledon (an embryonic leaf, or seed leaf).	 The seed has two cotyledons.
 Veins on the leaf are usually parallel.	 Veins on the leaf usually form a branched network.
 The vascular bundles are scattered throughout the stem.	 The vascular bundles are arranged in a ring.
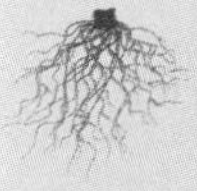 Roots are usually fibrous (no main central root).	 There is usually one main taproot from which the rest of the root system branches off.
 Flower parts (such as petals and stamens) are usually in multiples of three.	 Flower parts are usually in multiples of four or five.

THE OLDEST INDIVIDUAL TREE IS OVER 5,000 YEARS OLD

Invertebrates

Most animals are invertebrates—multicellular organisms lacking a vertebral column. About 1.3 million species have been described, but with new discoveries being made all the time the total is likely to be much higher.

Types of invertebrate

The invertebrate grouping is an assemblage of convenience. Some types of invertebrate are more closely related to vertebrates than they are to other invertebrates. They occupy almost every habitat on Earth from the frozen wastes of Antarctica to ocean hydrothermal vents. Reflecting this variety, they exhibit a bewildering array of lifestyles and body forms.

IT IS ESTIMATED THAT ABOUT **97% OF ALL ANIMALS** ON EARTH ARE **INVERTEBRATES**

KEY INVERTEBRATE GROUPS

There are 30 phyla of invertebrates (a phylum is a major division of a kingdom). Some of the largest and most interesting are listed below. By far the largest group is phylum Arthropoda, which includes all insects and spiders, and accounts for more than three out of four known species of animal.

GROUP	TAXONOMIC RANK	CHARACTERISTICS
Insects Over 1 million species	Phylum: Arthropoda Class: Insecta	Three pairs of legs; antennae; three-part body with head, thorax, and abdomen
Arachnids Over 100,000 species	Phylum: Arthropoda Class: Arachnida	Four pairs of legs; no wings or antennae; head and thorax fused
Crustaceans 67,000 species	Phylum: Arthropoda Subphylum: Crustacea	Flexible exoskeleton or shell; two pairs of antennae
Molluscs 85,000 species	Phylum: Mollusca	Body has head, muscular foot, and mantle (which may secrete a shell)
Sea urchins 7,000 species	Phylum: Echinodermata	Radially symmetrical; spiny skinned; exclusively marine
Annelid worms Over 15,000 species	Phylum: Annelida	Segmented body; respire through body surface

Dragonfly
Like other insects, a dragonfly has six legs and three main body sections—head, thorax, and abdomen. While dragonflies have four wings, flies have only two, and many insects have none.

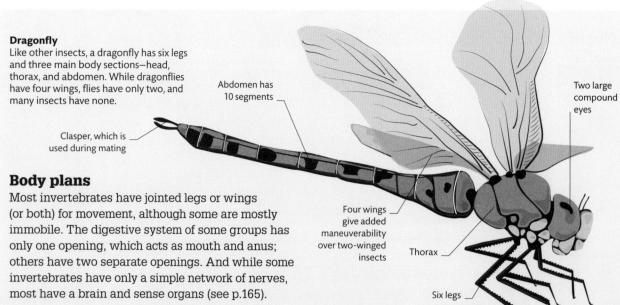

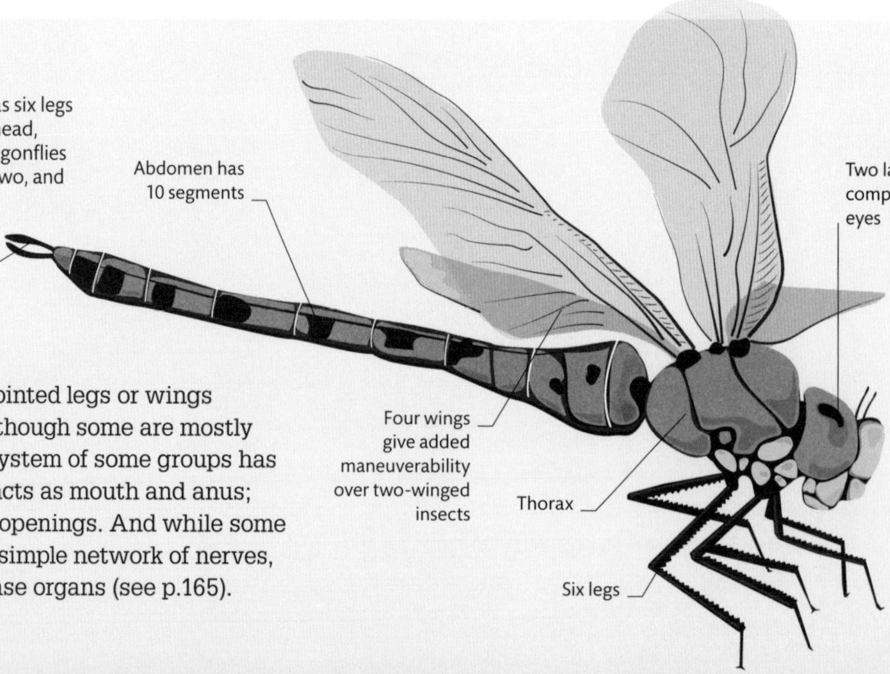
Abdomen has 10 segments
Two large compound eyes
Clasper, which is used during mating
Four wings give added maneuverability over two-winged insects
Thorax
Six legs

Body plans

Most invertebrates have jointed legs or wings (or both) for movement, although some are mostly immobile. The digestive system of some groups has only one opening, which acts as mouth and anus; others have two separate openings. And while some invertebrates have only a simple network of nerves, most have a brain and sense organs (see p.165).

Growth and development

While all invertebrates' bodies change as they grow, in some groups—such as spiders—the juveniles look like miniature adults. Transformations are much more dramatic in other groups. For example, insects undergo incomplete or complete metamorphosis. The latter involves changing from egg to larva, pupa, and finally adult. In incomplete metamorphosis, the insect develops from an egg into a nymph before becoming an adult.

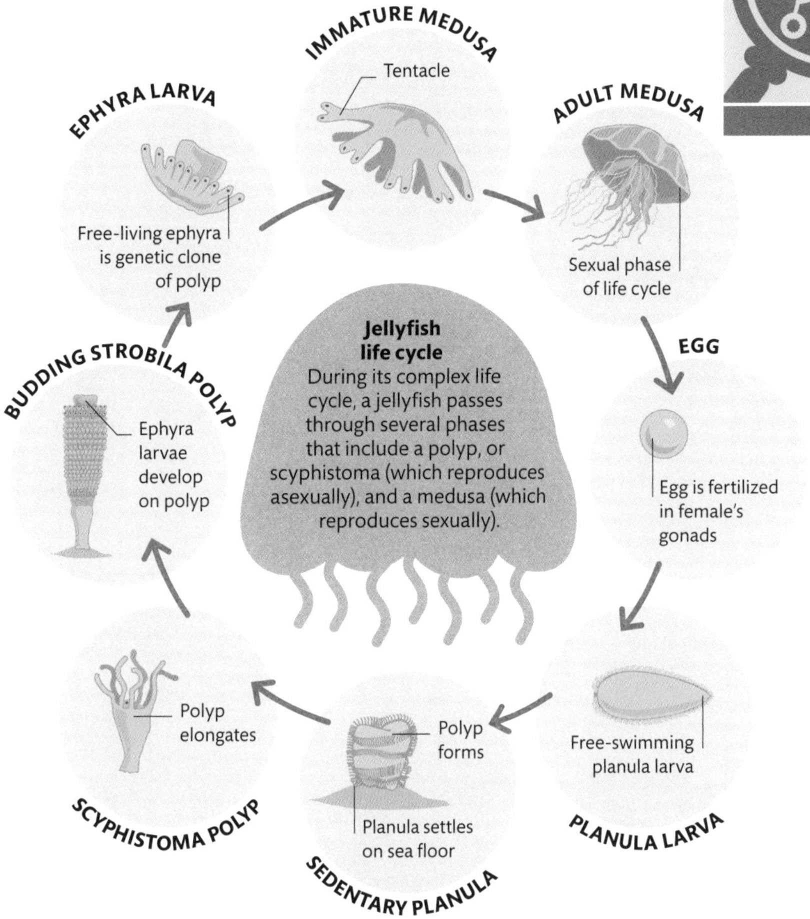
IMMATURE MEDUSA
Tentacle
ADULT MEDUSA
Sexual phase of life cycle
EGG
Egg is fertilized in female's gonads
PLANULA LARVA
Free-swimming planula larva
SEDENTARY PLANULA
Polyp forms
Planula settles on sea floor
SCYPHISTOMA POLYP
Polyp elongates
BUDDING STROBILA POLYP
Ephyra larvae develop on polyp
EPHYRA LARVA
Free-living ephyra is genetic clone of polyp
Jellyfish life cycle
During its complex life cycle, a jellyfish passes through several phases that include a polyp, or scyphistoma (which reproduces asexually), and a medusa (which reproduces sexually).

COLONIAL INVERTEBRATES

Some marine invertebrates, such as corals and sponges, live together in large groups called colonies. A coral colony is made up of polyps, which are sedentary for most of their life cycle.

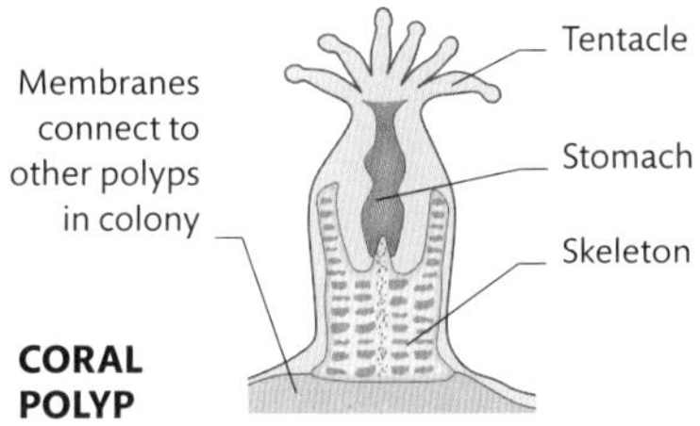
Tentacle
Membranes connect to other polyps in colony
Stomach
Skeleton
CORAL POLYP

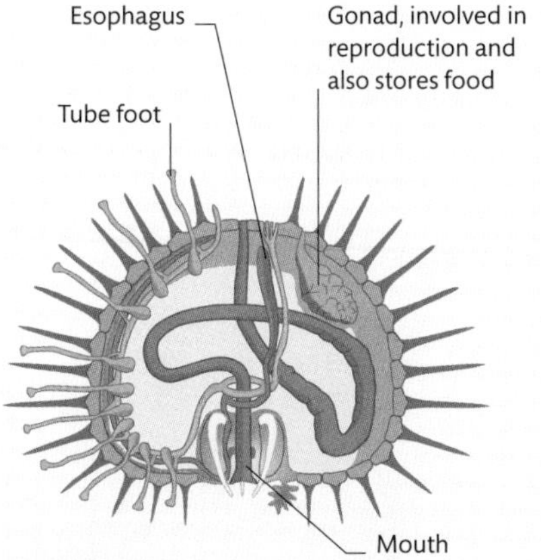
Esophagus
Gonad, involved in reproduction and also stores food
Tube foot
Mouth

Sea urchin
Roughly spherical, sea urchins lack arms but have five rows of tube feet that enable them to move slowly. Muscles pivot long spines, which give protection and help locomotion.

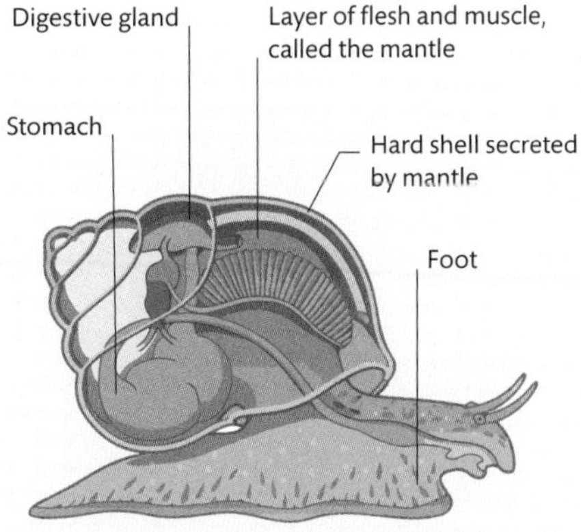
Digestive gland
Layer of flesh and muscle, called the mantle
Stomach
Hard shell secreted by mantle
Foot

Mollusk
Three-quarters of all mollusks, like this snail, are gastropods. Gastropods typically have a large foot with a flat sole, a protective shell, and a head with a pair of eyes and tentacles.

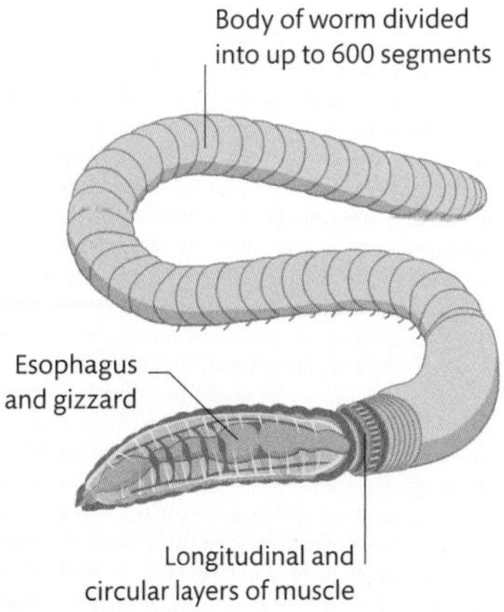
Body of worm divided into up to 600 segments
Esophagus and gizzard
Longitudinal and circular layers of muscle

Worm
The segments of annelid worms are surrounded by longitudinal muscle covered with circular muscle. Worms coordinate the contraction of muscle sets for locomotion.

Vertebrates

Almost all vertebrates have an internal skeleton that includes a spinal column and a skull. This group of animals gets its name from the chain of bones (vertebrae) that form the spine.

Chordates and vertebrates

Vertebrates evolved from chordates, the earliest-known fossils of which date from 530 million years ago (MYA). While all chordates have a notochord (a flexible rod running the length of the body), in the vertebrates this is replaced by a spinal column. The first vertebrates were jawless fish, which appear in the fossil record about 518 MYA. Vertebrates remained aquatic organisms for almost 150 million years, until the evolution of limbs in one group of fish prepared the way for the colonization of land. Land-dwelling vertebrates diversified into amphibians, reptiles (including the ancestors of dinosaurs and birds), and mammals.

KEY TO DEFINING CHARACTERISTICS

- Warm-blooded
- Cold-blooded
- Lays eggs
- Live birth
- Covered in feathers
- Covered in hair
- Covered in scales
- Scaleless
- Fish with jaws
- Fish with bony skeleton
- Fish with flexible cartilage skeleton
- Fish with cartilage skeleton, no collagen

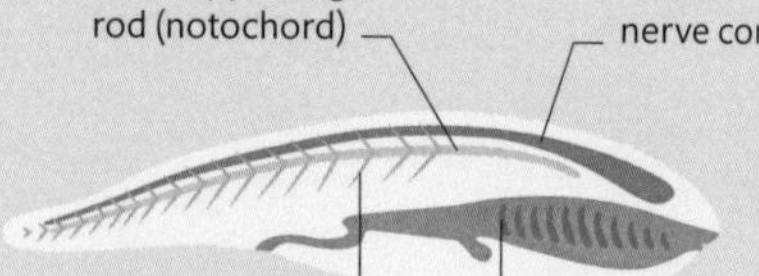
Flexible, supporting rod (notochord)
Brain and nerve cord
LANCELET
Muscle segment
Pharyngeal slits (filter-feeding organs)

Chordates
All chordates have a notochord and a hollow nerve cord. The chordates are categorized into three groups: vertebrates, tunicates (sea squirts and their close relatives), and cephalochordates (lancelets).

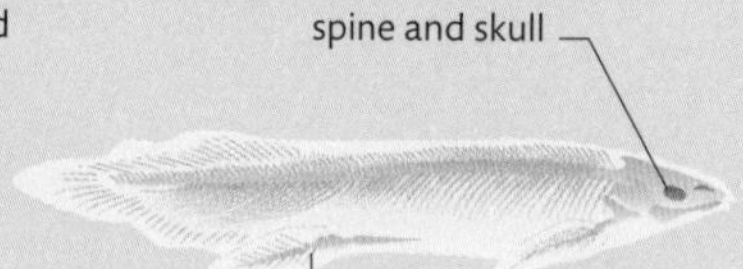
Internal skeleton with spine and skull
BONY FISH
Pelvic fin aids stability when swimming

Vertebrates
All vertebrates have a backbone with vertebrae apart from hagfish, which have only a primitive notochord. A vertebrate's mouth is at the front end of the animal, and the anus is just before the back end of the body.

WHAT IS THE MOST COMMON VERTEBRATE ON EARTH?

The tiny deep-sea fish known as bristlemouths are thought to be the most abundant vertebrates on Earth.

THE WORLD'S **SMALLEST VERTEBRATE**, THE **PAEDOPHYRNE FROG**, IS JUST 0.3 IN (7.7 MM) LONG

THE TETRAPOD EVOLVES

The earliest evidence of four-limbed vertebrates (tetrapods) is footprints dating from 390 million years ago. Tetrapods evolved from early lobe-finned fish living in shallow water. The earliest species had legs and lungs as well as gills, but they remained primarily aquatic.

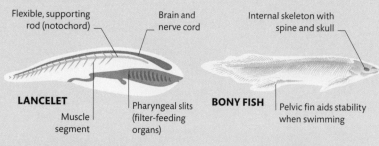

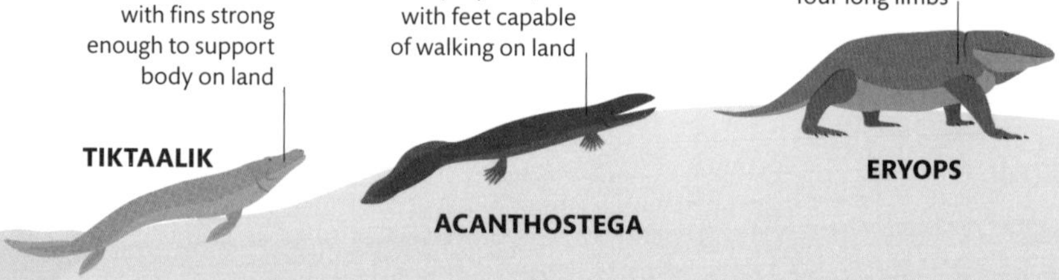
Swamp-dweller with fins strong enough to support body on land
TIKTAALIK
Mostly aquatic, but with feet capable of walking on land
ACANTHOSTEGA
Amphibious (comfortable on land and in water) with four long limbs
ERYOPS

Types of vertebrates

Vertebrates are usually divided into seven groups: the wholly aquatic jawless, cartilaginous, and bony fish; the partially aquatic amphibians; and the mostly terrestrial reptiles, mammals, and birds. Mammals and birds are endotherms (warm-blooded), meaning that they can maintain their body temperature by generating heat internally. Other vertebrates are ectotherms (cold-blooded), and are mostly dependent on external heat sources.

BIRDS

Birds evolved from dinosaurs around 160 million years ago. They have feathers, lay hard-shelled eggs, and most species are capable of powered flight.

REPTILES

The earliest examples of reptiles date from 320–310 million years ago. Reptiles have scales and are mostly terrestrial (live on land). Most species lay eggs, but some species give birth to live young.

MAMMALS

Mammals evolved from early reptiles more than 300 million years ago. They have hair, nourish their young with milk, and give birth to live offspring.

BONY FISH

Bony fish first appeared about 425 million years ago. They have a bony skeleton, are covered in scales, and have one pair of gill openings. Modern examples of bony fish are either ray-finned or lobe-finned.

AMPHIBIANS

Amphibians evolved from lobe-finned fish around 370 million years ago. They are scaleless and partly terrestrial, but most species lay their eggs in or near water. Modern examples include frogs and newts.

CARTILAGINOUS FISH

The earliest examples of cartilaginous fish date from 430 million years ago. They have a skeleton made of relatively soft, flexible cartilage. Modern examples include sharks and rays.

JAWLESS FISH

The first jawless fish appeared about 530 million years ago. They have a skeleton made of collagen-free cartilage. Modern examples of jawless fish include lampreys.

INVERTEBRATES ...

Adapting to land

Vertebrates include the heaviest animals ever to walk on land. In mammals, reptiles, and birds, the embryo grows inside a waterproof membrane called an amnion. This adaptation allowed early tetrapods to evolve outside bodies of water.

HOW PLANTS WORK

Seeds

Seeds form when pollen grains fertilize egg cells. The seed protects the plant's embryo until it germinates. It may also enable the embryo to survive cold or dry conditions, or aid dispersal over a long distance.

Structure of seeds

In flowering plants, seeds develop inside an ovary. These plants are referred to as angiosperms (literally, "clothed seeds"). In gymnosperms ("naked seeds"), seeds are held on scales within cones. In all seed-bearing plants, the seed holds an embryo (developing plant) and a nutrient-rich tissue that forms the embryo's food store. These structures are enclosed in a seed coat. In angiosperms, a double fertilization process occurs to produce the embryo and the food store, which is called the endosperm. The embryo, in turn, comprises a plumule (shoot), radicle (root), one or two seed leaves (cotyledons), and the first true leaves. Plants with one cotyledon are called monocotyledons; those with two are called dicotyledons.

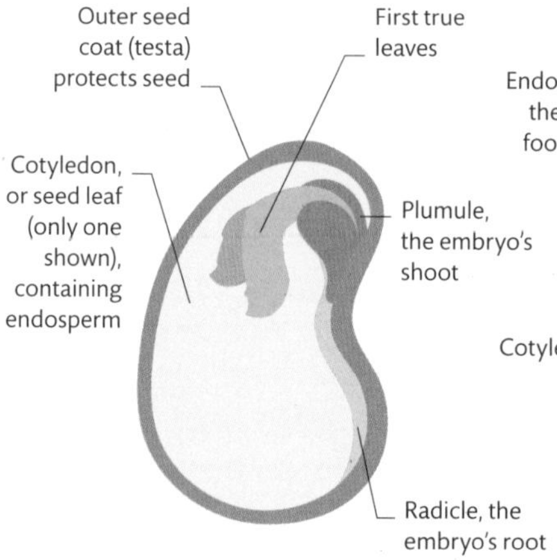

Outer seed coat (testa) protects seed
First true leaves
Cotyledon, or seed leaf (only one shown), containing endosperm
Plumule, the embryo's shoot
Radicle, the embryo's root

Dicotyledon (dicot) seed
Dicots have two cotyledons, which emerge as an opposing pair. The endosperm is contained inside the cotyledons, so these first leaves are fat and rounded.

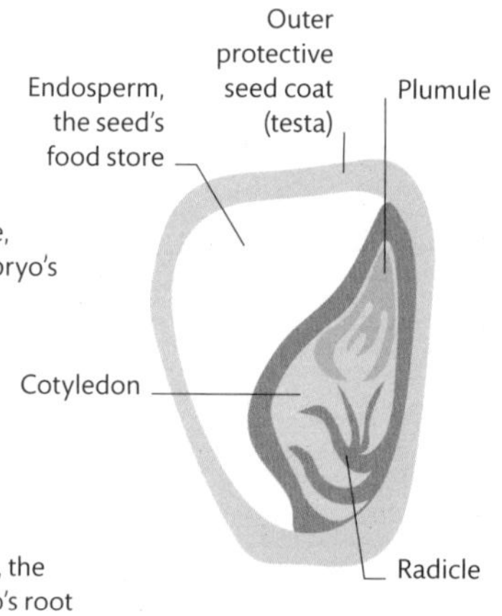

Endosperm, the seed's food store
Outer protective seed coat (testa)
Plumule
Cotyledon
Radicle

Monocotyledon (monocot) seed
Monocots, such as grasses and wheat, have one cotyledon, so the seedling has one vertical first leaf. The endosperm is not contained in the cotyledon, so the leaf is thin.

Germination and hormones

Germination is the process by which a seed breaks dormancy and an embryo plant emerges. Warmth, moisture, and oxygen are required for the seed to begin metabolising and the seedling to start to grow. The timing and process of germination are regulated by hormones in the seed. Abscisic acid (ABA) keeps the seed dormant, while gibberellins trigger germination. Auxins stimulate tropisms—growth changes in response to water and light sources. In shoots, high auxin concentrations stimulate growth, whereas in roots, high auxin concentrations inhibit growth. Ethylene also helps to stimulate germination, while hormones called cytokinins stimulate cell growth and division.

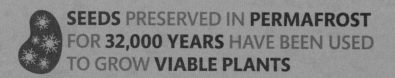

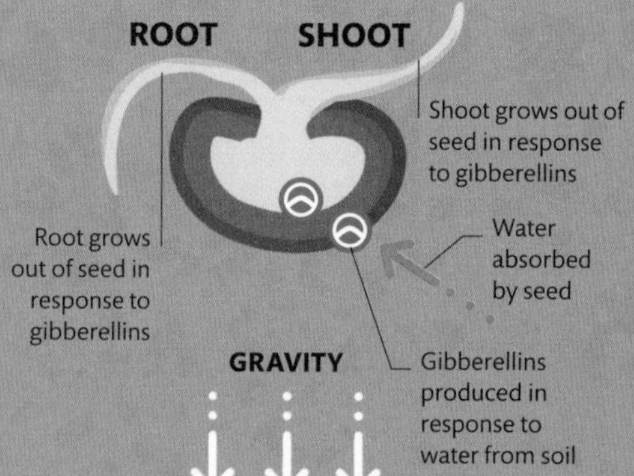

ROOT
SHOOT
Shoot grows out of seed in response to gibberellins
Root grows out of seed in response to gibberellins
Water absorbed by seed
GRAVITY
Gibberellins produced in response to water from soil

1 Seed germinates
When a seed absorbs water from the soil, gibberellins released from the embryo cause it to break dormancy and begin to germinate. Enzymes break down starch in the endosperm to release glucose for energy. This enables first the root and then the shoot to grow and emerge from the seed case.

SEEDS PRESERVED IN **PERMAFROST** FOR **32,000 YEARS** HAVE BEEN USED TO GROW **VIABLE PLANTS**

SUN

LIGHT

WHAT IS THE WORLD'S LARGEST SEED?

The coco-de-mer palm from the Seychelles produces the world's largest seeds. A single seed can weigh up to 55 lb (25 kg) and be 20 in (50 cm) long.

LIGHT

Plant grows toward light (positive phototropism)

Auxins accumulate on side that is shaded, stimulating cell elongation on that side

Shoot grows upward, toward light and away from gravity (positive phototropism and negative geotropism)

PLANT HORMONES

Plants rely on hormones to bring about germination and also to regulate growth and responses to the environment throughout their life cycle. Gibberellins, auxins, and cytokinins control stem growth and the development of buds and flowers. Other key hormones are ethylene, which controls fruit ripening, and abscisic acid, which controls leaf shedding in the fall.

ROOT SHOOT

Auxins pulled down by gravity stimulate cell elongation in the shoot

Auxins pulled down by gravity inhibit cell elongation, causing root to bend downwards

Abscisic acid buildup in leaves causes shedding

GRAVITY

Auxins build up on shaded side of roots, where they inhibit growth

GRAVITY

Roots bend away from light (negative phototropism)

LEAF SHEDDING

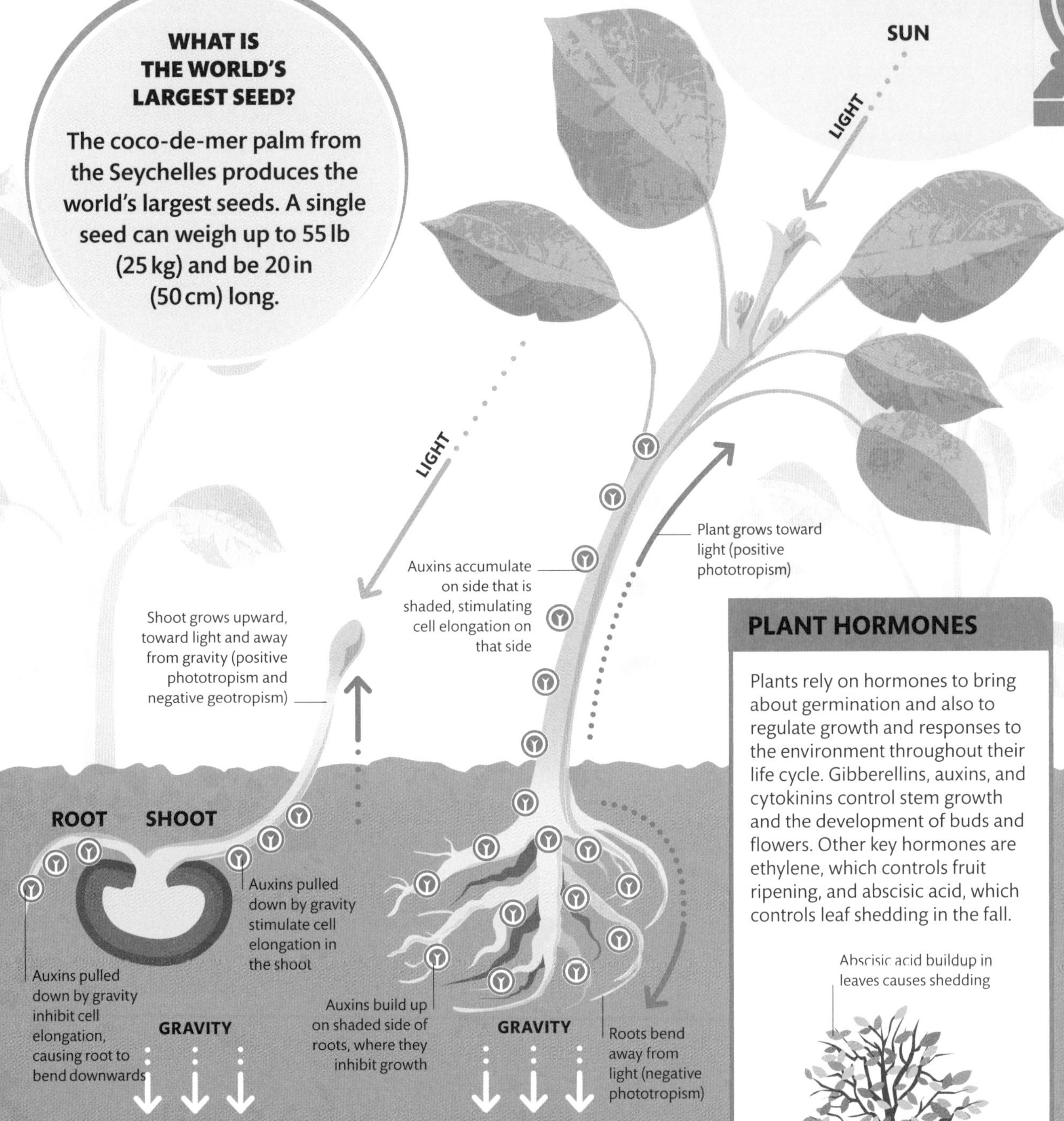

2 Auxins and gravity
The growth response to gravity is called geotropism. In roots, a buildup of auxins on the lower side inhibits cell growth, and this causes the root to bend downward. In the shoot, auxin buildup on the side facing the ground increases cell elongation, causing the shoot to bend upward.

3 Auxins and light
The growth response to light is called phototropism. In this response, auxins build up in parts of the plant that are shaded from light. In shoots, the cells on the shaded side grow longer, causing the shoot to bend toward the light. In roots, auxins have the opposite effect, so the roots angle away from light.

Roots

Roots anchor a plant in the ground and take up water and minerals from the soil. They can also store carbohydrates produced by photosynthesis. Plant roots, particularly grass roots, are key in minimizing soil erosion, allowing ecological communities to develop.

DO ROOTS STOP GROWING?

No, roots grow throughout the entire life of the plant. They grow from the tip, which is covered with a tough layer of dead cells known as the root cap.

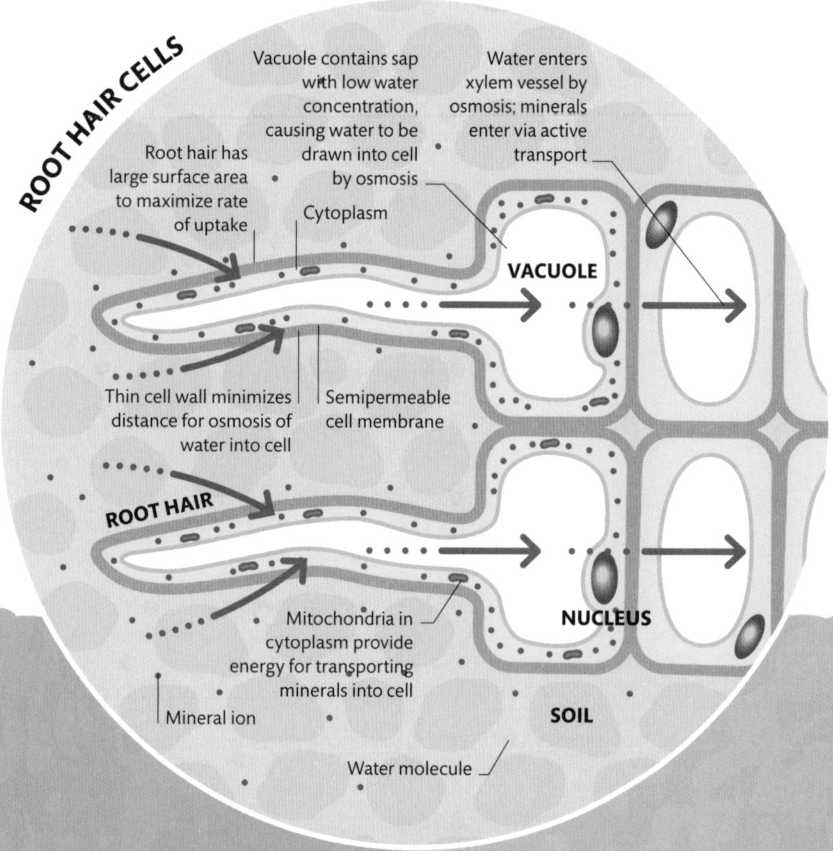

ROOT HAIR CELLS
Vacuole contains sap with low water concentration, causing water to be drawn into cell by osmosis
Water enters xylem vessel by osmosis; minerals enter via active transport
Root hair has large surface area to maximize rate of uptake
Cytoplasm
VACUOLE
Thin cell wall minimizes distance for osmosis of water into cell
Semipermeable cell membrane
ROOT HAIR
Mitochondria in cytoplasm provide energy for transporting minerals into cell
NUCLEUS
Mineral ion
SOIL
Water molecule

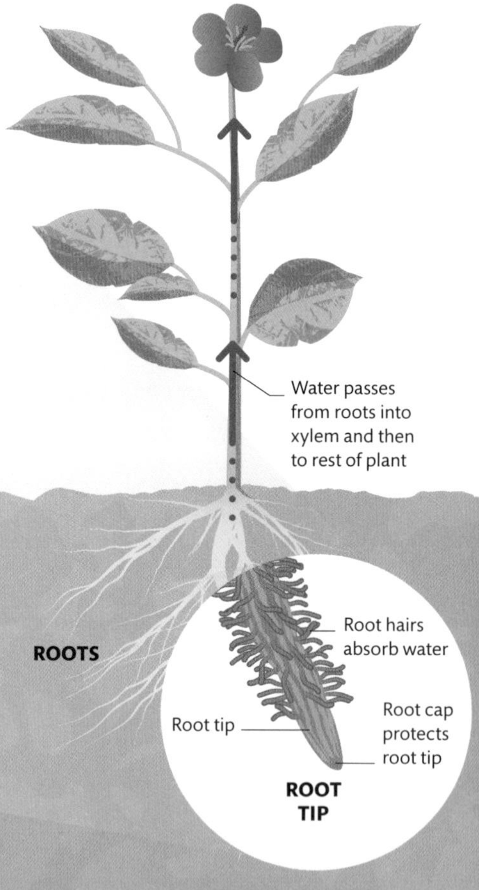

Water passes from roots into xylem and then to rest of plant
ROOTS
Root hairs absorb water
Root tip
Root cap protects root tip
ROOT TIP

The structure of a root system

A root network enables a plant to draw water and minerals out of the surrounding soil. Dicotyledons (many flowering plants including shrubs and trees) and gymnosperms (such as conifers) have a central taproot off which smaller roots branch. Monocotyledons (such as grasses) and ferns have shallow roots radiating from the stem base. Both systems end in fine roots covered with hairs, which hugely increase the surface area for absorption. In addition, most plants have a symbiotic relationship with fungi in the soil. The fungi act as root extensions, increasing total surface area for absorption. In return, the plant shares sugars with the fungi.

Root anatomy and function
Root hair cells absorb water by osmosis—the movement of water molecules from a high concentration area to a lower concentration area—through the semipermeable cell membranes. From there, water passes into the root cortex and then into the xylem for transport to the shoots and leaves.

Storage organs

As well as taking up water and minerals, some root systems store carbohydrates to provide energy for the plant. The taproot of dicotyledons and gymnosperms is often swollen with sugars and starch, as in carrots. Other plants store starch in special underground storage organs such as bulbs, tubers, and rhizomes. These are modified underground stems or shoots rather than swollen roots.

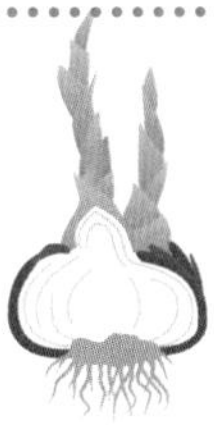

CROCUS

Corm
This is a short, swollen stem, protected by scale leaves formed from the previous year's foliage. It has no fleshy storage leaves.

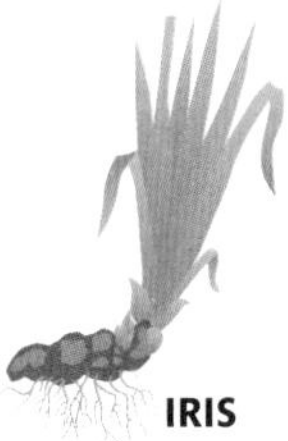

IRIS

Rhizome
A rhizome is a modified shoot that grows horizontally just below the soil surface. Vertical shoots arise from buds along its length.

ONION

Bulb
A bulb is a condensed shoot with many layers formed from the previous year's leaves, whose bases have swollen for storage.

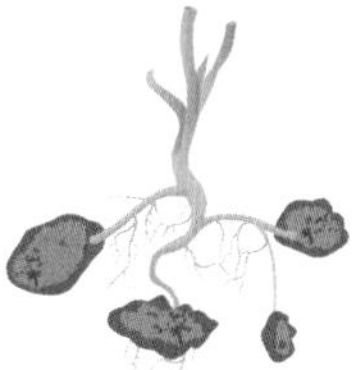

POTATO

Tuber
A swollen rhizome tip, adapted for storing lots of food, is called a tuber. Shoots grow from clusters of buds, called eyes, on the tuber.

Underground energy storage
In root vegetables such as carrots, sugars formed by photosynthesis accumulate in the taproot during the first growing season. The plant draws on these reserves when it produces flowers and seeds.

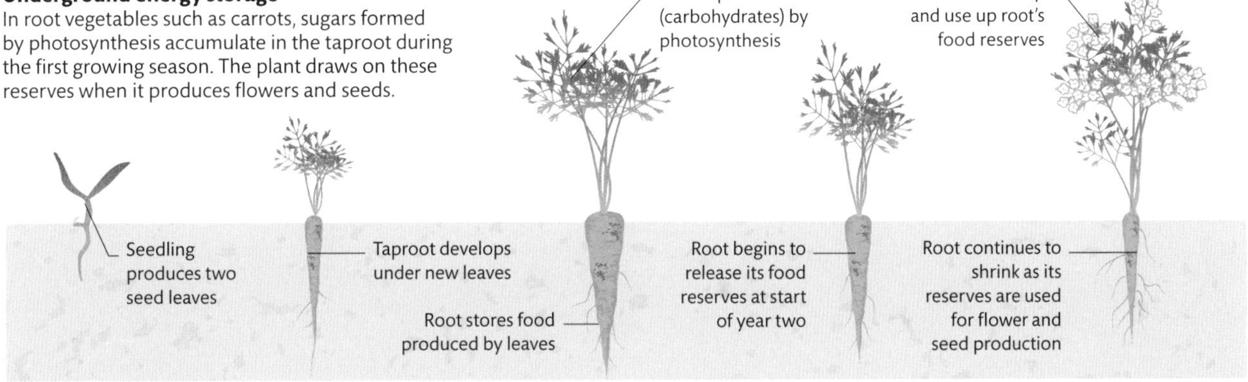
Leaves produce food (carbohydrates) by photosynthesis
Flowers develop and use up root's food reserves
Seedling produces two seed leaves
Taproot develops under new leaves
Root stores food produced by leaves
Root begins to release its food reserves at start of year two
Root continues to shrink as its reserves are used for flower and seed production

CARROT LIFE CYCLE

ROOT ARCHITECTURE

The root systems of dicotyledons and monocotyledons are differently organized (see left). In addition, xerophytic (desert) plants develop different types of root architecture as adaptations to dry conditions. Cacti have wide, shallow root systems, which absorb overnight condensation and any rainwater before it sinks into the soil. In contrast, acacia trees have narrow, deep root systems that obtain groundwater from below the water table. Both systems are extensive to maximize uptake.

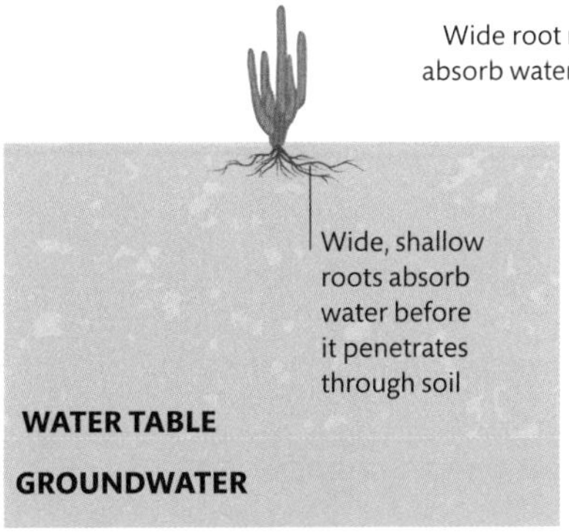
Wide, shallow roots absorb water before it penetrates through soil
WATER TABLE
GROUNDWATER

CACTUS

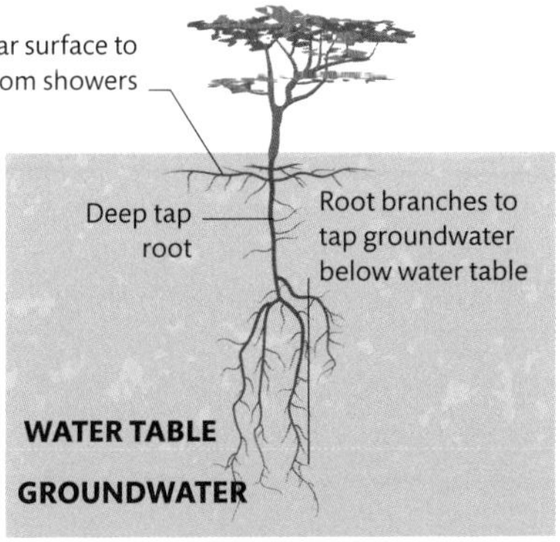
Wide root near surface to absorb water from showers
Deep tap root
Root branches to tap groundwater below water table
WATER TABLE
GROUNDWATER

ACACIA TREE

Plant stems

The stem of a plant has two functions. It supports the above-ground parts of the plant, such as flowers and leaves, allowing them to reach for light. The stem also houses a transport system, in which long strands called vascular bundles carry water and nutrients around the plant (see pp.148–49).

Stem structure

The stem is protected by the epidermis, an impermeable outer layer that prevents water from escaping. The cortex, which helps maintain the plant's shape, is strengthened by an outer ring of collenchyma and sclerenchyma, and filled with packing tissue (parenchyma). The pith at the core provides further support. Within the cortex, vascular bundles have xylem on the inside, phloem on the outside, and a layer of cambium in between that makes new xylem and phloem.

FLOWERING PLANT

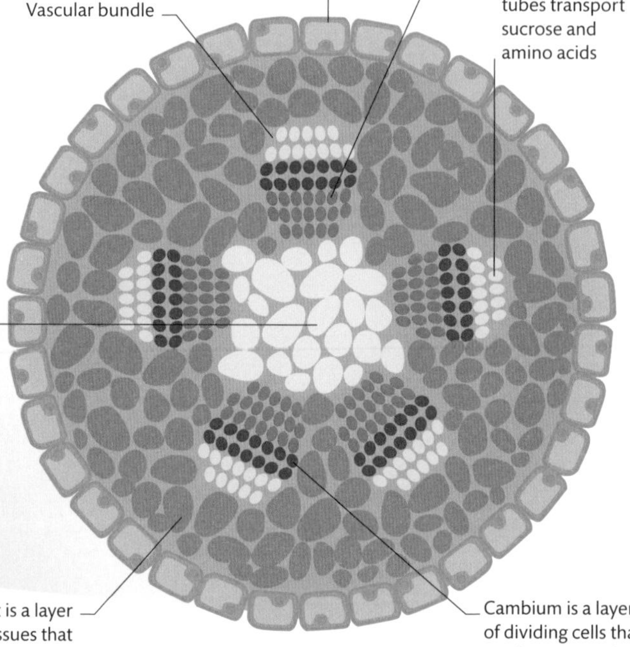
Epidermis forms external layer; it is one cell thick, except for a few stomata
Xylem vessels transport water and minerals
Vascular bundle
Phloem sieve tubes transport sucrose and amino acids
Pith is an inner layer of packing tissue (parenchyma)
Cortex is a layer of tissues that give support and shape to stem
Cambium is a layer of dividing cells that produce new xylem and phloem

CROSS SECTION OF PLANT STEM

Inside a plant stem
In the interior of a stem, xylem and phloem are packed together into vascular bundles, surrounded by other tissues to provide additional support and structure to the stem. An outer water-resistant epidermis layer prevents water loss.

The vascular system

Like animals, plants have vascular systems that transport essential nutrients and fluids from one part to another. Instead of blood vessels, plant vascular systems consist of two specialized tissues: xylem vessels and phloem sieve tubes. Each of these tissues has a different structure and purpose, but together they form part of vascular bundles. As a plant grows, new xylem and phloem are produced by the cambium, a region of actively dividing cells in the center of the vascular bundle.

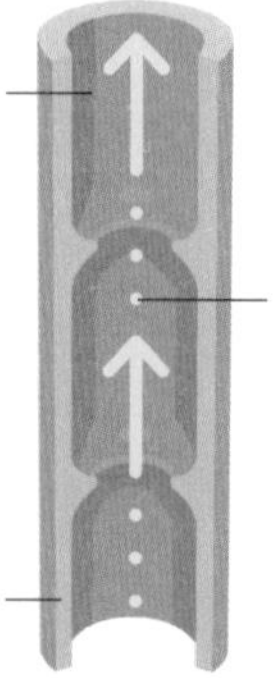
Cell wall is made of lignin, which provides support to plant
Upward flow of water and mineral ions (atoms or molecules with electrical charge)
Xylem vessel is formed of dead cells joined together

SECTION THROUGH XYLEM

Xylem vessel
Xylem vessels transport water and minerals from the roots to the rest of the plant. They consist of dead, hollow cells connected to form a tube. Xylem walls are waterproofed with a tough substance called lignin.

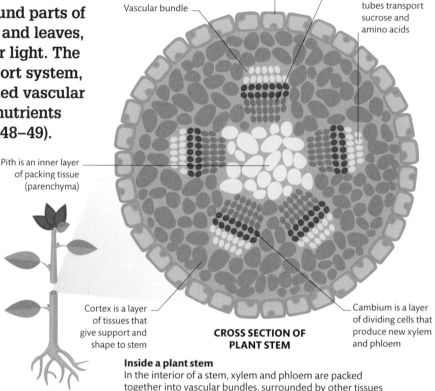

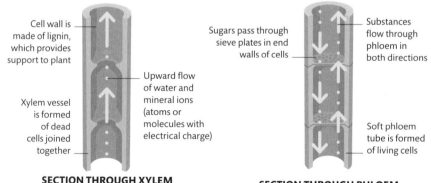

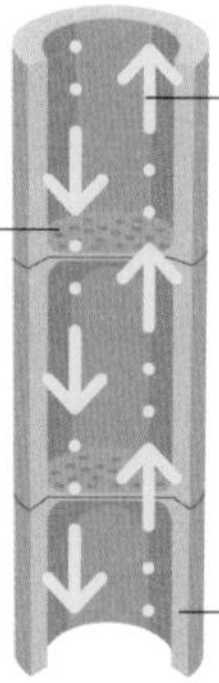
Sugars pass through sieve plates in end walls of cells
Substances flow through phloem in both directions
Soft phloem tube is formed of living cells

SECTION THROUGH PHLOEM

Phloem sieve tube
Phloem sieve tubes transport sucrose (converted glucose from photosynthesis) and amino acids up and down the plant in a process called translocation. The direction of travel depends on where sugar is needed.

Making wood in stems

As a plant grows taller, its stem must also widen, to support the plant and to meet its increased demand for water and minerals. It does so by producing more xylem and phloem. This thickening, or outward growth, is called secondary growth (primary plant growth is upward). Instead of staying in separate vascular bundles, the secondary xylem and phloem form complete rings, and the stem—or trunk—becomes increasingly woody.

WHICH PLANT HAS THE TALLEST STEM?

The world's tallest tree is a coast redwood in California, named Hyperion. At 380 ft (116 m), this towering tree is taller than a soccer field is long.

TUBERS, SUCH AS POTATOES, ARE UNDERGROUND STEMS THAT HAVE BECOME ADAPTED FOR STORING STARCH

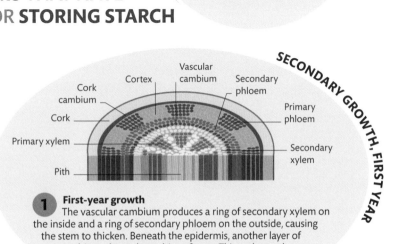

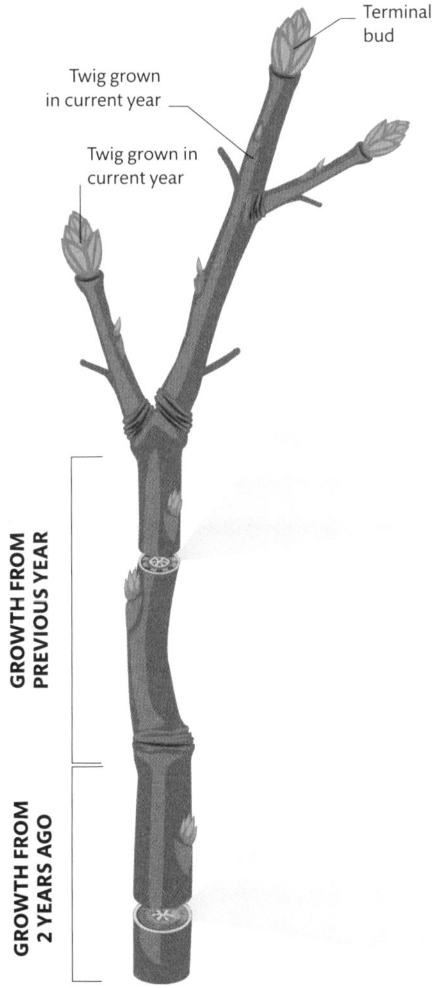
Terminal bud
Twig grown in current year
Twig grown in current year
GROWTH FROM PREVIOUS YEAR
GROWTH FROM 2 YEARS AGO

SECONDARY GROWTH, FIRST YEAR

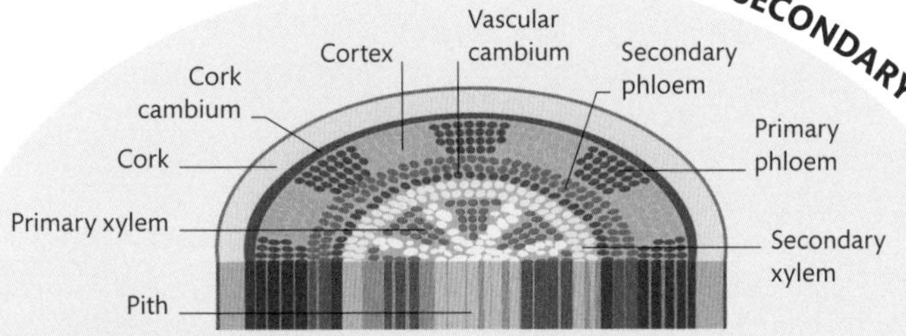
Cork cambium
Cortex
Vascular cambium
Secondary phloem
Primary phloem
Cork
Primary xylem
Secondary xylem
Pith

1 First-year growth
The vascular cambium produces a ring of secondary xylem on the inside and a ring of secondary phloem on the outside, causing the stem to thicken. Beneath the epidermis, another layer of cambium, the cork cambium, forms. This makes cork, which replaces the epidermis and becomes the stem's outer layer of bark.

SECONDARY GROWTH, SECOND YEAR

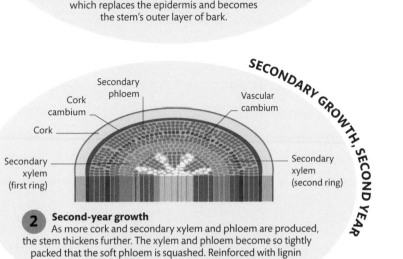

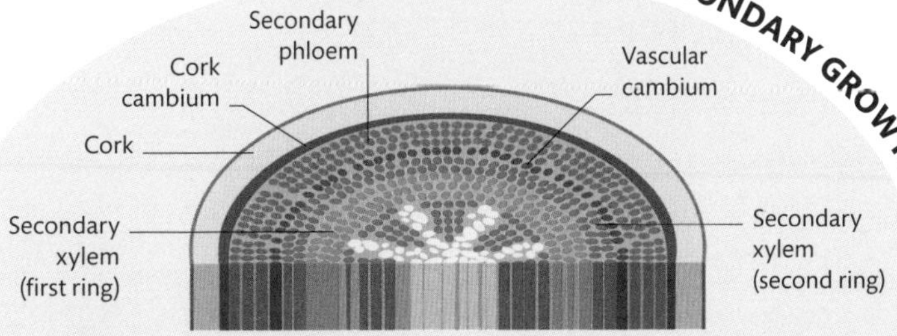
Secondary phloem
Cork cambium
Vascular cambium
Cork
Secondary xylem (first ring)
Secondary xylem (second ring)

2 Second-year growth
As more cork and secondary xylem and phloem are produced, the stem thickens further. The xylem and phloem become so tightly packed that the soft phloem is squashed. Reinforced with lignin and cellulose, the hard xylem increasingly dominates the space inside the stem, so the stem becomes woody throughout.

Leaves

While animals simply eat to obtain energy, plants make their own food. They do so through the process of photosynthesis (see pp.46–47). Leaves are a plant's main photosynthetic organs. They capture light energy using chlorophyll, the green pigment in chloroplasts, and use this energy to convert carbon dioxide and water into sugars (glucose) and oxygen.

The structure of a leaf

A typical leaf consists of a flattened blade, which is attached to the plant stem by a stalk called a petiole. A network of veins, made up of xylem and phloem (see pp.144–145), carries water, minerals, and nutrients around the leaf. The large central vein is called the midrib. Leaves are made from several layers of cells, including the epidermis, palisade layer, and spongy mesophyll layer. Each type of cell has a different function, with specific adaptations that allow it to perform that function.

Upper side of leaf
A waterproof waxy cuticle covers the leaf's upper surface, reducing water loss by transpiration (evaporation). Epidermal cells beneath form a protective layer. They are transparent, so light passes through them.

Midrib vein runs along center of leaf and provides support to blade
Veins have xylem and phloem tubes running through them to transport water, minerals, and sugars
Blade is flat part of leaf

THE **LEAVES** OF THE **GIANT WATERLILY *VICTORIA AMAZONICA*** ARE UP TO 10 FT (3 M) WIDE

NON-PHOTOSYNTHETIC LEAVES

In some plants, leaves are modified into other structures whose main function is not photosynthesis. Cactus leaves have become protective spines, and photosynthesis occurs in the stems. In peas, some leaves have become tendrils to help the plant climb. The red "petals" (bracts) of a poinsettia are actually leaves that mimic flowers to attract pollinators.

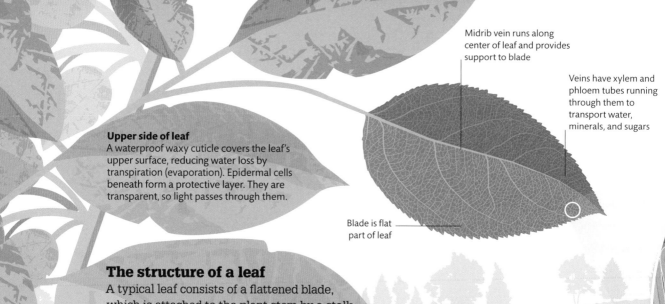

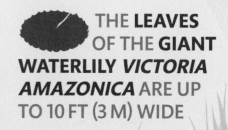

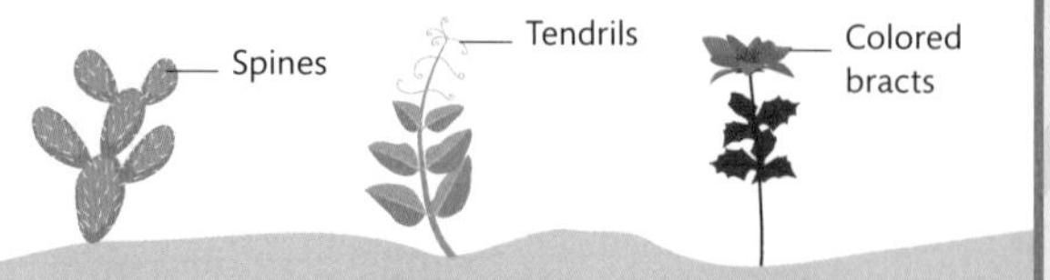

Spines
Tendrils
Colored bracts
CACTUS
PEA PLANT
POINSETTIA

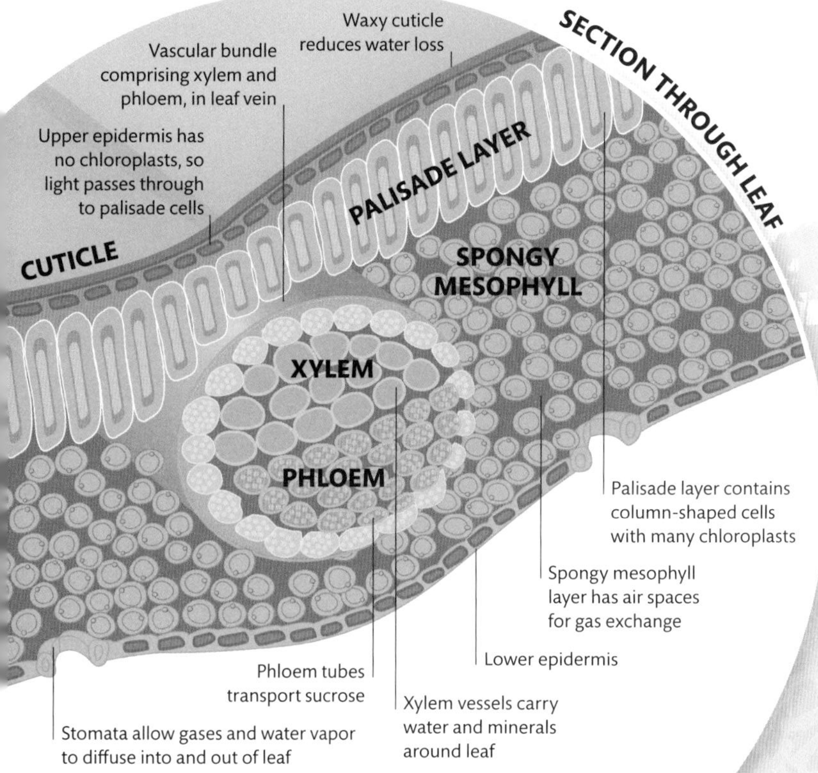
SECTION THROUGH LEAF
Waxy cuticle reduces water loss
Vascular bundle comprising xylem and phloem, in leaf vein
Upper epidermis has no chloroplasts, so light passes through to palisade cells
CUTICLE
PALISADE LAYER
SPONGY MESOPHYLL
XYLEM
PHLOEM
Palisade layer contains column-shaped cells with many chloroplasts
Spongy mesophyll layer has air spaces for gas exchange
Lower epidermis
Phloem tubes transport sucrose
Xylem vessels carry water and minerals around leaf
Stomata allow gases and water vapor to diffuse into and out of leaf

Inside a leaf
Most photosynthesis occurs in the palisade layer, which has many chloroplasts to trap light energy. Spongy mesophyll cells also photosynthesize, and air spaces between cells allow oxygen and carbon dioxide to diffuse. Xylem and phloem facilitate transport.

WHY DO LEAVES CHANGE COLOR?

When temperature and light levels drop in the fall, chlorophyll, the green pigment in leaves, breaks down first, leaving behind other pigments, such as yellow carotenes and red or pink anthocyanins.

Stomata

On the underside of the leaf are openings called stomata (singular: stoma). These allow carbon dioxide to diffuse into the leaf for photosynthesis, and allow the oxygen and water vapor produced by the process to diffuse out. Stomata are usually open during the day, to enable photosynthesis, and closed at night. If a plant lacks water, it closes its stomata to reduce water loss—so a wilted plant cannot photosynthesize. Stomatal opening and closing is controlled by a pair of sausage-shaped guard cells.

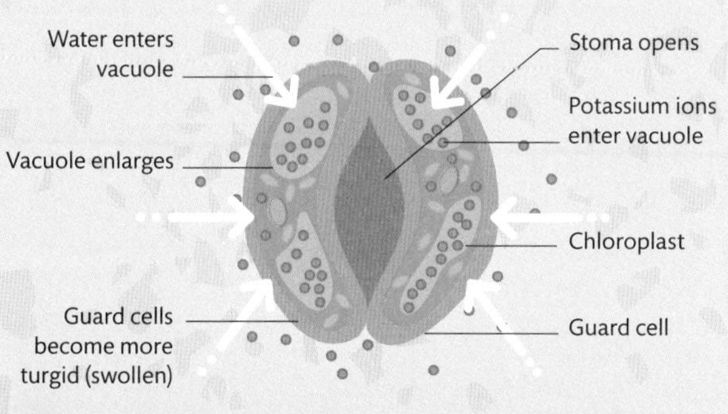
Water enters vacuole
Stoma opens
Potassium ions enter vacuole
Vacuole enlarges
Chloroplast
Guard cells become more turgid (swollen)
Guard cell

1 Stoma opens
Light stimulates guard cells to accumulate potassium ions. The build-up of potassium causes water to enter the cells by osmosis (see pp.64–65). Guard cells' inner walls are thicker than their outer walls, so when they swell with water, they bend, opening the stoma.

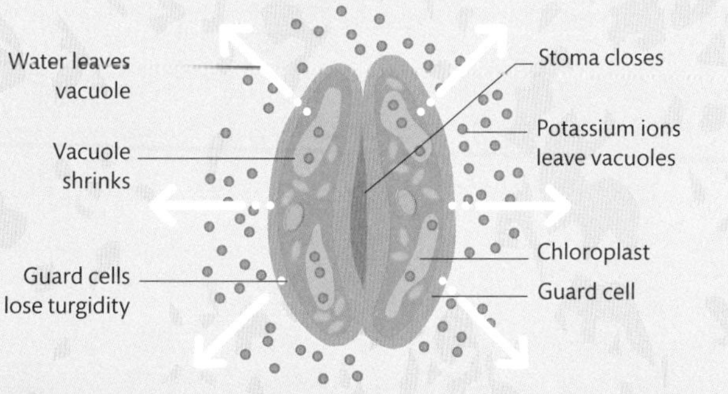
Water leaves vacuole
Stoma closes
Potassium ions leave vacuoles
Vacuole shrinks
Chloroplast
Guard cells lose turgidity
Guard cell

2 Stoma closes
At night, or when water- or heat-stressed, a plant produces the stress hormone abscisic acid. Abscisic acid binding causes potassium ions to leave the guard cells. Water follows by osmosis, so the guard cells shrink and become flaccid. This causes the stoma to close.

SAP CAN **MOVE** AT UP TO **150 FT (45 M) PER HOUR** IN LARGE TREES

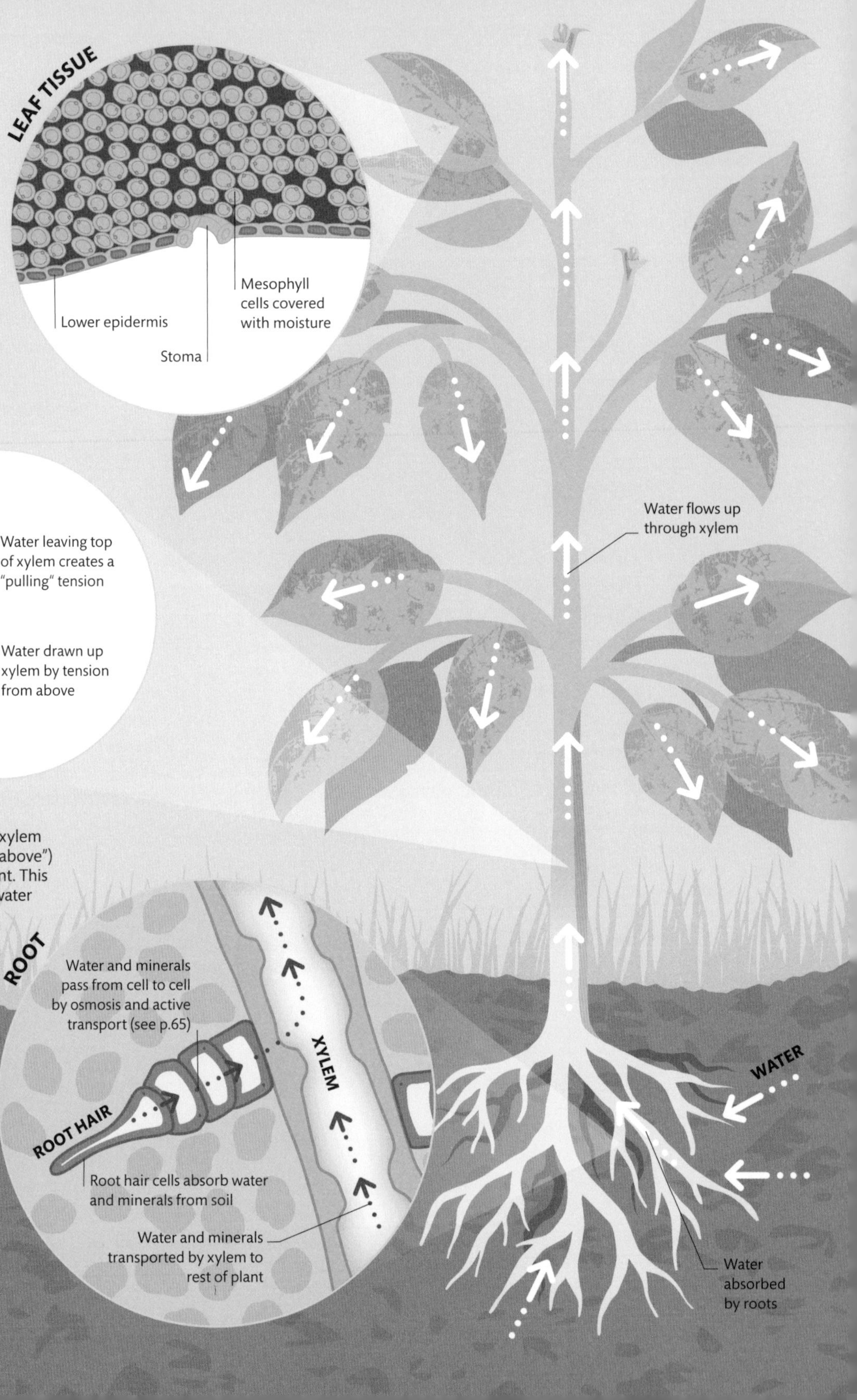
LEAF TISSUE
Lower epidermis
Stoma
Mesophyll cells covered with moisture
Water flows up through xylem
ROOT
Water and minerals pass from cell to cell by osmosis and active transport (see p.65)
XYLEM
ROOT HAIR
Root hair cells absorb water and minerals from soil
Water and minerals transported by xylem to rest of plant
WATER
Water absorbed by roots

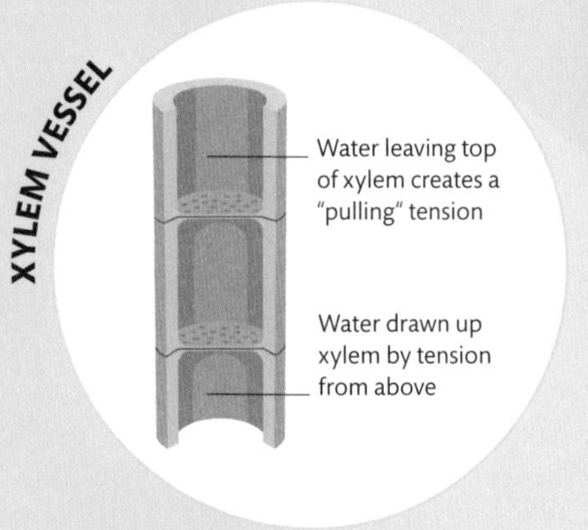
XYLEM VESSEL
Water leaving top of xylem creates a "pulling" tension
Water drawn up xylem by tension from above

2 Water moves up xylem vessels

Water leaving the top of the xylem creates tension (a "pull from above") and a water potential gradient. This draws columns of cohesive water molecules up the xylem.

3 Water absorbed by roots

Water and minerals are absorbed from the soil into the root hair cells. The base of the xylem has a lower water potential than the root cells; this water potential gradient causes the water to be drawn into the cortex (inner layer) of the roots and then into the xylem.

Transport in plants

While animals have a heart or similar pumping system to carry substances around the body, plants use chemical and biological processes to transport water and nutrients to cells. These processes are driven by differences in pressure and fluid concentrations.

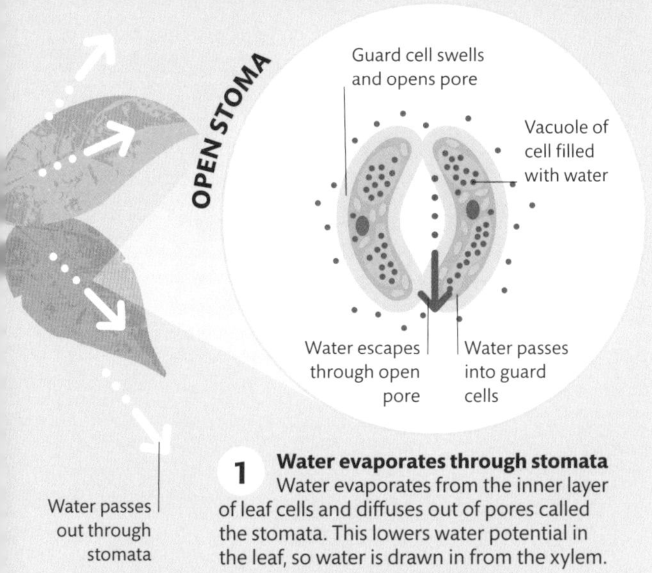
OPEN STOMA
Guard cell swells and opens pore
Vacuole of cell filled with water
Water escapes through open pore
Water passes into guard cells
Water passes out through stomata

1 Water evaporates through stomata Water evaporates from the inner layer of leaf cells and diffuses out of pores called the stomata. This lowers water potential in the leaf, so water is drawn in from the xylem.

Transpiration and root pressure

Transpiration is the evaporation of water from leaves. This creates water potential gradients in the plant. Water potential is the tendency of water to move from areas of high water potential (with a high level of water molecules) to areas of low water potential (low levels of water). Transpiration causes water to be drawn in through the roots and carried up the plant in the xylem (see p.144). At night, when the stomata are closed, the plant takes on water by root pressure. Minerals are still being drawn from the roots into the xylem, lowering water potential in the xylem. A buildup of water at the base of the xylem creates pressure, forcing water up the stem.

Translocation

Sucrose and amino acids are transported in phloem (see p.144) by translocation—movement from areas of production (sources) to areas where they are used in respiration or growth (sinks). These nutrients are carried in a fluid called sap. Water from nearby tissues is drawn into the concentrated sap at the source, pushing it toward the sink. Depending on the season, the source may be leaves or storage organs such as tubers, while the sink is shoots, buds, flowers, fruits, seeds, or storage organs.

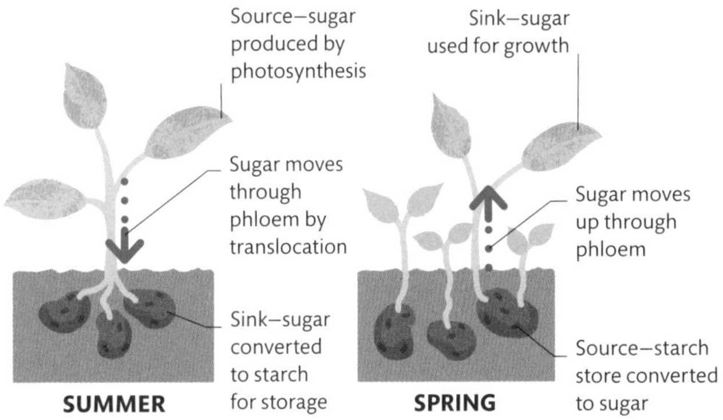
Source–sugar produced by photosynthesis
Sugar moves through phloem by translocation
Sink–sugar converted to starch for storage
SUMMER
Sink–sugar used for growth
Sugar moves up through phloem
Source–starch store converted to sugar
SPRING

Translocation in a potato
In summer, the leaves are the source: glucose is produced there by photosynthesis and converted to sucrose. The growing tubers are the sink, where sucrose is converted to starch for storage. The leaves die down over winter, leaving the tubers as the only source. In spring, the tubers start to sprout. These shoots become the sink.

WILTING

In a hydrated plant, the outward pressure of water inside cells (turgor pressure) presses the cell membrane against the cell wall and presses cells against each other. These tightly packed cells keep the plant upright. Wilting occurs when transpiration exceeds water uptake. As cells lose water, they become flaccid. The cell membrane shrinks away from the cell wall (plasmolysis). Flaccid cells do not press against each other, so the plant wilts.

Leaves and stems flaccid due to water loss
WILTED PLANT

WIND-POLLINATED FLOWER	INSECT-POLLINATED FLOWER
No nectar produced	Nectar (food) often produced to attract insects
No scent produced	Sweet scent often produced to attract insects
If present, petals are small and bland—often green	Colorful petals, often with patterns visible only to insects
Pendulous stigma and stamens hang outside the flower so they are exposed to the wind	Stigma and stamens are inside the flower to rub pollen onto or off an insect
Stigma has large surface area	Stigma has small surface area
Stigma is feathery to catch pollen from the air	Stigma is not feathery but is often sticky
Small, light, smooth pollen grains are easily carried by the wind	Pollen grains are large and barbed or spiky to attach to insects
Large amounts of pollen produced, increasing chance of pollen reaching its target	Small amounts of pollen produced as insects ensure accurate transfer of pollen between flowers

Flowers

Flowering plants, or angiosperms, are the dominant form of land plants, being much more numerous and diverse than gymnosperms such as conifers. As well as their flowers, they are notable for their various ways of spreading pollen and seeds.

WHICH PLANT HAS THE LARGEST FLOWER?

Individual blooms of the corpse flower *Rafflesia arnoldii*, a parasitic plant that is indigenous to the Indonesian rainforest, grow up to about 3 ft (1 m) across.

Flower structure

Flowers are a plant's reproductive system. The stamens are the male organs, producing pollen. The female organs—collectively called the pistil—are made up of the ovary, style, and stigma. Many flowers also have brightly colored petals and give off scent to attract pollinators such as insects. Most flowers have both female and male organs, although some plants, such as zucchini, have separate male and female flowers. Certain species, such as holly, have entirely separate male and female plants, which have all male or all female flowers.

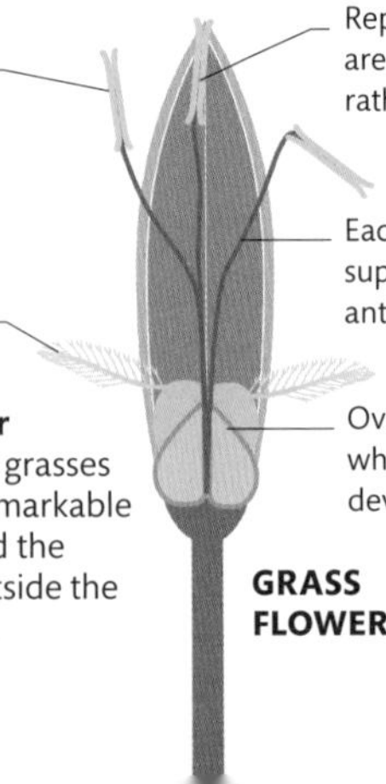

Anthers (part of stamen) produce pollen
Flower has two or three stigmas, which are feathery in order to capture pollen grains from wind
Reproductive parts are enclosed in bracts rather than petals
Each filament supports an anther
Ovary is site where seeds develop
GRASS FLOWER

Wind-pollinated flower
Flowers of plants such as grasses are small and often unremarkable in color. The stamens and the feathery stigma hang outside the flower to catch the wind.

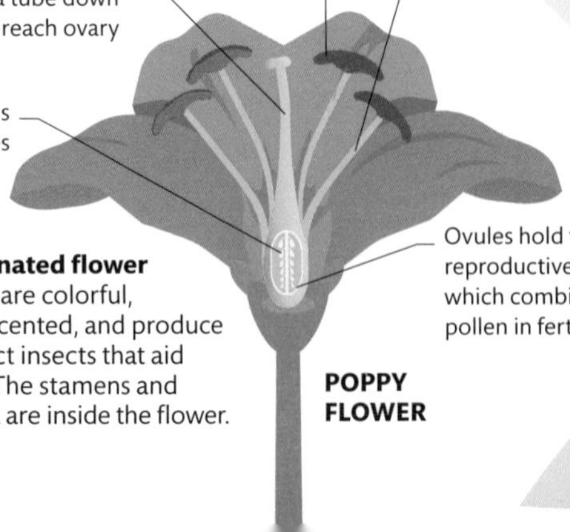

Sticky stigma traps pollen rubbed off by insects
Stamens (anthers plus filaments) are located inside flower, where insects can easily brush against them
Pollen grains must grow a tube down style to reach ovary
Ovary contains ovules
Ovules hold female reproductive cells, which combine with pollen in fertilization
POPPY FLOWER

Insect-pollinated flower
Flowers that are colorful, sometimes scented, and produce nectar, attract insects that aid pollination. The stamens and sticky stigma are inside the flower.

CONIFER CONES

Gymnosperms do not have flowers. Their seeds are held in cones instead of fruits. Conifers have small male cones and large female cones. Male cones produce large quantities of pollen, which is dispersed by the wind. Once it lands on a female cone and fertilizes it, the papery seeds take up to three years to form. Then the female cone opens and releases seeds into the wind.

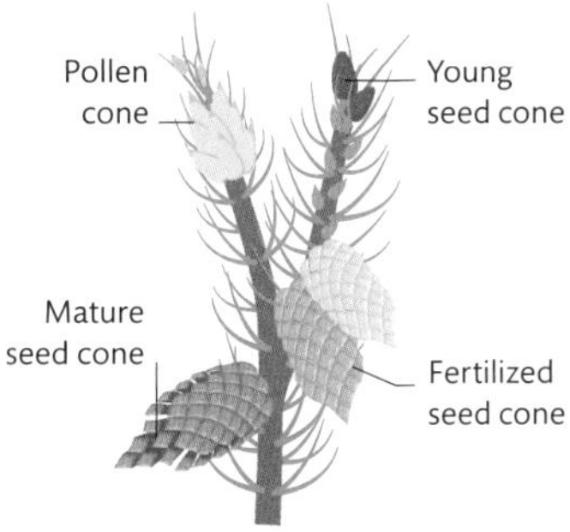
Pollen cone
Young seed cone
Mature seed cone
Fertilized seed cone

THE AUSTRALIAN WESTERN UNDERGROUND ORCHID LIVES AND FLOWERS ENTIRELY UNDERGROUND

Insect cross-pollination
Over 80 percent of animal pollination is carried out by insects—bees, butterflies, moths, and flies. Other pollinators include birds and bats.

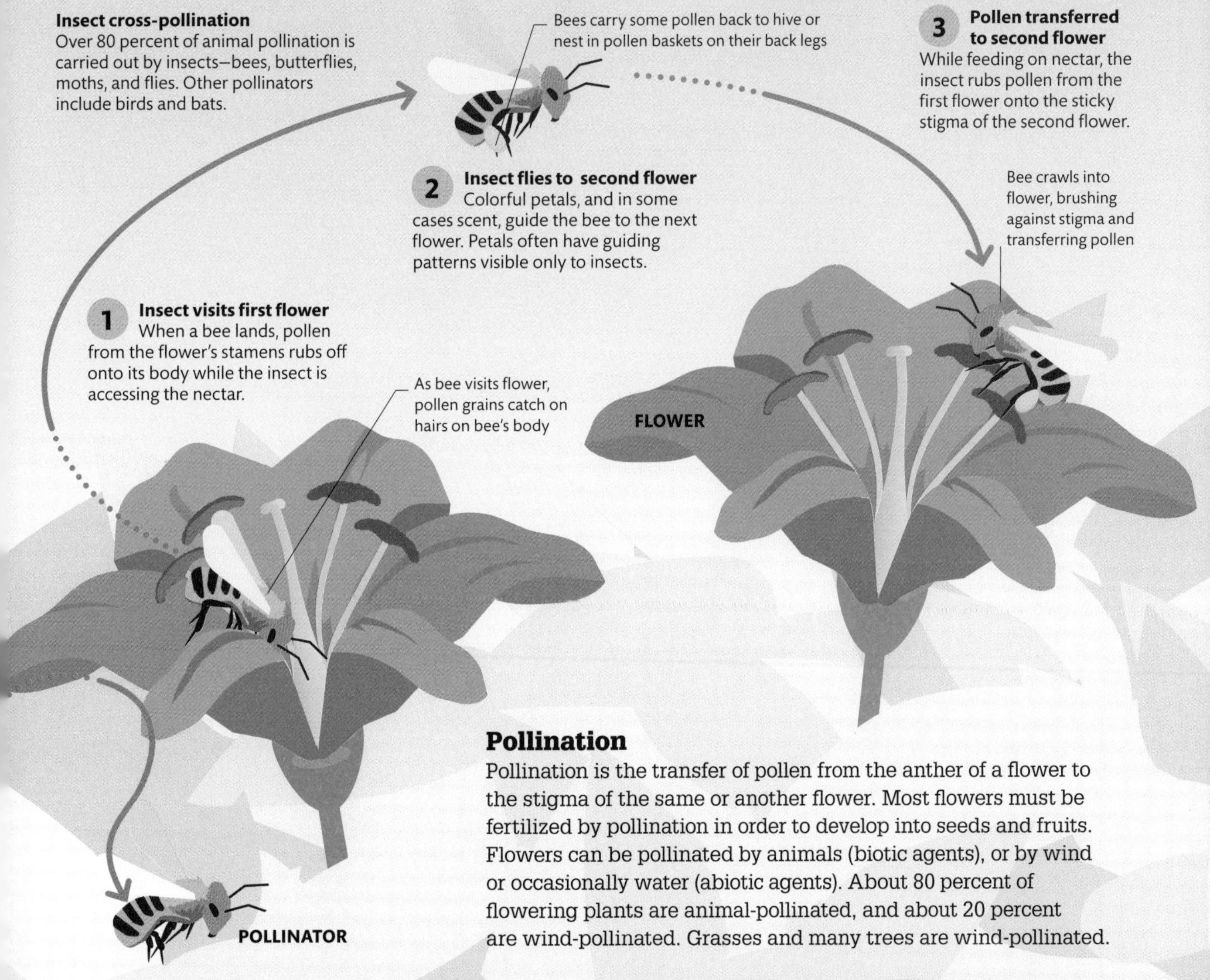
Bees carry some pollen back to hive or nest in pollen baskets on their back legs
3 Pollen transferred to second flower
While feeding on nectar, the insect rubs pollen from the first flower onto the sticky stigma of the second flower.
2 Insect flies to second flower
Colorful petals, and in some cases scent, guide the bee to the next flower. Petals often have guiding patterns visible only to insects.
Bee crawls into flower, brushing against stigma and transferring pollen
1 Insect visits first flower
When a bee lands, pollen from the flower's stamens rubs off onto its body while the insect is accessing the nectar.
As bee visits flower, pollen grains catch on hairs on bee's body
FLOWER
POLLINATOR

Pollination

Pollination is the transfer of pollen from the anther of a flower to the stigma of the same or another flower. Most flowers must be fertilized by pollination in order to develop into seeds and fruits. Flowers can be pollinated by animals (biotic agents), or by wind or occasionally water (abiotic agents). About 80 percent of flowering plants are animal-pollinated, and about 20 percent are wind-pollinated. Grasses and many trees are wind-pollinated.

Fruits

Every flowering plant produces fruits. In botanical terms, a fruit is a mature ovary containing seeds. The term covers not just familiar fruits, such as apples, but also tomatoes, peppers, and zucchini; nuts; and poppy heads. Fruits have a dual function: they protect the seeds, and they aid dispersal of those seeds.

Fruit development

A flower must be fertilized in order to develop into a fruit. During this process, sperm cells from the male pollen grains fuse with the female egg cells contained in ovules inside the ovary. Separate pollen grains are required to fertilize each ovule. After fertilization, each ovule develops into a seed, and the ovary itself develops into a fruit. As it transforms into a fruit, the ovary grows in size, and its wall thickens to become the fleshy outer layers of the fruit (pericarp). As the fruit ripens, sugars accumulate inside it and the wilted petals and other flower parts are shed. Fruit development and ripening is controlled by hormones.

Double fertilization

Angiosperms (flowering plants) are unique in employing double fertilization. Two sperm from the pollen grain fertilize separate nuclei within each ovule. One fusion creates a zygote, which becomes the plant embryo. The other fusion forms the endosperm, which is the food store for the embryo.

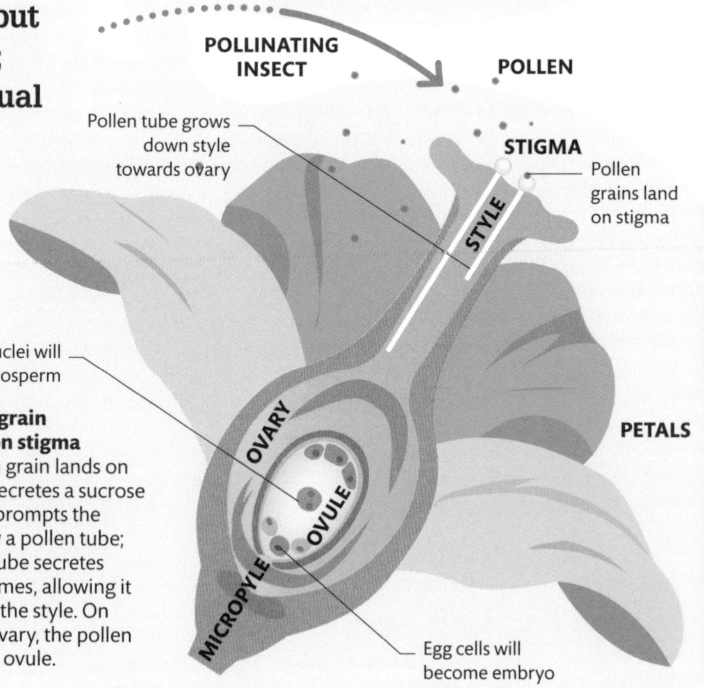
POLLINATING INSECT
POLLEN
Pollen tube grows down style towards ovary
STIGMA
Pollen grains land on stigma
STYLE
Polar female nuclei will become endosperm
PETALS
OVARY
OVULE
MICROPYLE
Egg cells will become embryo within seed

1 Pollen grain lands on stigma
When a pollen grain lands on it, the stigma secretes a sucrose solution. This prompts the pollen to grow a pollen tube; the tip of the tube secretes digestive enzymes, allowing it to grow down the style. On reaching the ovary, the pollen tube enters an ovule.

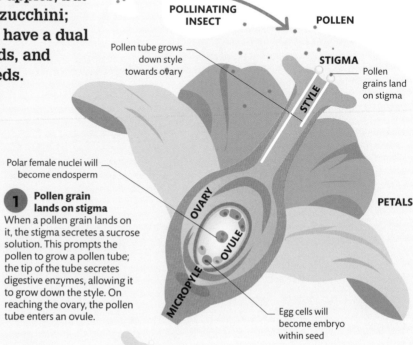

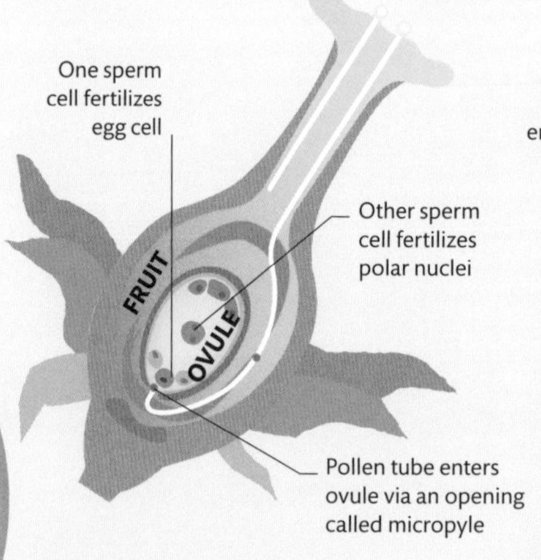
One sperm cell fertilizes egg cell
Other sperm cell fertilizes polar nuclei
FRUIT
OVULE
Pollen tube enters ovule via an opening called micropyle

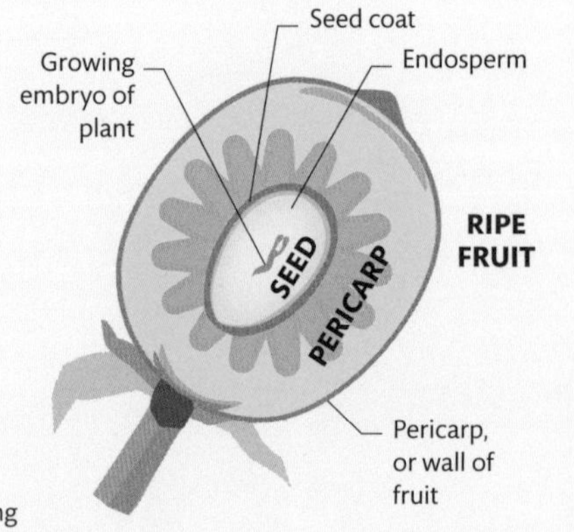
Seed coat
Growing embryo of plant
Endosperm
RIPE FRUIT
SEED
PERICARP
Pericarp, or wall of fruit

2 Fertilization of ovary
Two sperm from the pollen grain travel down the tube into the ovule. One fuses with the egg cell to form a zygote, which grows into an embryo. The second sperm fuses with two other female nuclei to form the endosperm (food store).

3 Fruit grows
Following fertilization, the ovules become seeds and the ovary a fruit. The ovary wall becomes the pericarp (fleshy wall of the fruit). The flower petals are shed. The fruit grows, accumulates sugars, and ripens as the seeds mature.

WHICH FRUIT IS THE SMELLIEST?

The durian, a large, spiky edible fruit indigenous to southeast Asia, has a putrid stench that is so unpleasant that the fruit is banned from public transportation in some countries.

MAIN METHODS OF SEED AND FRUIT DISPERSAL

ANIMAL DISPERSAL

Tomato
Fruits are sweet and brightly colored to attract animals and birds. Seeds are then dispersed in droppings.

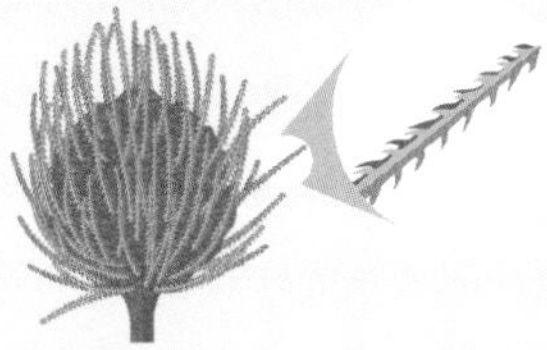

Burr
Fruits have hooked spines pointing in all directions; these hooks readily attach themselves to animal fur.

WATER DISPERSAL

Coconut
The fruits are buoyant and surrounded by a thick, fibrous husk, allowing them to survive months or even years at sea.

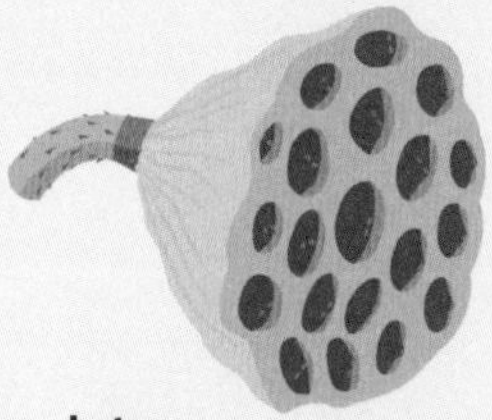

Lotus
Lotuses grow in water. Seeds fall out of the fruit and are swept away by the current, later germinating in mud.

WIND DISPERSAL

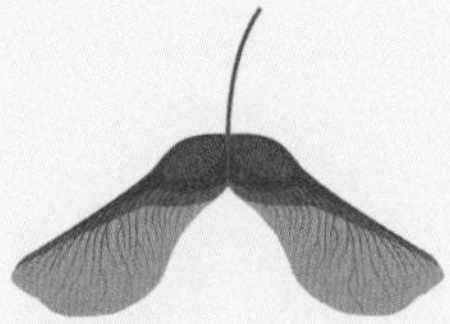

Sycamore
The fruits have stiff wings, so they can fly. The wings are twisted and balanced so that the fruit spins.

Dandelion
Each fruit, containing just one lightweight seed, has a feathery parachute to carry it long distances on the breeze.

SELF-DISPERSAL

Squirting cucumber
Pressure buildup inside the fruit makes it rocket off the stem and expel a jet of seeds as it falls to the ground.

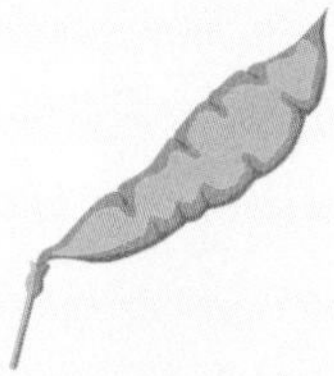

Pride of Barbados
The pericarp (seed pod) dries out on the plant and then violently twists open, ejecting the seeds.

Methods of dispersal

Once the seeds are ripe, fruits act to help disperse them. Dispersal provides a means for plants to establish themselves in new areas; it also prevents seedlings from competing with the parent plant and with each other for space, light, water, and nutrients. Fruits are highly adapted for dispersal, with different plant species employing a variety of methods. Some species utilize animals (biotic dispersal); others spread their seeds by abiotic means such as wind, water, or mechanical self-dispersal.

THE TROPICAL SANDBOX TREE HAS EXPLODING FRUITS THAT CAN EJECT SEEDS AT SPEEDS OF UP TO 160 MPH (257 KPH)

TYPES OF FRUIT

Fruits are classified in various ways. One is based on how they grow. Simple fruits, such as lemons, grow from one flower. Aggregate fruits, such as raspberries, form a cluster of small fruits, also from a single flower. In false fruits, such as apples, tissue other than the ovary makes up most of the fruit. More broadly, fruits may be dry or fleshy. Dry fruits, such as nuts and pea pods, have a hard or papery pericarp. Fleshy fruits, such as cherries, have a soft, pulpy pericarp.

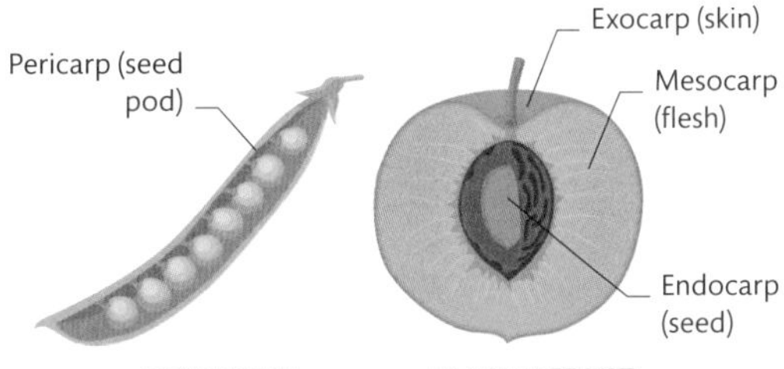
Pericarp (seed pod)
Exocarp (skin)
Mesocarp (flesh)
Endocarp (seed)

DRY FRUIT **FLESHY FRUIT**

HOW ANIMALS WORK

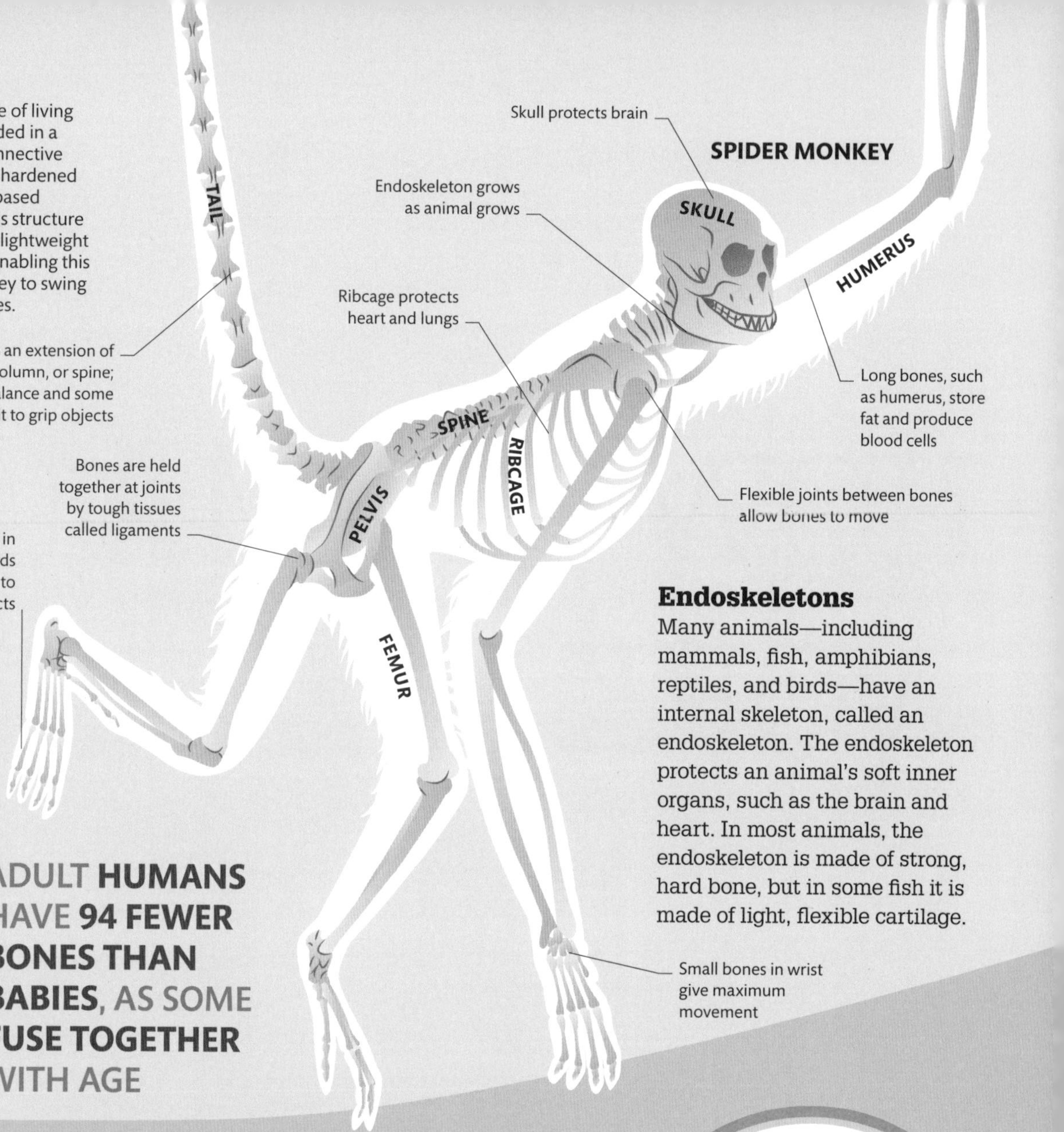
SPIDER MONKEY
Skull protects brain
SKULL
Endoskeleton grows as animal grows
HUMERUS
Ribcage protects heart and lungs
Long bones, such as humerus, store fat and produce blood cells
TAIL
Tail is an extension of vertebral column, or spine; it aids balance and some animals use it to grip objects
SPINE
RIBCAGE
Flexible joints between bones allow bones to move
Bones are held together at joints by tough tissues called ligaments
PELVIS
Many bones in feet and hands give flexibility to grasp objects
FEMUR
Small bones in wrist give maximum movement

Bone
Bone is made of living cells embedded in a matrix of connective tissue that is hardened by calcium-based minerals. This structure makes bone lightweight but strong, enabling this spider monkey to swing from the trees.

Endoskeletons
Many animals—including mammals, fish, amphibians, reptiles, and birds—have an internal skeleton, called an endoskeleton. The endoskeleton protects an animal's soft inner organs, such as the brain and heart. In most animals, the endoskeleton is made of strong, hard bone, but in some fish it is made of light, flexible cartilage.

ADULT **HUMANS** HAVE **94 FEWER BONES THAN BABIES**, AS SOME **FUSE TOGETHER** WITH AGE

Support and movement

Like the poles of a tent, a skeleton provides the essential framework that gives an animal shape, protects its vital organs from harm, and forms the rigid structure on which muscles can pull to create movement.

DO TURTLES HAVE AN ENDOSKELETON AND AN EXOSKELETON?

No, the hard outer shell of a turtle is not a true exoskeleton, but an adaptation of its endoskeleton.

HYDROSTATIC SKELETONS

Some primitive animals, such as earthworms, have a flexible, fluid-filled chamber surrounded by muscles to provide support, shape, and movement. This is called a hydrostatic skeleton.

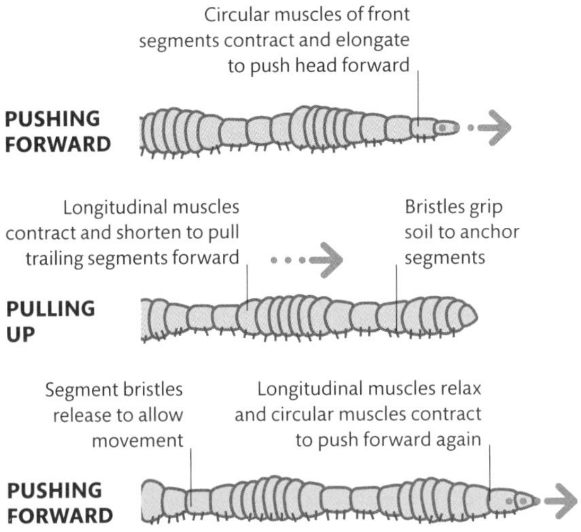
Circular muscles of front segments contract and elongate to push head forward
PUSHING FORWARD
Longitudinal muscles contract and shorten to pull trailing segments forward
Bristles grip soil to anchor segments
PULLING UP
Segment bristles release to allow movement
Longitudinal muscles relax and circular muscles contract to push forward again
PUSHING FORWARD

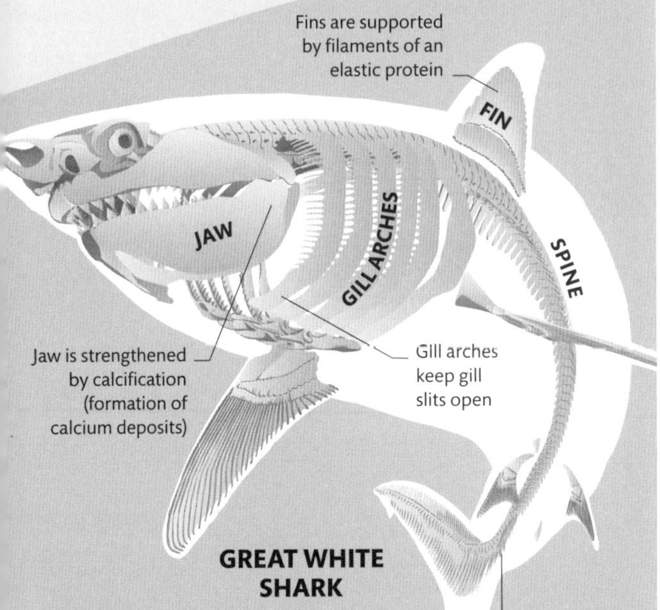
Fins are supported by filaments of an elastic protein
FIN
JAW
GILL ARCHES
SPINE
Jaw is strengthened by calcification (formation of calcium deposits)
Gill arches keep gill slits open
GREAT WHITE SHARK
Flexible skeleton allows shark to make tight turns

Cartilage
Sharks are cartilaginous fish, meaning their endoskeletons are made of a connective tissue called cartilage. Cartilage is much lighter than bone, helping keep sharks buoyant in water.

Exoskeletons

Many invertebrates have armorlike external skeletons. Like an endoskeleton, an exoskeleton provides the structure to support movement, protects the soft inner tissues of the animal and, in animals such as insects, it also prevents an animal from drying out. Some exoskeletons do not grow, so must be molted as the animal grows.

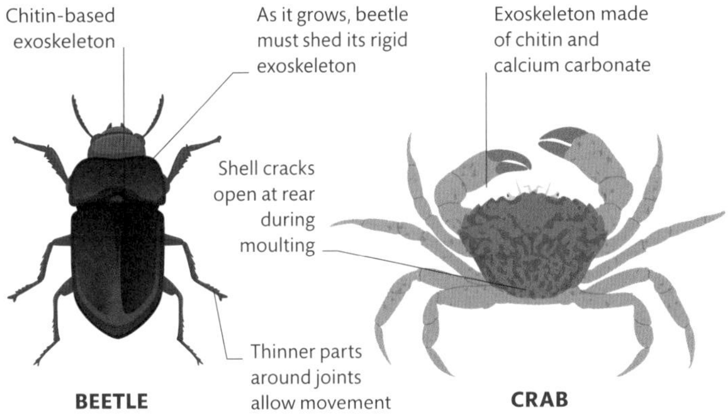
Chitin-based exoskeleton
As it grows, beetle must shed its rigid exoskeleton
Exoskeleton made of chitin and calcium carbonate
Shell cracks open at rear during moulting
Thinner parts around joints allow movement
BEETLE
CRAB

How muscles work

A skeleton is moved by skeletal muscles pulling on its bones. In endoskeletons, these muscles are attached to the bones by tough, fibrous tissues called tendons, but in exoskeletons the muscles attach directly to the skeleton itself. Muscles in joints work in opposing pairs—while one muscle contracts to cause movement, the other relaxes (see pp.72–73).

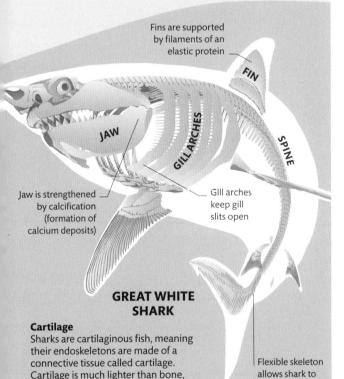

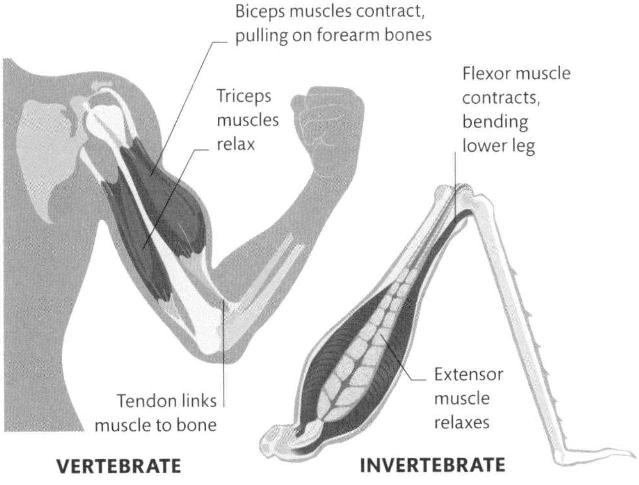
Biceps muscles contract, pulling on forearm bones
Triceps muscles relax
Flexor muscle contracts, bending lower leg
Tendon links muscle to bone
Extensor muscle relaxes
VERTEBRATE
INVERTEBRATE

Breathing

Animals need oxygen to live, but different groups have evolved their own methods of getting oxygen into their bodies, depending on their size, shape, and whether they breathe in air or water.

Lungs

Mammals, birds, reptiles, and some amphibians and fish use lungs to breathe. Lungs are like a suction pump, expanding to lower the air pressure within them, which in turn draws air into the body. As well as being powerful pumps, lungs contain the surfaces where oxygen can enter the bloodstream and carbon dioxide can leave.

WHY DO WE SNORE?

Snoring happens when soft tissues at the back of the roof of the mouth relax and flap when breathing, or when airways in the nose narrow and vibrate during sleep.

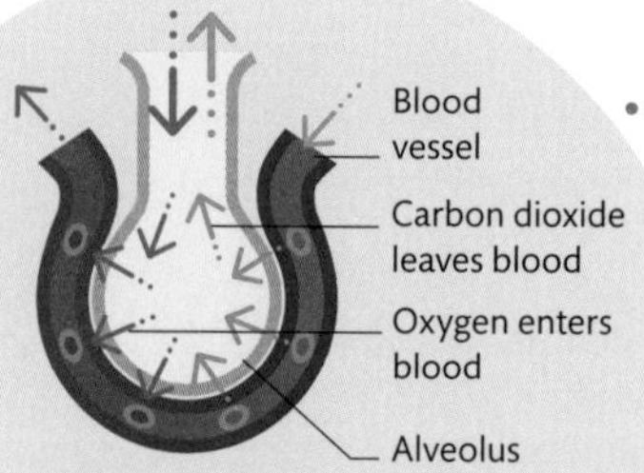
Blood vessel
Carbon dioxide leaves blood
Oxygen enters blood
Alveolus

Gas exchange
Lungs are filled with millions of tiny, thin-walled air sacs called alveoli. Blood vessels surround each sac, allowing gases to diffuse into and out of the bloodstream.

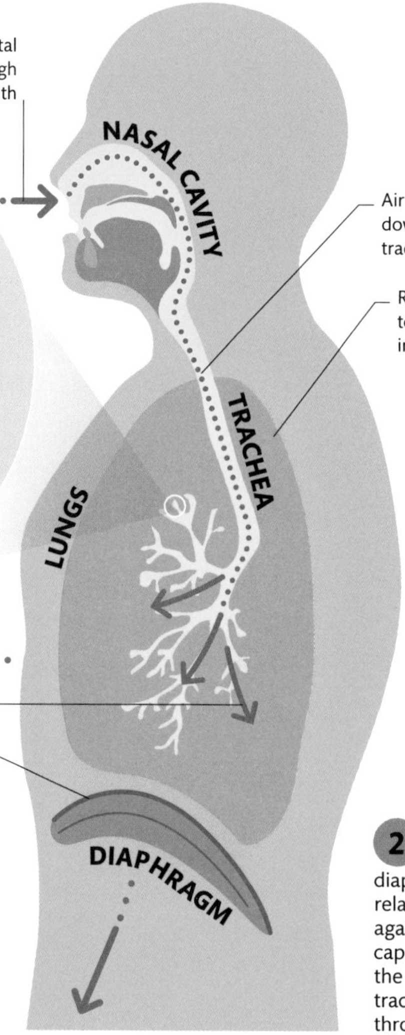
Air containing vital oxygen drawn in through nose and mouth
NASAL CAVITY
Air travels down trachea
Rib muscles contract to pull ribs up and out, increasing chest space
TRACHEA
LUNGS
Chest expands
Air fills lungs
Diaphragm (a domed muscle) contracts and flattens, expanding lungs
DIAPHRAGM

1 Inhaling
When a person inhales, the diaphragm contracts and pulls down, and the rib muscles contract to expand the chest. This causes the lungs to expand, drawing air in through the nose and mouth, down the trachea, and into the lungs.

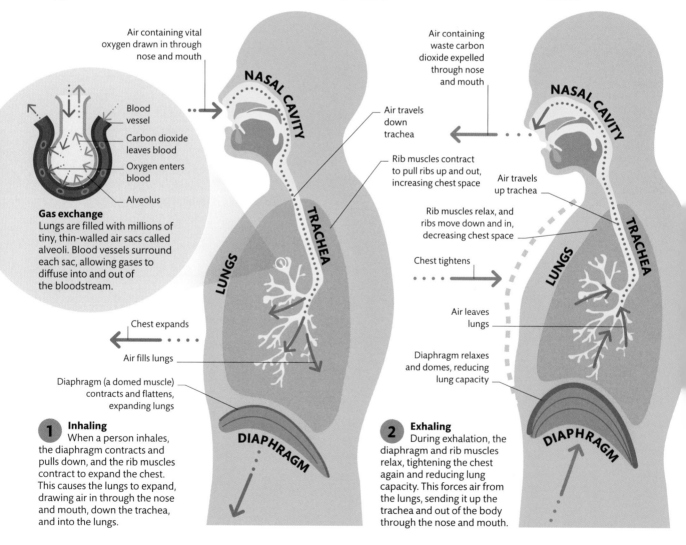

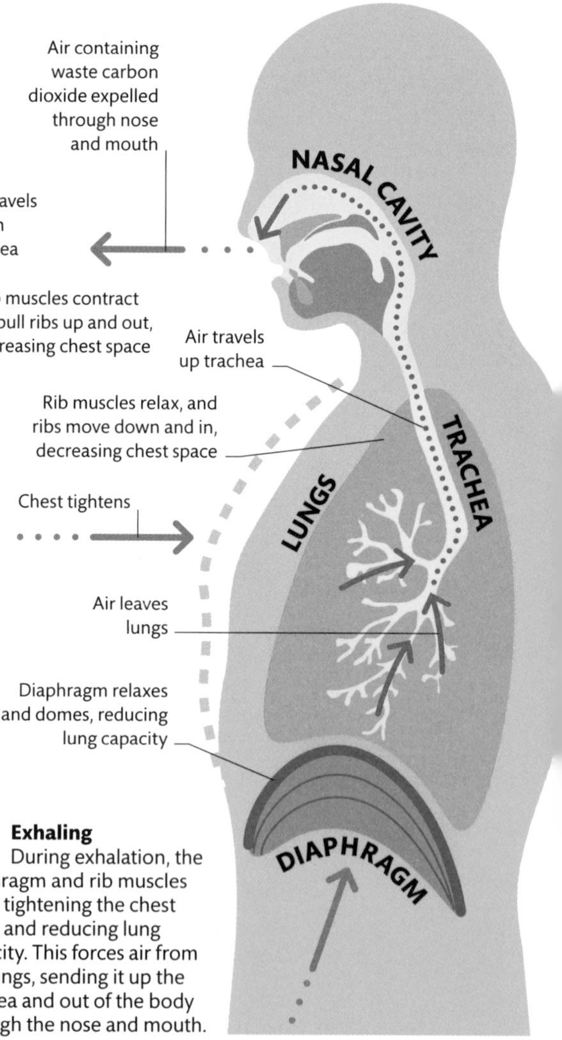
Air containing waste carbon dioxide expelled through nose and mouth
NASAL CAVITY
Air travels up trachea
Rib muscles relax, and ribs move down and in, decreasing chest space
TRACHEA
LUNGS
Chest tightens
Air leaves lungs
Diaphragm relaxes and domes, reducing lung capacity
DIAPHRAGM

2 Exhaling
During exhalation, the diaphragm and rib muscles relax, tightening the chest again and reducing lung capacity. This forces air from the lungs, sending it up the trachea and out of the body through the nose and mouth.

Gills

Fish, crabs, mollusks, and some larval stages of other animals that begin life in water breathe using gills. Gills are a series of filaments, shaped to provide as large a surface area as possible for gas exchange. Like the air sacs of lungs, these filaments are surrounded by tiny blood vessels that absorb oxygen into the blood and allow carbon dioxide to leave. Gills must be kept wet to prevent them from drying out and collapsing.

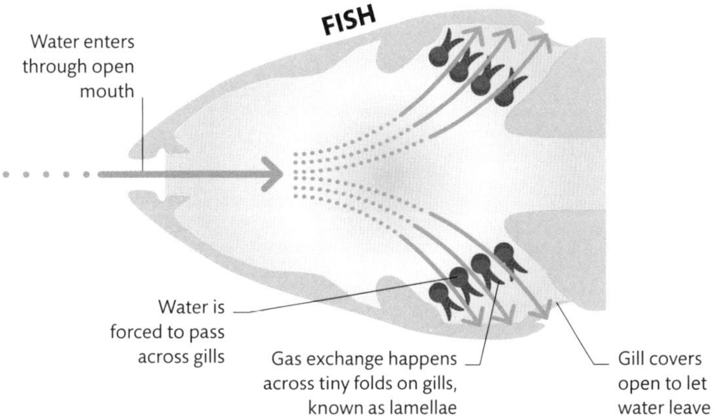
FISH
Water enters through open mouth
Water is forced to pass across gills
Gas exchange happens across tiny folds on gills, known as lamellae
Gill covers open to let water leave

Tracheae and spiracles

The breathing system of insects, such as cockroaches, is completely separate from the circulatory system, so their blood does not carry gases. Instead, they have a tracheal system, in which air is taken in through pores in the body and carried through a network of air tubes called tracheae. This tracheal system delivers vital oxygen for respiration directly to the body's tissues and removes waste carbon dioxide.

A HUMMINGBIRD TAKES ABOUT **250 BREATHS PER MINUTE,** WHILE HUMANS TAKE **AN AVERAGE OF 12**

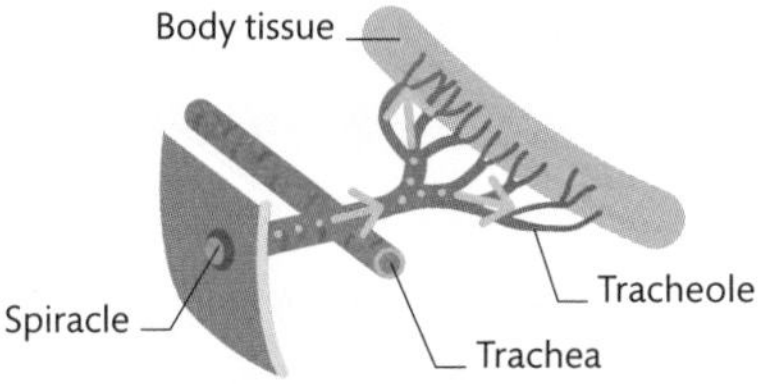
Body tissue
Spiracle
Tracheole
Trachea

Tube network
Spiracles open to let air into the tracheal system, which branches out all over the body to deliver oxygen directly to the body's tissues.

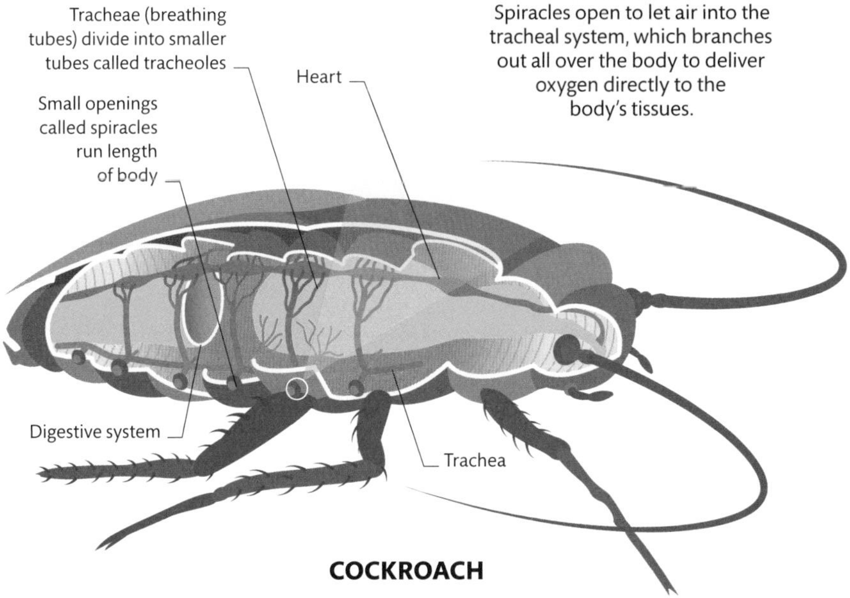
Tracheae (breathing tubes) divide into smaller tubes called tracheoles
Heart
Small openings called spiracles run length of body
Digestive system
Trachea
COCKROACH

SKIN BREATHING

Some animals—including sponges, corals, jellyfish, and worms—are able to breathe entirely through their skin. Amphibians, such as frogs, use skin breathing in combination with either gills (as tadpoles) or lungs (as adults). The skin must be moist and thin for gases to pass through this barrier—a process called diffusion.

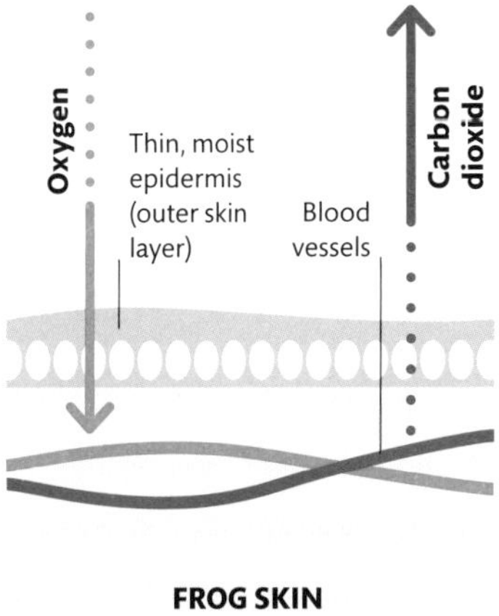
Oxygen
Thin, moist epidermis (outer skin layer)
Blood vessels
Carbon dioxide
FROG SKIN

Circulatory systems

The circulatory system is an essential transport system, responsible for supplying every cell in an animal's body with the nutrients and immune cells that it needs to function, and removing waste.

HOW MANY HEARTS DOES AN EARTHWORM HAVE?

No chambered organs like the human heart, but five pairs of heartlike organs called aortic arches that pump blood around the body.

Life-support system

All complex animals need a system for moving nutrients and waste around the body. Much like a network of highways and trucks, animals have circulatory systems with networks of blood vessels that are able to carry blood to every cell in the body. The heart is the powerful pump that lies at the center of the system, keeping it moving.

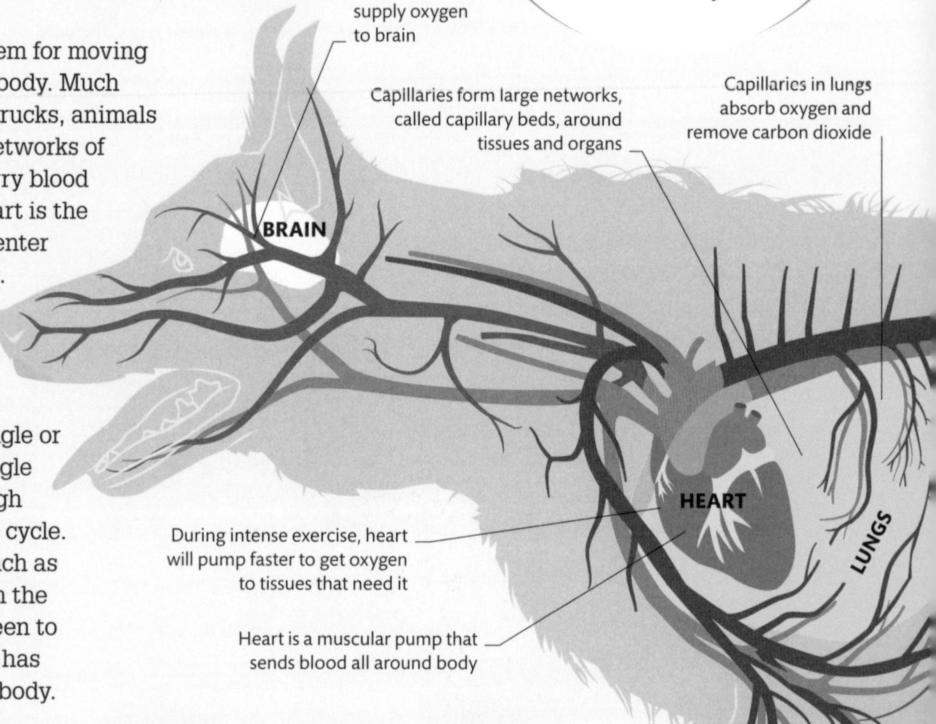

Arteries in head supply oxygen to brain
Capillaries form large networks, called capillary beds, around tissues and organs
Capillaries in lungs absorb oxygen and remove carbon dioxide
BRAIN
HEART
LUNGS
During intense exercise, heart will pump faster to get oxygen to tissues that need it
Heart is a muscular pump that sends blood all around body
Blood is made up of red blood cells, white blood cells, and platelets, in a fluid called plasma

Single and double circulation

Vertebrates can have either a single or double circulatory system. In single systems, the blood passes through the heart just once in a complete cycle. In double circulation systems, such as in humans, blood passes through the heart twice—once after it has been to the lungs and then again after it has been sent around the rest of the body.

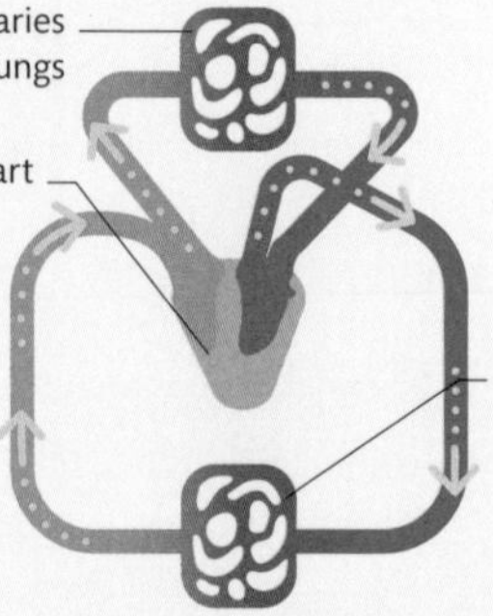

Capillaries in lungs
Heart
Capillaries in body

Mammals and birds
Mammals and birds have a double circulation system. Blood is pumped to the lungs to be oxygenated before being pumped to the body.

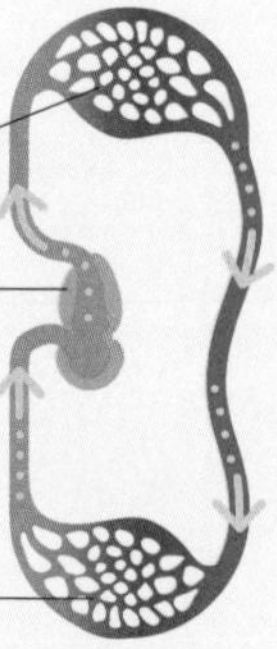

Capillaries in gills
Two-chambered heart
Capillaries in body

Fish
In fish, blood is pumped in a single loop from the heart to the gills where it picks up oxygen on its way to the rest of the body.

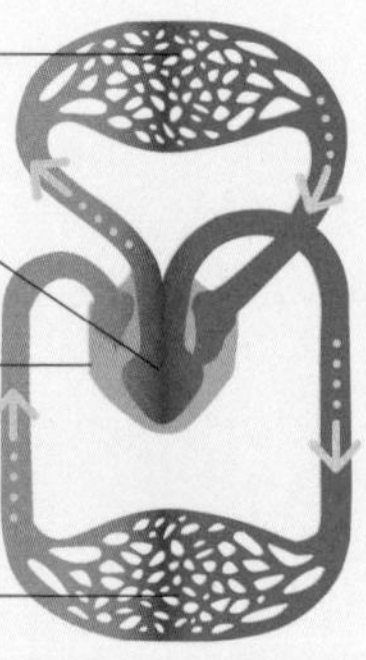

Amphibians
Amphibians have a double system, but oxygen-poor and oxygen-rich blood mixes in the heart before being pumped around the body.

OPEN CIRCULATORY SYSTEMS

Small, simple animals, such as earthworms and insects, have an open system in which a fluid called hemolymph is pumped directly into the body cavity by their long, tubular heart to exchange chemicals between the fluid and cells.

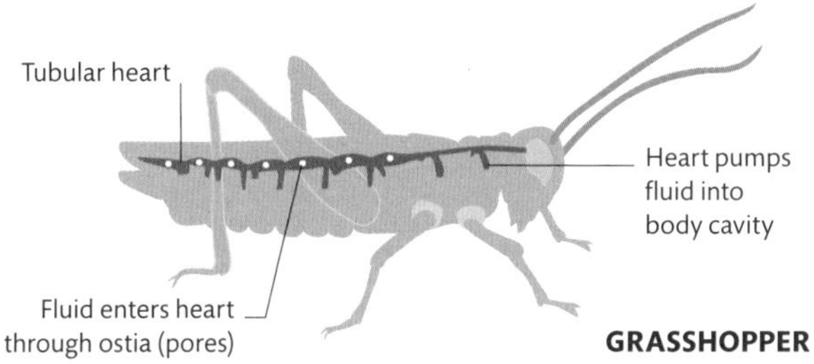
Tubular heart
Heart pumps fluid into body cavity
Fluid enters heart through ostia (pores)
GRASSHOPPER

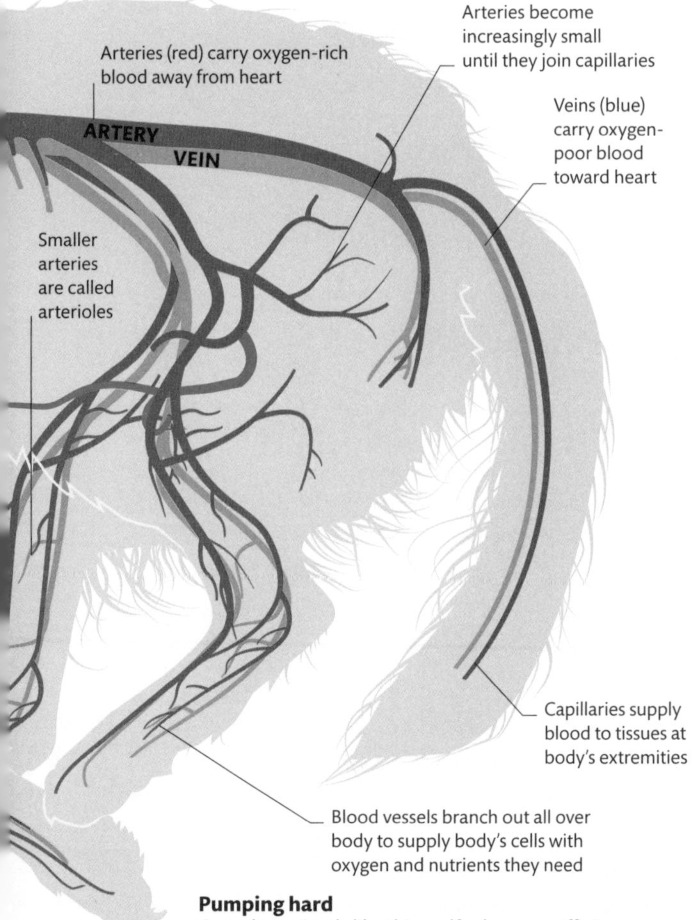
Arteries (red) carry oxygen-rich blood away from heart
Arteries become increasingly small until they join capillaries
ARTERY
VEIN
Veins (blue) carry oxygen-poor blood toward heart
Smaller arteries are called arterioles
Capillaries supply blood to tissues at body's extremities
Blood vessels branch out all over body to supply body's cells with oxygen and nutrients they need

Pumping hard
Complex animals like this wolf rely on an efficient circulatory system to quickly provide their muscles with the energy-giving chemicals they need for action.

Types of blood vessel

Three main types of blood vessels—arteries, veins, and capillaries—carry blood around a mammal's body. Arteries carry blood away from the heart and can widen or narrow to control the flow of blood. Veins carry blood back to the heart. Capillaries are where the exchange of nutrients and waste substances in tissues occurs.

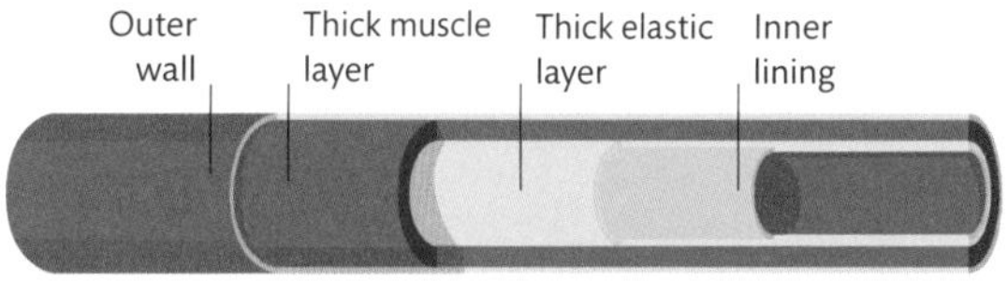
Outer wall
Thick muscle layer
Thick elastic layer
Inner lining

Artery
Arteries are surrounded by a layer of connective tissue and have thick, muscular, elastic walls that allow them to cope with surges in pressure as blood is pumped through them.

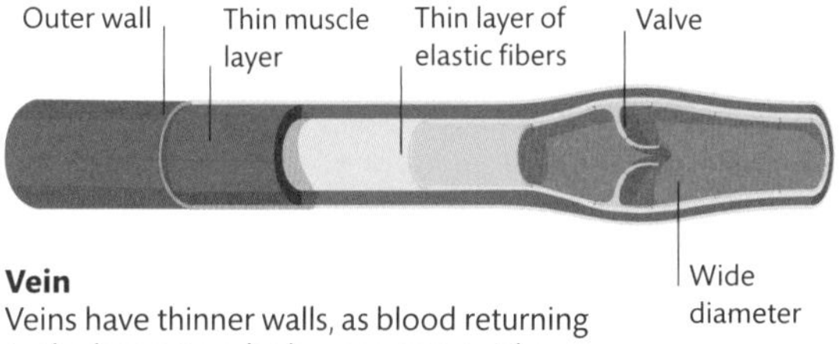
Outer wall
Thin muscle layer
Thin layer of elastic fibers
Valve
Wide diameter

Vein
Veins have thinner walls, as blood returning to the heart is under lower pressure. They contain one-way valves that prevent blood from flowing backward.

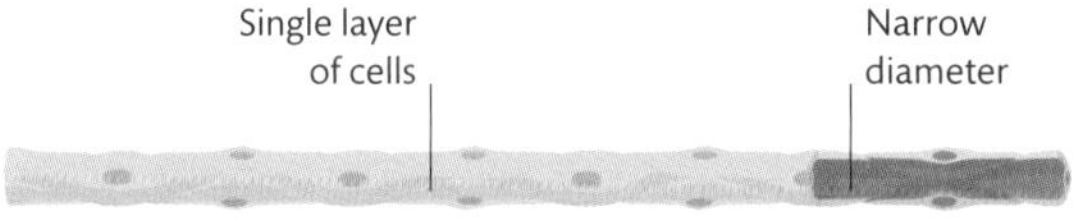
Single layer of cells
Narrow diameter

Capillary
Capillaries have very thin walls that allow nutrients, gases, and other molecules to travel between the blood and tissues. Some have gaps or pores that allow larger molecules through.

A BLUE WHALE'S **HEART** IS **5 FT (1.5 M) LONG, 4 FT (1.2 M) WIDE**, AND **WEIGHS 400 LB (181 KG)**

Digestive systems

Animals are fueled by food that they ingest. The digestive system gradually breaks down food so that an animal can absorb the nutrients it needs to survive. The system varies between animals depending on their nutritional needs.

From food to feces

The journey of food for all animals starts at the mouth. For many animals this is the start of food's passage through the long gastrointestinal (digestive) tract, which uses muscular contractions and chemicals to break down food into its component nutrients, before it can be absorbed across the gut lining and into the bloodstream. Everything that is left—the indigestible components—is then pooped out as feces.

2 Esophagus
The esophagus contracts in wavelike motions, called peristalsis, to squeeze food and liquid from the mouth to the stomach.

3 Stomach
Muscles in the stomach wall contract to churn the food with acidic gastric juices and enzymes, breaking it down into a digestive slurry called chyme. It is then released into the small intestine.

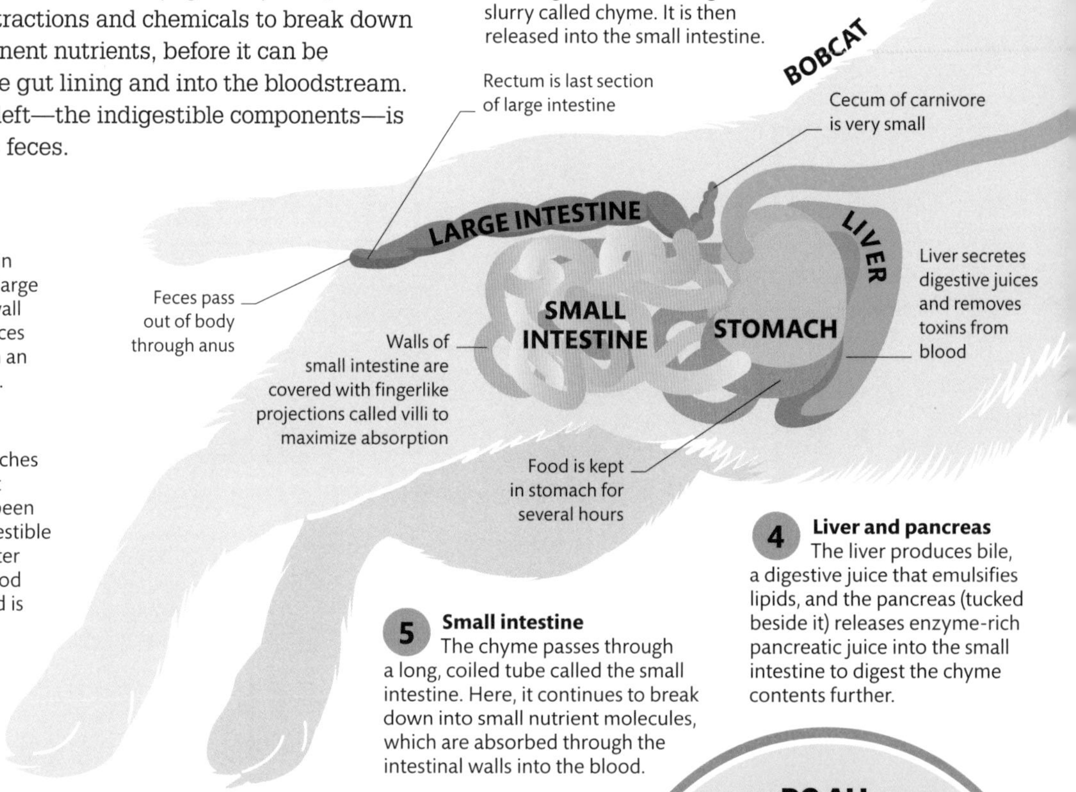
BOBCAT
Rectum is last section of large intestine
Cecum of carnivore is very small
LARGE INTESTINE
LIVER
Liver secretes digestive juices and removes toxins from blood
Feces pass out of body through anus
SMALL INTESTINE
STOMACH
Walls of small intestine are covered with fingerlike projections called villi to maximize absorption
Food is kept in stomach for several hours

8 Rectum
Feces are stored in the rectum (part of the large intestine). The rectum wall contracts to push the feces out of the body through an opening called the anus.

7 Large intestine
When chyme reaches the large intestine, most nutrients have already been absorbed, leaving indigestible food and water. The water is absorbed into the blood and the undigested food is pressed into feces.

4 Liver and pancreas
The liver produces bile, a digestive juice that emulsifies lipids, and the pancreas (tucked beside it) releases enzyme-rich pancreatic juice into the small intestine to digest the chyme contents further.

5 Small intestine
The chyme passes through a long, coiled tube called the small intestine. Here, it continues to break down into small nutrient molecules, which are absorbed through the intestinal walls into the blood.

Meat eaters
Animals that primarily eat meat are called carnivores. Meat eaters have a short digestive tract because meat is dense in nutrients that can be extracted relatively easily. Carnivore stomachs are large and highly acidic to break down meat.

6 Cecum
In carnivores, the cecum is a small chamber at the start of the large intestine that absorbs salts and minerals. In herbivores, the cecum is larger and more developed to cope with their plant-heavy diet.

DO ALL ANIMALS HAVE A DIGESTIVE SYSTEM?

Parasitic tapeworms have no digestive system. They absorb nutrients through their skin directly from the digestive tract of their host.

WOMBATS ARE THE ONLY ANIMALS IN THE WORLD TO PRODUCE CUBIC POOP

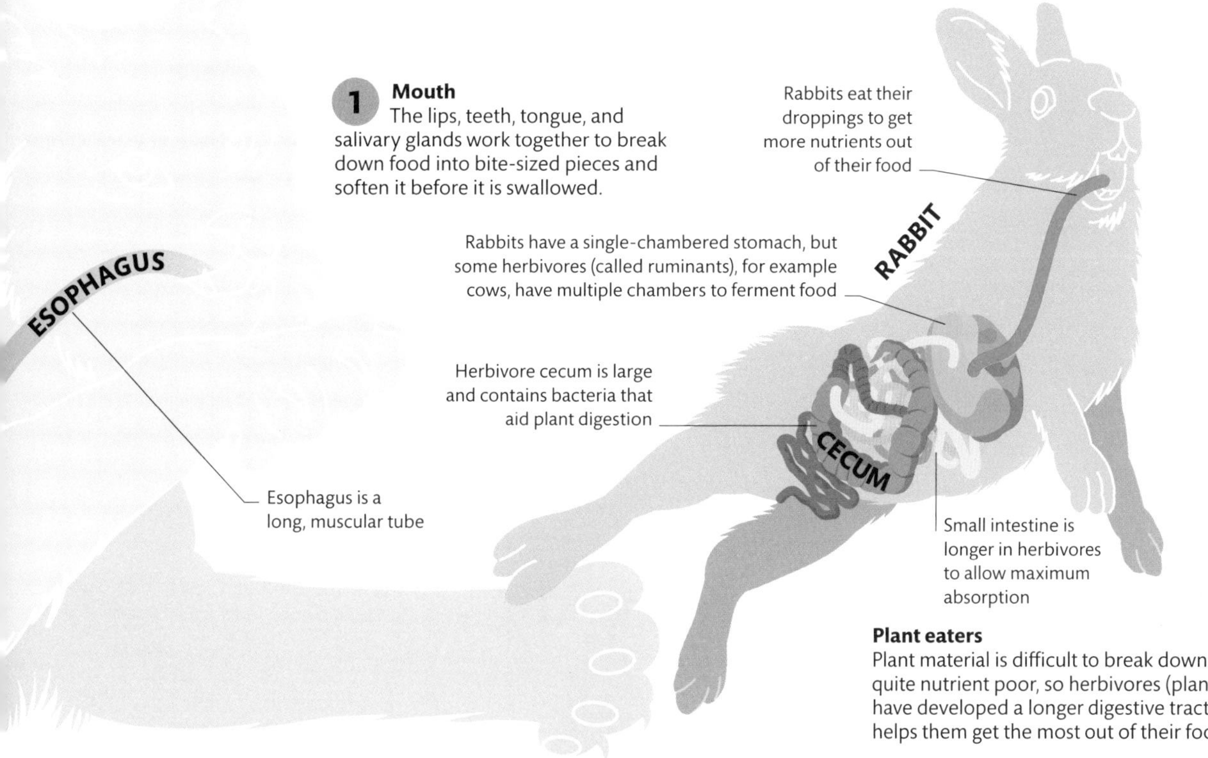
ESOPHAGUS
Esophagus is a long, muscular tube
RABBIT
Rabbits eat their droppings to get more nutrients out of their food
Rabbits have a single-chambered stomach, but some herbivores (called ruminants), for example cows, have multiple chambers to ferment food
Herbivore cecum is large and contains bacteria that aid plant digestion
CECUM
Small intestine is longer in herbivores to allow maximum absorption

1 Mouth
The lips, teeth, tongue, and salivary glands work together to break down food into bite-sized pieces and soften it before it is swallowed.

Plant eaters
Plant material is difficult to break down and quite nutrient poor, so herbivores (plant eaters) have developed a longer digestive tract that helps them get the most out of their food.

Digestive diversity

Although the function of the digestive tract is universal, there is significant variation in the form it takes across different groups of animals. Digestive systems have evolved based on what animals eat, how they eat, and where they live. Systems range from the advanced, multichambered gastrointestinal tract of the vertebrates shown above to the primitive cavity seen in some invertebrates, such as sea anemones.

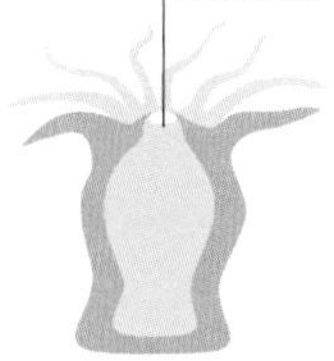
Mouth and anus

Single opening
Sea anemones eat and excrete through the same opening. They pass food through this opening into a central stomach and excrete their waste from the same opening.

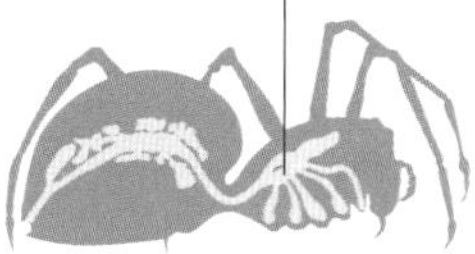
Sucking stomach

External digestion
Spiders begin digestion outside the body by regurgitating digestive enzymes onto their prey to soften it, before using a special sucking stomach to slurp it up.

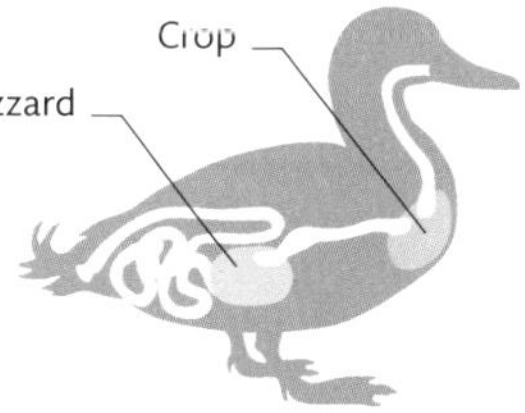
Crop
Gizzard

Grinding chamber
Birds do not have teeth but have evolved a muscular pouch (crop) that stores and moistens food, as well as a gizzard, where food is mechanically ground up with grit and small stones.

TEETH

From a carnivore's piercing canines to an herbivore's grinding molars, teeth are an important tool in acquiring and preparing food. As a result, they have become specialized depending on an animal's diet.

Carnivores kill prey and rip flesh with long pointed front teeth

CARNIVORE

Herbivores grind vegetation with flat ridged teeth at back

HERBIVORE

Omnivores that eat meat and plants have both grinding and piercing teeth

OMNIVORE

Brains and nervous systems

The brain and nervous system of an animal function a little like a command center and fiberoptic network: the brain processes information from its external environment and coordinates the body using a network of nerve fibers.

GOLDFISH BRAIN

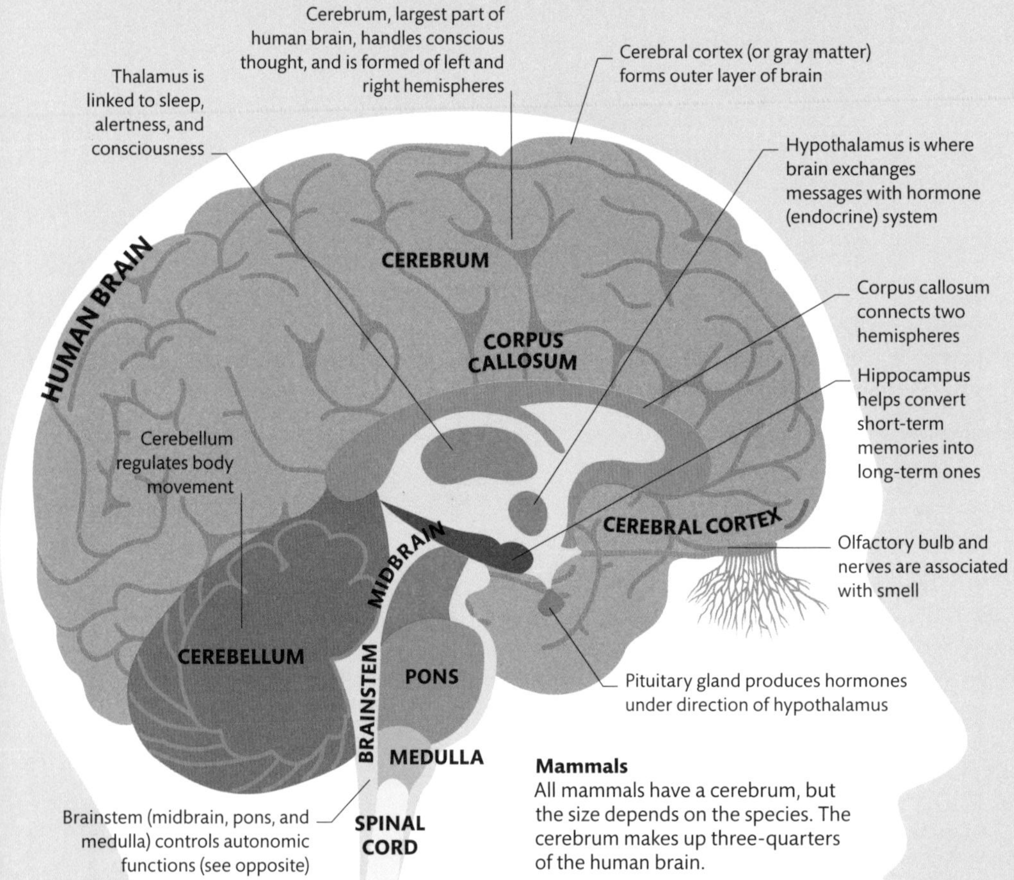

Fish
A large proportion of a fish brain is dedicated to processing visual information (optic lobe), and the cerebrum is relatively tiny.

BULLFROG BRAIN

Amphibians
Relative to body size, amphibian brains are minute compared to human brains. The proportions of the regions suggest they rely largely on reflex movements.

QUAIL BRAIN

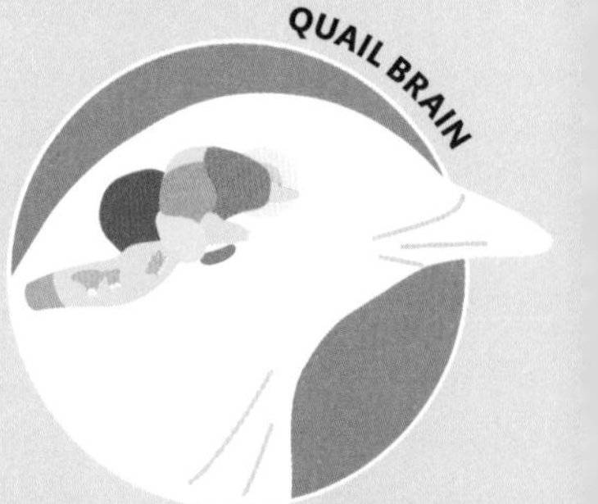

Birds
A significant proportion of a bird's brain is dedicated to olfaction (smell). The cerebellum and cerebrum are also large compared to other parts of their brains.

Thalamus is linked to sleep, alertness, and consciousness

Cerebrum, largest part of human brain, handles conscious thought, and is formed of left and right hemispheres

Cerebral cortex (or gray matter) forms outer layer of brain

Hypothalamus is where brain exchanges messages with hormone (endocrine) system

Corpus callosum connects two hemispheres

Hippocampus helps convert short-term memories into long-term ones

Olfactory bulb and nerves are associated with smell

Cerebellum regulates body movement

Pituitary gland produces hormones under direction of hypothalamus

Brainstem (midbrain, pons, and medulla) controls autonomic functions (see opposite)

HUMAN BRAIN

CEREBRUM

CORPUS CALLOSUM

CEREBRAL CORTEX

MIDBRAIN

CEREBELLUM

BRAINSTEM

PONS

MEDULLA

SPINAL CORD

Mammals
All mammals have a cerebrum, but the size depends on the species. The cerebrum makes up three-quarters of the human brain.

How brains work

The brain is constantly processing information, comparing it with stored information, and coordinating the body's responses. A vertebrate's brain is made up of regions, each with billions of connected nerve cells, or neurons, working to perform specific functions. Brains use a lot of energy, and the brains of individual species have evolved to maximize the functional areas they need.

KEY

- Cerebellum
- Optic lobe
- Cerebrum
- Pituitary gland
- Medulla
- Olfactory bulb

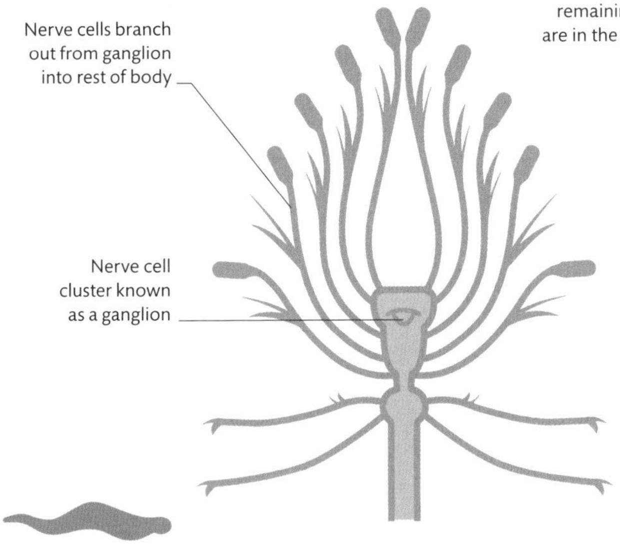
Nerve cells branch out from ganglion into rest of body
Nerve cell cluster known as a ganglion
LEECH
LEECH BRAIN

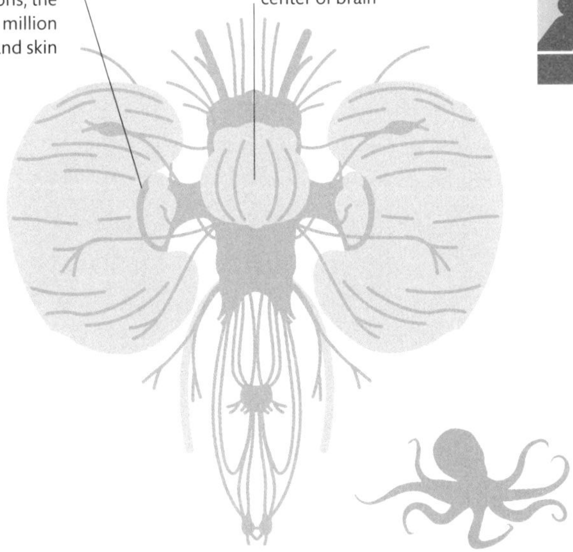
Brain contains about 180 million neurons; the remaining 320 million are in the arms and skin
Esophagus runs through center of brain
OCTOPUS BRAIN
OCTOPUS

The invertebrate nervous system

Brain complexity and shape are far more variable among invertebrates, which account for as much as 97 percent of the animal kingdom, than they are in vertebrates. This variation ranges from sponges with no nerve cells at all to octopuses, which have a donut-shaped brain and around 500 million neurons. Many invertebrates have what are known as ganglia, which are clusters of nerve cells linked together, but they are not organized to the level of a brain.

A SQUID'S **ESOPHAGUS** RUNS THROUGH A **HOLE** IN ITS **DONUT-SHAPED BRAIN**

THE PERIPHERAL NERVOUS SYSTEM

The nervous system is divided into the central nervous system (the brain and spinal cord) and the peripheral nervous system (everything else). The spinal cord is the main highway for information to pass between the brain and the body, and it is from here that pairs of spinal nerves branch out to connect the spinal cord with the rest of the body. Some of these peripheral nerves carry out voluntary actions and gather sensory information (the somatic nervous system). The rest carry out involuntary actions in the body, such as digestion or breathing (the autonomic nervous system).

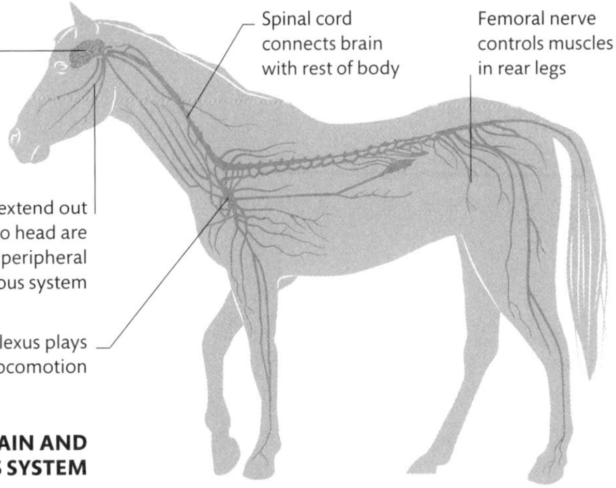
Brain
Spinal cord connects brain with rest of body
Femoral nerve controls muscles in rear legs
Nerves that extend out from brain into head are also part of peripheral nervous system
Brachial plexus plays a role in locomotion

HORSE BRAIN AND NERVOUS SYSTEM

RODS AND CONES

Electrical signal sent to brain

Rod cell

Cone cell

Light

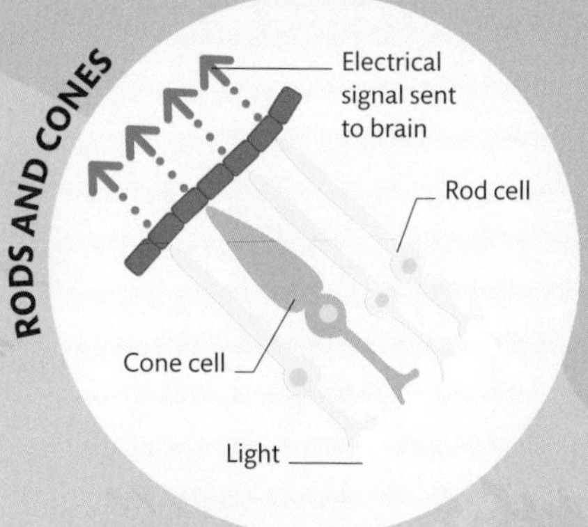

Photoreceptor cells
There are two types of light-sensitive cells in the retina. Cones are sensitive to color in high light conditions, whereas rods make it possible to see images (but not color) in low light conditions.

Vision

Animals see using their sense of vision, which converts light into electrical signals that are processed by the brain. Simple visual systems such as eyespots detect light or dark, while complex eyes allow animals to spot prey from afar.

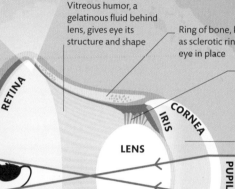

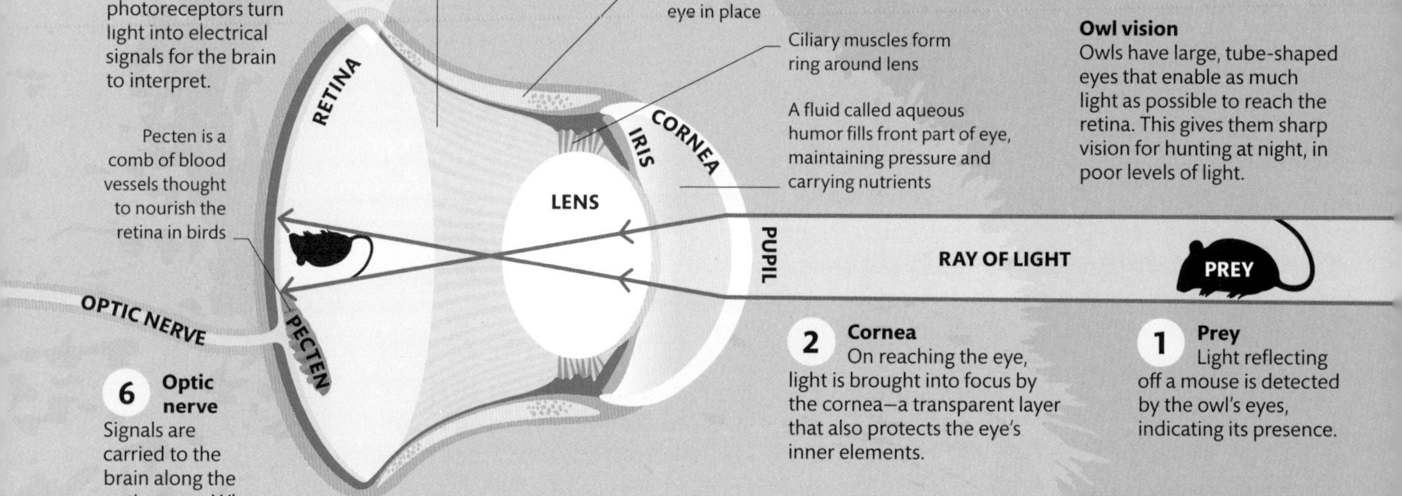

5 Retina
An upside-down image forms on the retina, where photoreceptors turn light into electrical signals for the brain to interpret.

Pecten is a comb of blood vessels thought to nourish the retina in birds

Vitreous humor, a gelatinous fluid behind lens, gives eye its structure and shape

Ring of bone, known as sclerotic ring, holds eye in place

Ciliary muscles form ring around lens

A fluid called aqueous humor fills front part of eye, maintaining pressure and carrying nutrients

RETINA

IRIS

CORNEA

LENS

PUPIL

RAY OF LIGHT

PREY

OPTIC NERVE

PECTEN

Owl vision
Owls have large, tube-shaped eyes that enable as much light as possible to reach the retina. This gives them sharp vision for hunting at night, in poor levels of light.

2 Cornea
On reaching the eye, light is brought into focus by the cornea—a transparent layer that also protects the eye's inner elements.

1 Prey
Light reflecting off a mouse is detected by the owl's eyes, indicating its presence.

6 Optic nerve
Signals are carried to the brain along the optic nerve. Where this connects to the retina, there are no photoreceptor cells, creating a blind spot.

4 Lens
The flexible lens fine-tunes the focus of the image. Ciliary muscles pull on the lens, changing its shape to create sharper focus.

3 Iris
The muscular iris controls the size of the pupil to allow more or less light to pass through. In low light, for example, the iris widens the pupil.

UV VISION IN RAPTORS HELPS THEM DETECT URINE LEFT ON THE PATHS OF THEIR RODENT PREY

The vertebrate eye

In a vertebrate eye, particles of light pass through a clear front layer (the cornea) and enter the eye through the pupil. A single lens focuses the light onto a layer of light-sensitive tissue at the back of the eye (the retina), where it is converted into electrical signals. The brain then interprets this information.

WHY ARE CAT PUPILS VERTICAL?

Vertical pupils are common in ambush predators, such as cats, because they optimize depth perception. This helps a predator estimate the distance to its prey.

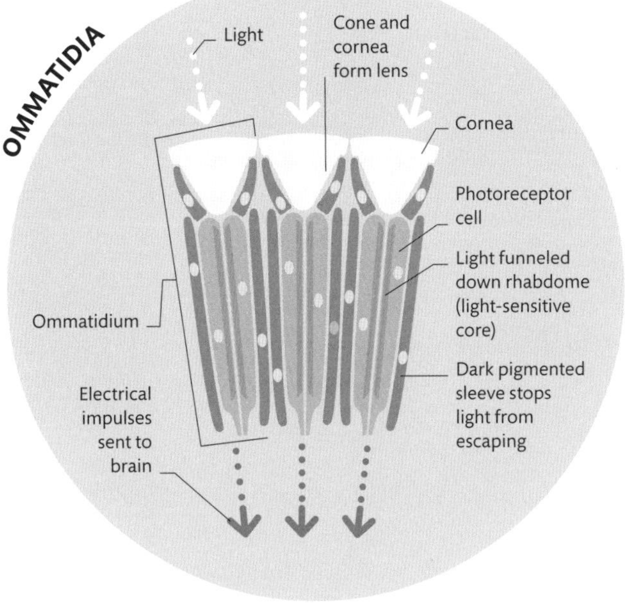
OMMATIDIA
Light
Cone and cornea form lens
Cornea
Photoreceptor cell
Light funneled down rhabdome (light-sensitive core)
Ommatidium
Dark pigmented sleeve stops light from escaping
Electrical impulses sent to brain

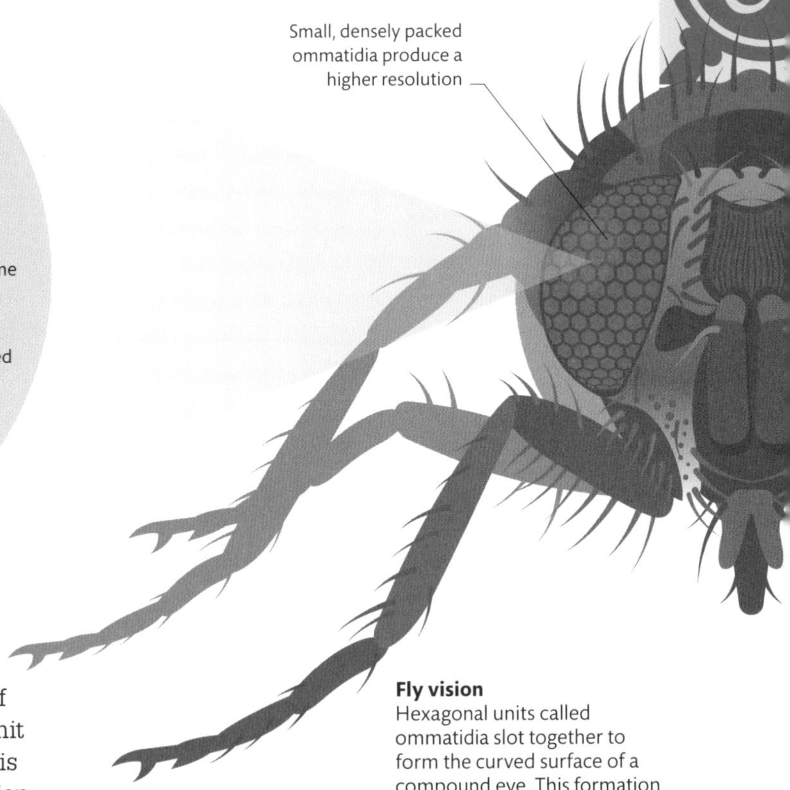
Small, densely packed ommatidia produce a higher resolution

Compound eyes

Most insects and crustaceans (such as crabs) have compound eyes, which are made up of thousands of small, independent units called ommatidia. Each unit includes a cornea, lens, and photoreceptor cells. This system gives a wide view but poorer image resolution compared with a vertebrate eye. The compound eye is an adaptation that helps animals detect fast movement.

Fly vision
Hexagonal units called ommatidia slot together to form the curved surface of a compound eye. This formation gives a fly an almost 360° view of its surroundings; each unit captures a fragment of the image.

Monocular and binocular vision

In monocular (one-eyed) vision, each eye creates a separate image, while in binocular (both-eyed) vision, the eyes work together. Predators usually have more binocular vision as it produces a clearer image and better depth perception. Prey animals tend to have more monocular vision as it allows them to scan for danger.

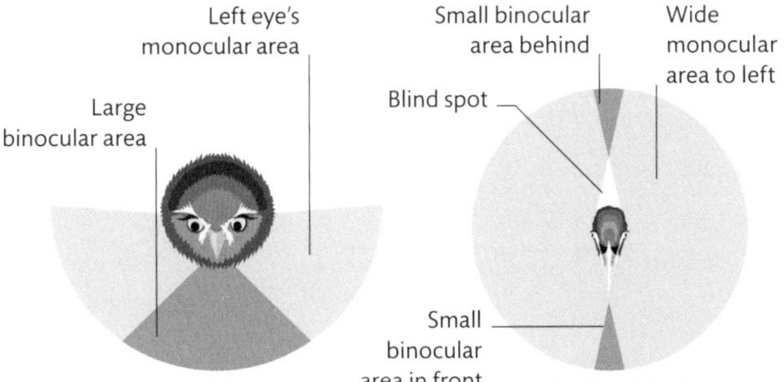
Left eye's monocular area
Large binocular area
Small binocular area behind
Wide monocular area to left
Blind spot
Small binocular area in front

OWL **SHOREBIRD**

BEYOND THE VISIBLE SPECTRUM

Many animals can see beyond the range of light wavelengths visible to humans. Bumblebees, for example, have photoreceptors that are sensitive to ultraviolet (UV) light. This allows them to detect seemingly invisible markings on flowers that direct them to sources of nectar, much like lights on a runway.

HUMAN VIEW **BUMBLEBEE VIEW**

Hearing

Sound can provide a vital layer of information about the environment and can sometimes mean the difference between life and death for an animal hunting prey, dodging a predator, or seeking a mate. Animals have evolved different ways of detecting sound vibrations—from simple hairs on the antennae of an insect, to the complex mammalian ear.

How do mammals hear?

All mammals have three sections to their ear: the outer, middle, and inner ear. The outer ear collects sound waves and funnels them into the middle ear, where the sound waves are amplified and transmitted to the inner ear. The inner ear then converts the mechanical stimulation of sound waves into electrical signals that are sent to the brain, where they can be interpreted and acted upon.

1 Outer ear
The outer ear captures sound waves. It usually includes a pinna (the visible "ear" part), the ear canal, and muscles to move the pinna.

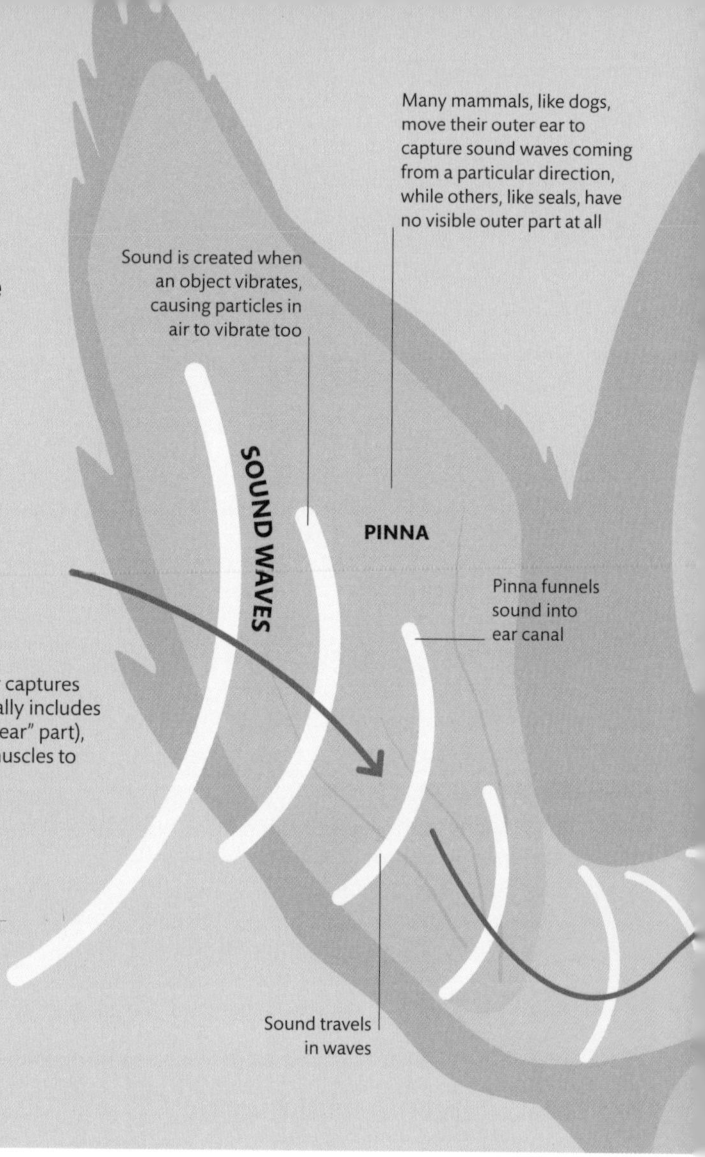

Many mammals, like dogs, move their outer ear to capture sound waves coming from a particular direction, while others, like seals, have no visible outer part at all
Sound is created when an object vibrates, causing particles in air to vibrate too
SOUND WAVES
PINNA
Pinna funnels sound into ear canal
Sound travels in waves

Do insects have ears?

The ability to hear has evolved in insects multiple times, which means that a range of adaptations for hearing can be seen in different types of insect. Some insects have hearing organs on their antennae, and others have them in their forelegs, wings, or even their mouths. Avoiding predators seems like an obvious benefit of hearing, but it seems that finding mates must also be important, because only species of cicada that sing to attract mates have evolved hearing.

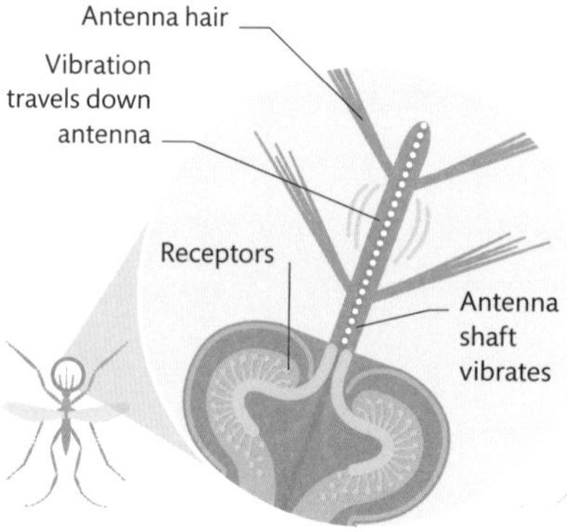

Antenna hair
Vibration travels down antenna
Receptors
Antenna shaft vibrates

Mosquito
Mosquitoes detect sound vibrations through tiny hairs on their antennae. Males are particularly sensitive to the sound of female wing vibrations.

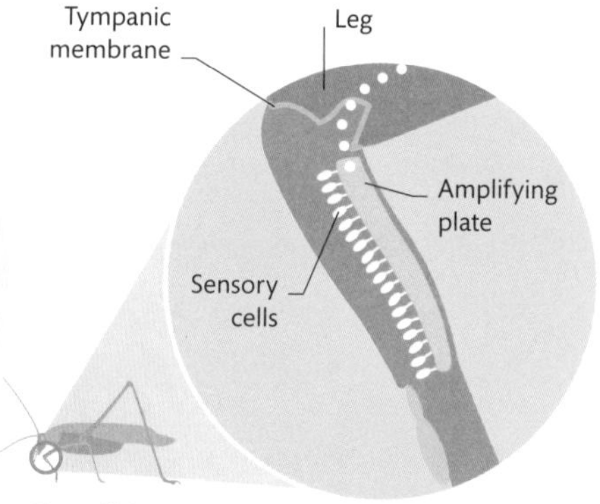

Tympanic membrane
Leg
Amplifying plate
Sensory cells

Katydid
A tympanic membrane on each of a katydid's (or bush cricket's) front legs transmits sound vibrations to an amplifying plate that then passes them to a cochlea-like organ.

OWLS HAVE NO OUTER EAR– THEY USE THEIR "FACE DISK" TO FUNNEL SOUND

3 Inner ear
The inner ear includes the semicircular canals, which are involved in balance, and the cochlea, which converts sound waves into electrical signals.

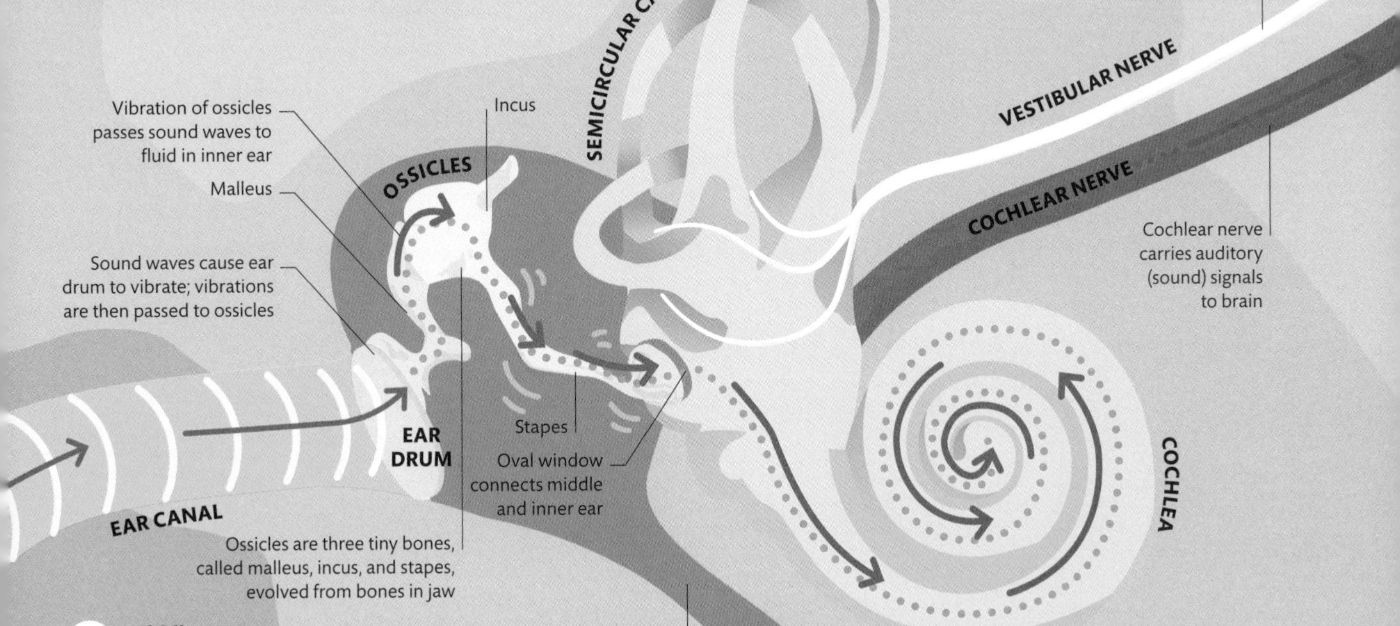
Fluid-filled semicircular canals contain tiny hairs that bend as fluid moves over them in response to movement; this motion is converted into electrical signals that pass to brain
SEMICIRCULAR CANALS
Vestibular nerve carries information about balance to brain
VESTIBULAR NERVE
COCHLEAR NERVE
Cochlear nerve carries auditory (sound) signals to brain
Vibration of ossicles passes sound waves to fluid in inner ear
Incus
OSSICLES
Malleus
Sound waves cause ear drum to vibrate; vibrations are then passed to ossicles
EAR DRUM
Stapes
Oval window connects middle and inner ear
COCHLEA
EAR CANAL
Ossicles are three tiny bones, called malleus, incus, and stapes, evolved from bones in jaw
Eustachian tube connects middle ear with throat, drains fluid, and equalizes air pressure in ear
EUSTACHIAN TUBE
Fluid-filled cochlea houses thousands of tiny hairs that convert sound waves into electrical signals

2 Middle ear
The middle ear amplifies sounds. Three small bones (malleus, incus, and stapes) work together to receive, amplify, and transmit the sound waves from the ear drum to the inner ear.

CAN SOUND BE HEARD IN SPACE?

No, sound cannot be heard in space because there is no air in space, so there is no medium for the sound waves to travel through.

BEYOND HUMAN HEARING

The number of vibrations per second made by the source of a sound is called the frequency. Some animals make and hear sounds at frequencies below (infrasound) or above (ultrasound) the human range of hearing. Elephants hear low-frequency sounds by sensing vibrations through their feet. Bats use a high-frequency sound detection called echolocation (see p.175).

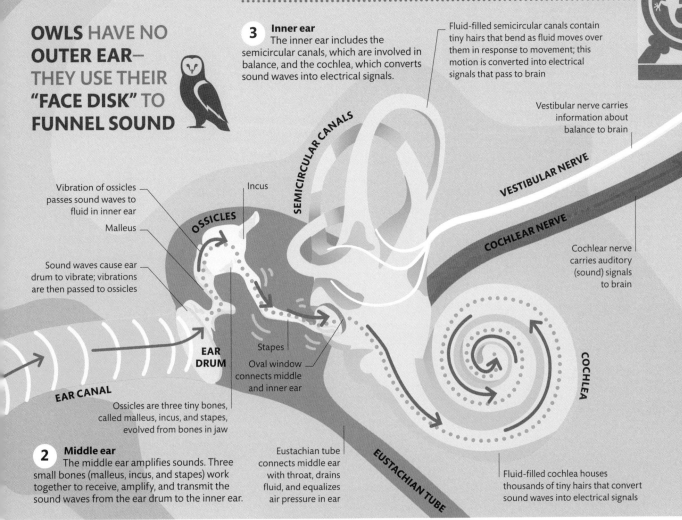

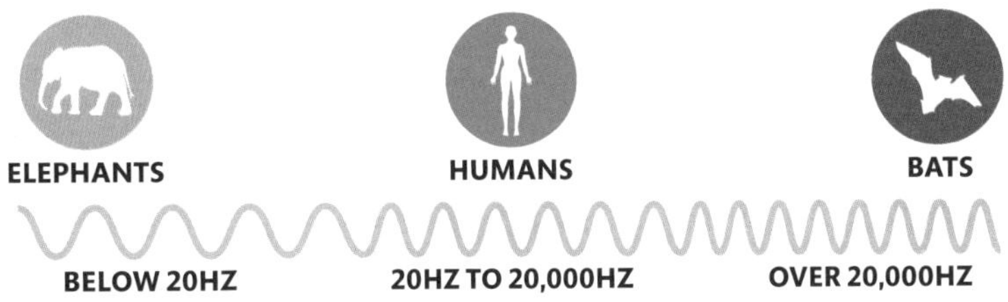
ELEPHANTS
HUMANS
BATS
BELOW 20HZ
20HZ TO 20,000HZ
OVER 20,000HZ

Sensing chemicals

Being able to detect chemicals through systems such as smell and taste can be a matter of life or death for animals, who rely on them to evade predators and locate food.

WHAT ANIMAL HAS THE BEST SENSE OF SMELL?

African Bush Elephants have 2,000 olfactory sensors, 2.5 times as many as well-known sniffers like blood hounds, and 50 times more than humans.

The mammalian nose

The system of smell is called the olfactory system. Mammals use it to detect scent chemicals, called odorants, from the air to gain useful information about the environment around them. When they enter the nose, odorants are picked up by sensory cells that send signals to the brain's olfactory bulb. The size of the olfactory bulb and the number and types of sensor differ considerably between mammals.

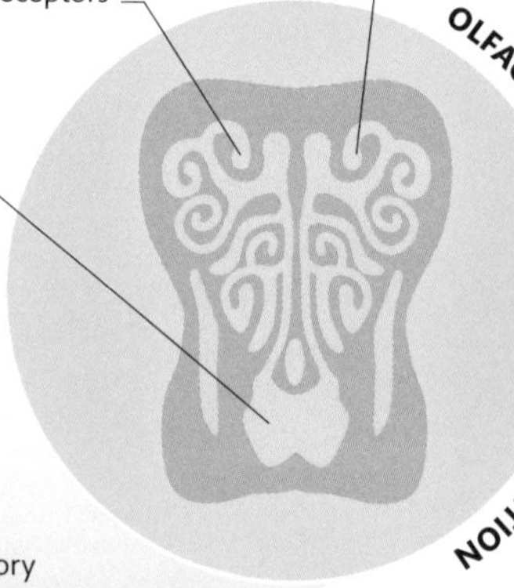

Structure creates large surface area for olfactory receptors
Thin bones are folded like scrolls
Epithelium-lined cavity
OLFACTORY EPITHELIUM CROSS SECTION

The olfactory epithelium
The back of the nasal passage contains a labyrinth of thin bones (the olfactory recess) covered by a thin layer of tissue called the olfactory epithelium.

3 Inside the brain
Signals from the olfactory receptor neurons are relayed through clusters of nerves called glomeruli to mitral cells located in an area of the brain called the olfactory bulb. The mitral cells then carry these signals to different areas of the brain.

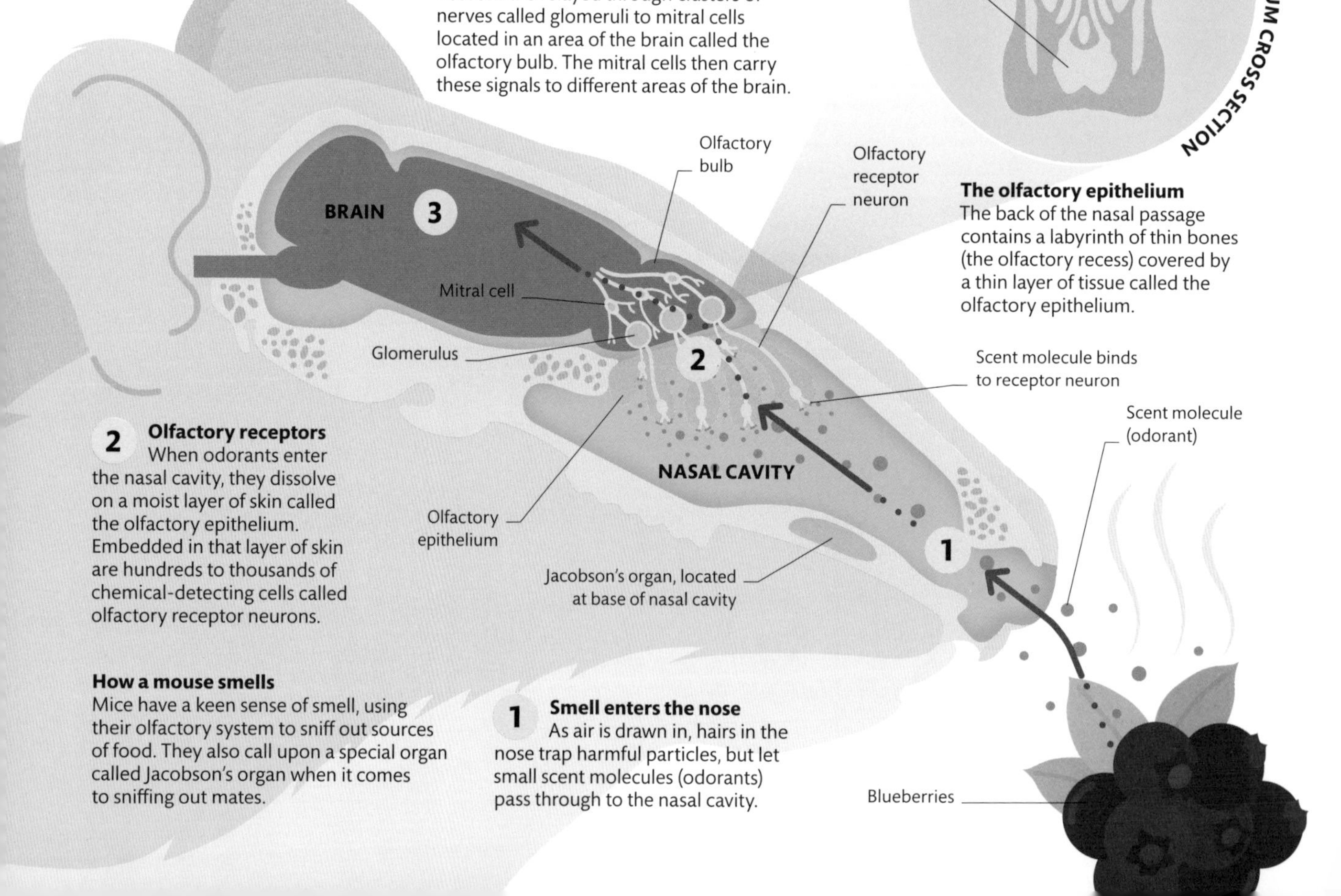

BRAIN
3
Olfactory bulb
Olfactory receptor neuron
Mitral cell
Glomerulus
2
Scent molecule binds to receptor neuron
Scent molecule (odorant)
NASAL CAVITY
Olfactory epithelium
1
Jacobson's organ, located at base of nasal cavity
Blueberries

2 Olfactory receptors
When odorants enter the nasal cavity, they dissolve on a moist layer of skin called the olfactory epithelium. Embedded in that layer of skin are hundreds to thousands of chemical-detecting cells called olfactory receptor neurons.

How a mouse smells
Mice have a keen sense of smell, using their olfactory system to sniff out sources of food. They also call upon a special organ called Jacobson's organ when it comes to sniffing out mates.

1 Smell enters the nose
As air is drawn in, hairs in the nose trap harmful particles, but let small scent molecules (odorants) pass through to the nasal cavity.

1 Tongue
The muscular tongue pushes food around the mouth, but it is also covered in a mucous membrane that dissolves chemicals for tasting, helping the brain determine if the food should be eaten.

Tongue

The mammalian tongue

Hosting an array of taste detectors, the mammalian tongue works in tandem with the nose to help an animal determine how nutritious or safe food is. Most mammals have five types of taste receptor cells: sweet, sour, bitter, salty, and umami (earthy flavors). Cats, however, have lost sensitivity to sweet things because they are strict carnivores.

COVERED HEAD TO FIN IN TASTE SENSORS, CATFISH ARE SWIMMING TONGUES

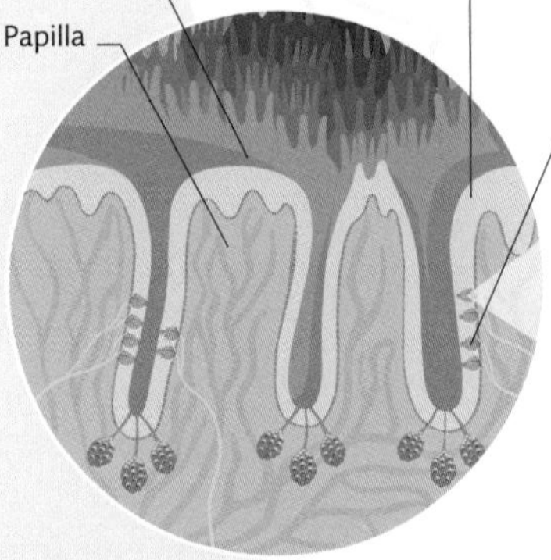
Tongue surface
Mucous membrane
Papilla

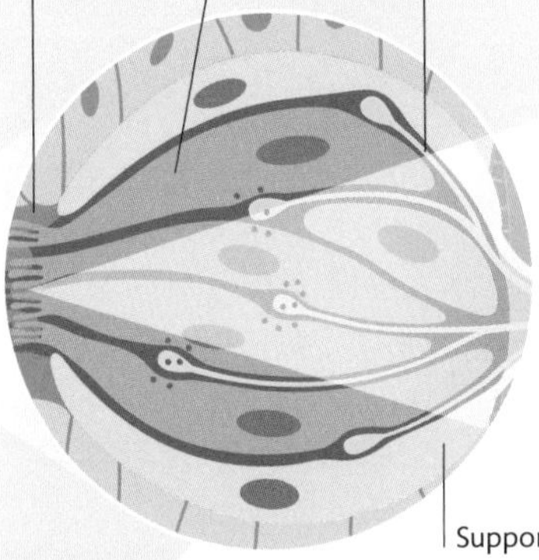
Taste bud
Taste receptor cell
Taste pore
Nerve fiber
Supporting cell

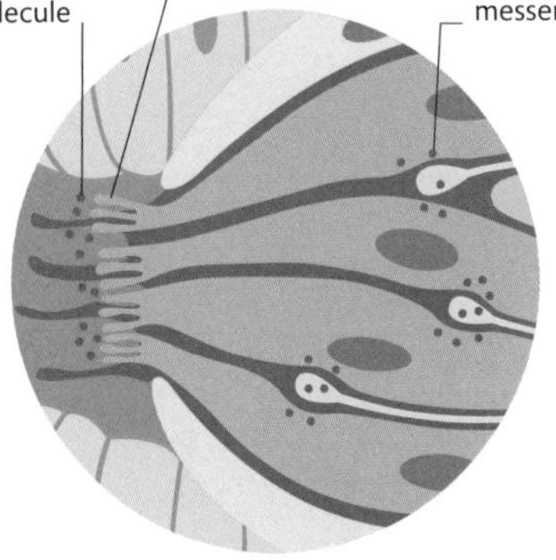
Microvilli containing receptors that bind with chemicals in food
Flavor molecule
Chemical messenger

2 Papillae
The small bumpy structures on the top of the tongue are papillae. They are thought to increase the surface area of the tongue, bringing more mucous membrane into contact with the food.

3 Taste buds
A taste bud is a collection of taste receptor cells, nerves, and supporting cells. At one end of the bud is a taste pore, where the tips of the sensory cells protrude out into the mucous membrane.

4 Taste receptor cells
Each taste receptor cell is specialized to bind to certain chemicals in food, for example, sugars or salts. When these chemicals bind to the receptor cells, a signal is sent to the brain along nerve fibers.

FORKED TONGUES AND JACOBSON'S ORGANS

A snake's tongue carries odorants to a specialized pouch called Jacobson's organ. The molecules bind to the organ's receptors, which send information to the brain. The snake can tell the direction of the odor based on how many molecules land on each fork of its tongue.

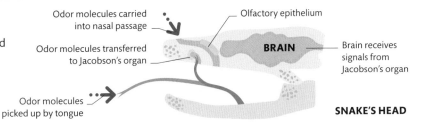

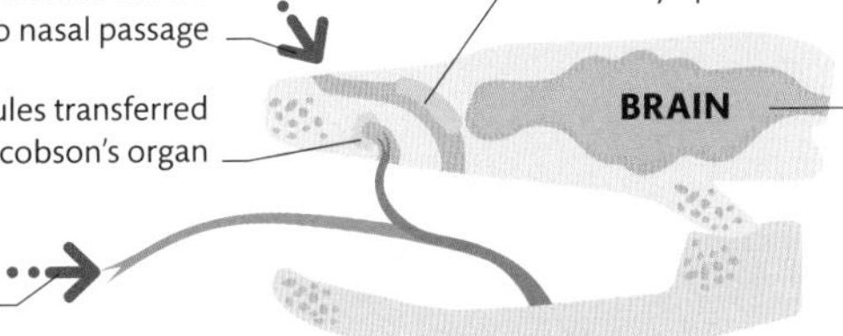
Odor molecules carried into nasal passage
Olfactory epithelium
Odor molecules transferred to Jacobson's organ
BRAIN
Brain receives signals from Jacobson's organ
Odor molecules picked up by tongue
SNAKE'S HEAD

Touch

Touch is one of the oldest and most basic sensory systems in the animal kingdom. Even simple animals without complex sense organs, such as eyes or ears, usually have some way to detect and respond to touch.

Touch receptors

The sense of touch is constructed from signals received from several types of receptor. Touch receptors are specialized structures at the ends of sensory nerves. Most are located in the skin; some lie near the surface to detect very light contact, and others are found in deeper layers and need more stimulation to activate. When physical information—such as heat, pressure, cold, stretch, and contact—disrupts or distorts the receptor, it translates this information into a nerve impulse. The signal travels to the brain, which generates a response to the information.

WHICH ANIMAL IS MOST SENSITIVE TO TOUCH?

The star-nosed mole has a nose designed for touch rather than smell, with 25,000 minute touch sensors on its star-shaped appendages.

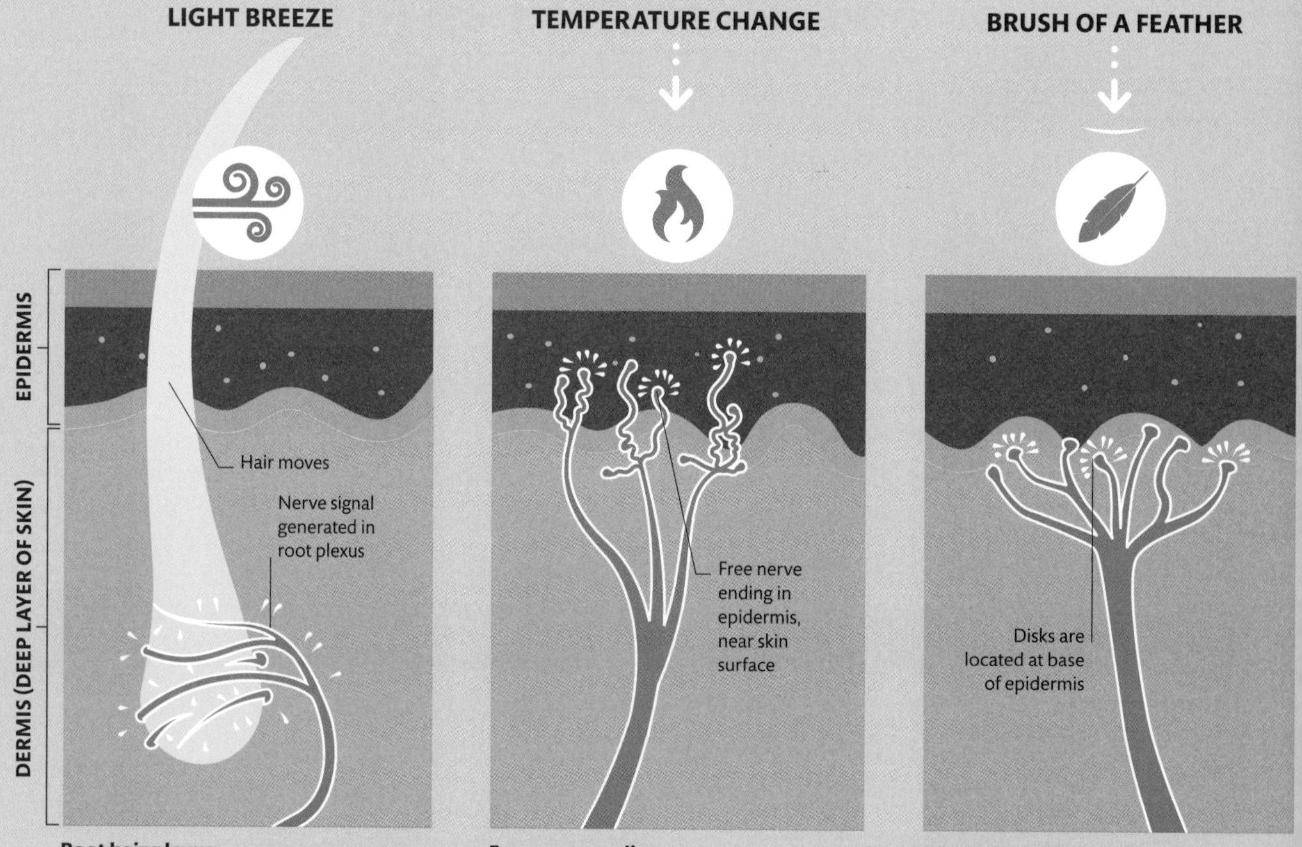

LIGHT BREEZE
TEMPERATURE CHANGE
BRUSH OF A FEATHER
EPIDERMIS
DERMIS (DEEP LAYER OF SKIN)
Hair moves
Nerve signal generated in root plexus
Free nerve ending in epidermis, near skin surface
Disks are located at base of epidermis

Root hair plexus
A network of nerve endings surrounds the root of a hair or whisker. When the hair moves or touches something, it triggers the plexus to send a signal down the nerve.

Free nerve endings
Some nerves have no special structures at their ends. These free ends extend into the surface layer of the skin and are sensitive to hot, cold, pain, and itching.

Merkel's disks
These nerve endings are sensitive to very light touch and help detect shapes and edges of objects. They are very dense in areas such as fingertips.

THE LATERAL LINE

Fish detect changes in water pressure and movement using a sensory system called the lateral line. Water enters the lateral-line canal through pores along the fish's sides. Specialized nerve endings called neuromasts bend in response to changes in water pressure or flow and convert that mechanical bending into an electrical signal that is transmitted to the brain.

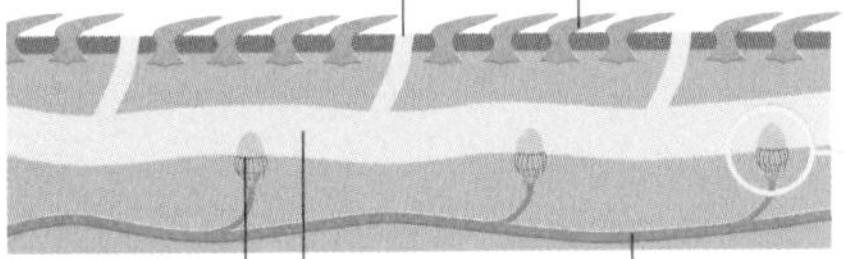
Pore in skin
Scale
Neuromast
Lateral-line canal
Sensory nerve

CROSS SECTION OF FISH BODY SURFACE

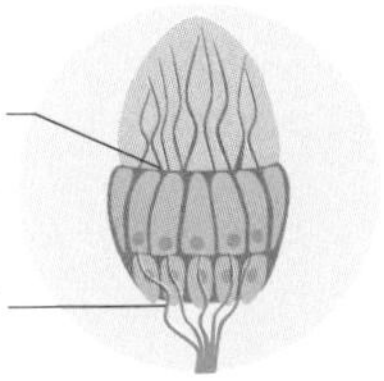
Sensory hairs embedded in jellylike cone
Sensory hair cell

NEUROMAST

SEALS CAN USE THEIR **WHISKERS** TO **DETECT A FISH** SWIMMING **330 FT (100 M) AWAY**

GENTLE TOUCH
Corpuscles are located in dermis, in bumps called papillae

Meissner's corpuscles
Deeper in the skin are endings sensitive to light pressure and vibration. These detect shapes and textures. Most are found in hairless areas such as fingertips and palms.

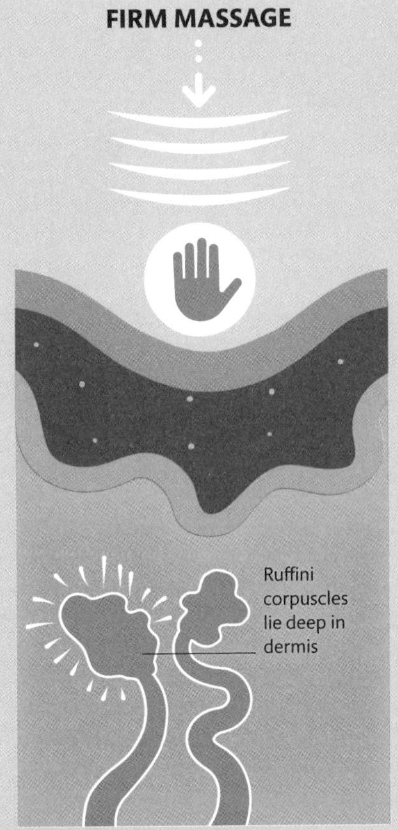
FIRM MASSAGE
Ruffini corpuscles lie deep in dermis

Ruffini endings
These endings sense sustained pressure and stretching. They also detect changes in the angles of joints, aiding awareness of body position and movement.

VIBRATION
Pacinian corpuscles are located at base of dermis

Pacinian corpuscles
Deep in the skin are large endings that are sensitive to vibration. They are found in low densities in the skin, but are also found in the intestines and joints.

Special sensors

From fire chaser beetles that can detect a fire from 80 miles (130 km) away to platypuses that use their bills like metal detectors, many animals have evolved finely tuned super sensors that help them survive in their environments.

WHY DO FIRE CHASER BEETLES NEED TO DETECT FIRE?

Their larvae feed only on burned wood, so they use sensors to detect infrared radiation emitted by heat to locate fires.

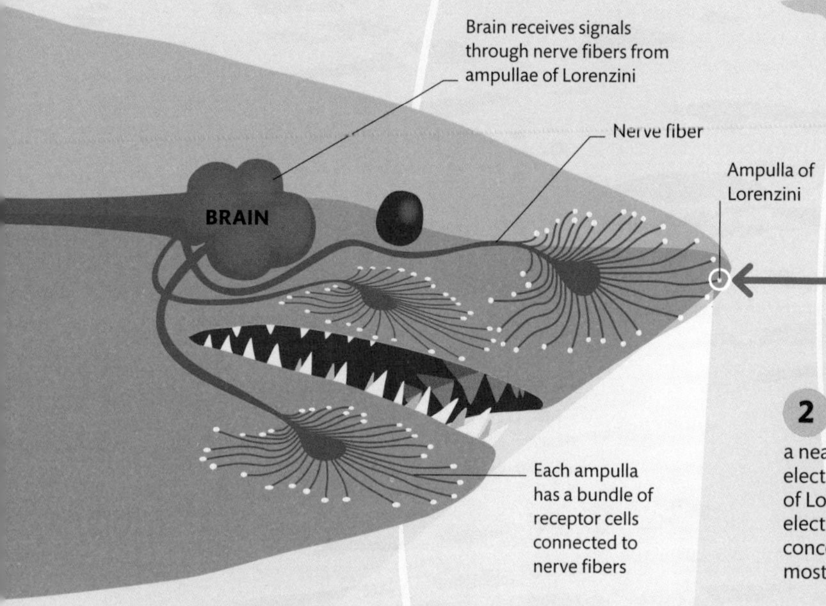
Brain receives signals through nerve fibers from ampullae of Lorenzini
Nerve fiber
Ampulla of Lorenzini
BRAIN
Each ampulla has a bundle of receptor cells connected to nerve fibers
Electrical signals propagate through surrounding water

2 Shark
As it swims through the water, a nearby shark uses bundles of electroreceptors called ampullae of Lorenzini to detect the fish's electric field. The visible pores are concentrated all around its head, mostly on the snout and jaw.

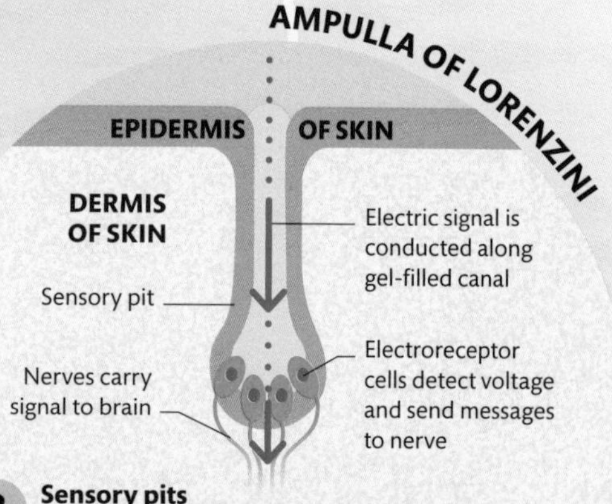
AMPULLA OF LORENZINI
EPIDERMIS OF SKIN
DERMIS OF SKIN
Electric signal is conducted along gel-filled canal
Sensory pit
Electroreceptor cells detect voltage and send messages to nerve
Nerves carry signal to brain

3 Sensory pits
The pores of the ampullae of Lorenzini open up into gel-filled canals. The conductive gel transfers electric signals in the water to electroreceptor cells at the base of the sensory pit. The signals are then sent to the brain, and the shark prepares to strike.

Electrical sensors

Some animals are able to detect the weak electric signals generated by muscle movements in other animals. This sense can be particularly useful for animals that live in low-light environments, such as murky rivers, and those that hunt at night or look for prey buried in sand. Some animals even generate a weak electric signal and use distortions in the field created by the signal to detect objects, such as rocks, that do not generate electric signals themselves. This sense, known as electroreception, is most common in aquatic environments because water carries electrical signals far better than air. Wired for hunting, sharks have a network of hundreds to thousands of electroreceptors that help them sense the location of nearby prey and line up to attack with precision.

BIRDS NAVIGATE DURING MIGRATION BY SENSING EARTH'S MAGNETIC FIELD

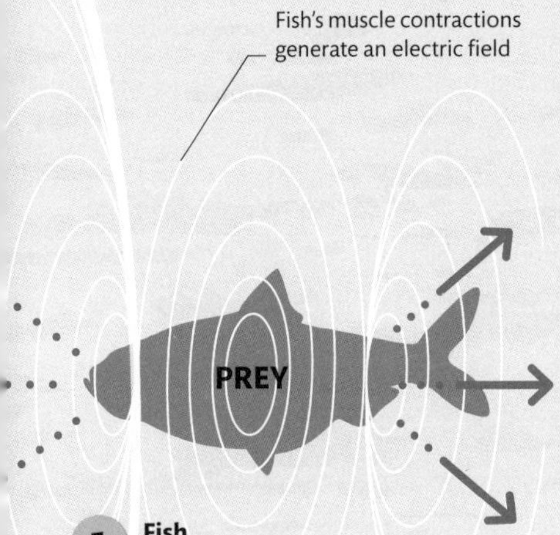
Fish's muscle contractions generate an electric field
PREY

1 Fish
As a fish moves through the water, its muscle contractions generate weak electric signals that create an electric field all around the fish. Even if a fish is perfectly still, its beating heart generates an electric field.

ECHOLOCATION

Some animals, including dolphins and bats, use high-frequency sound waves to navigate the world around them. Animals can detect their surroundings and prey using echoes that are produced when the sound waves bounce off objects. Called echolocation, this capability can help animals determine an object's distance (how quickly the echo comes back), size (how large the echo is), and direction (the strength of the echo in each ear).

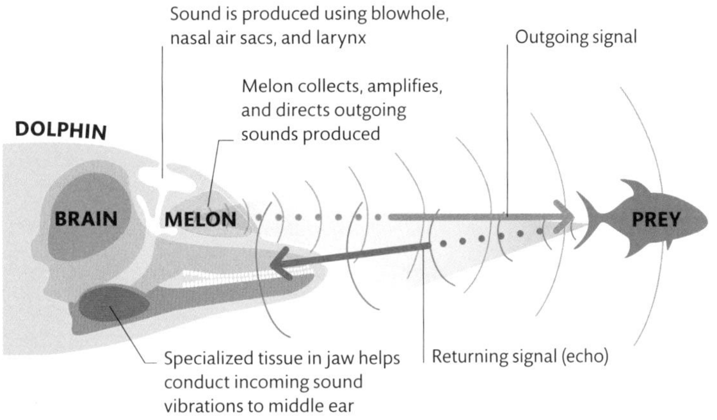
Sound is produced using blowhole, nasal air sacs, and larynx
Outgoing signal
Melon collects, amplifies, and directs outgoing sounds produced
DOLPHIN
BRAIN
MELON
PREY
Specialized tissue in jaw helps conduct incoming sound vibrations to middle ear
Returning signal (echo)

Dolphin echolocation
Dolphins send out a directed beam of clicks that resonate out through a fatty area of their head called the melon. When sound waves hit a fish, they bounce back, creating an echo.

Antennae

Most arthropods (including insects and crustaceans) have antennae that sense a wide variety of information, from smell, touch, taste, wind speed, heat, and moisture, to sound waves. Many species have combinations of these sensors, but they are predominantly for smell. Animals have evolved differently shaped antennae depending on how they use them. An ant smelling its way along a trail has elbow-shaped antennae to touch the ground, while a moth has featherlike antennae for detecting odor molecules in the air.

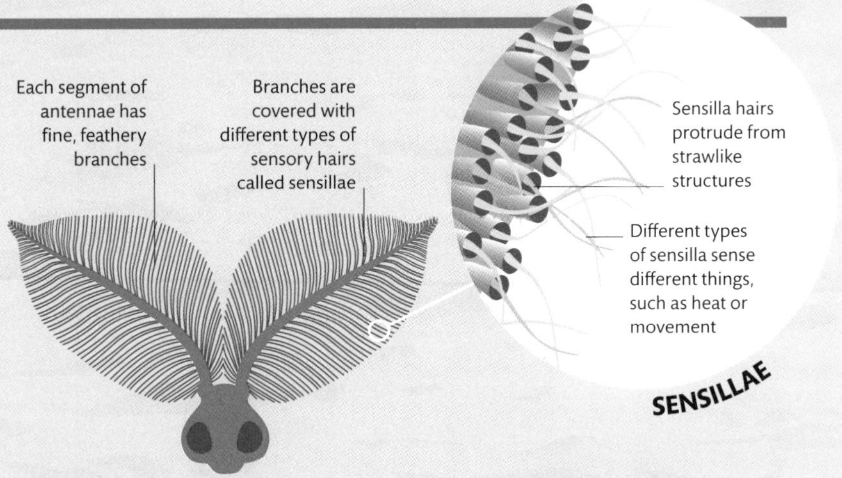
Each segment of antennae has fine, feathery branches
Branches are covered with different types of sensory hairs called sensillae
Sensilla hairs protrude from strawlike structures
Different types of sensilla sense different things, such as heat or movement
SENSILLAE

Moth antennae
A moth's featherlike antennae provide a large surface area for sensory cells (many of which are for smell). The moth can detect the direction of an odor based on where odor molecules land on its antennae.

ANTENNAE POSITION

IT CAN TAKE **20–60 MINUTES** FOR THE BODY TO **RECOVER FROM STRESS**

Fight or flight?
The presence of a threat instantly triggers a chain of hormonal and physical responses to release energy and prepare the muscles for action. This enables an animal or person to fend off or flee a threat.

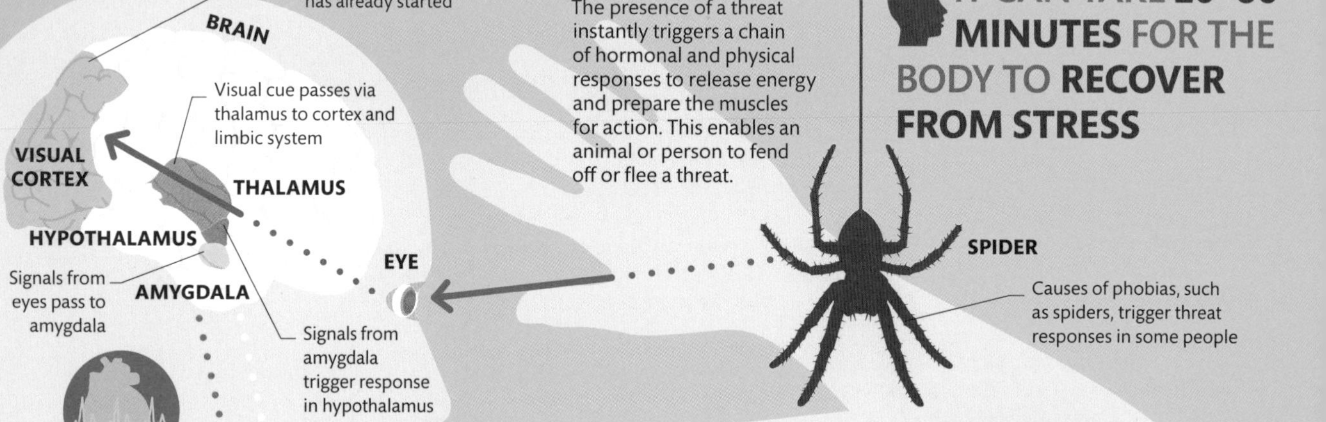
Image of spider is processed after fight-or-flight response has already started
BRAIN
Visual cue passes via thalamus to cortex and limbic system
VISUAL CORTEX
THALAMUS
HYPOTHALAMUS
EYE
SPIDER
Signals from eyes pass to amygdala
AMYGDALA
Signals from amygdala trigger response in hypothalamus
Causes of phobias, such as spiders, trigger threat responses in some people

1 Detecting the threat
Certain cues—notably visual cues such as the shape, position, or movement of a threat—trigger an unconscious or even instinctive response.

2 Sending signals
Before a conscious perception has formed, the amygdala sends a signal to the hypothalamus. This triggers the sympathetic nervous system to respond and the pituitary to release a hormone called ACTH.

NERVOUS SIGNAL

HORMONES

Heart
Heart rate increases to supply oxygen and energy to body.

Breathing
Airways expand and breathing quickens, to increase oxygen intake.

Pupils dilate
Pupils dilate, allowing more light to reach the retina.

Blood flows to muscles
Blood is diverted from other areas of the body to the muscles.

Fat used for energy
A rich source of energy, fat is released, ready to power muscles.

Hormonal and nervous signals trigger responses in adrenal glands

Stress hormone cortisol is produced in cortex (outer layer) of adrenal glands

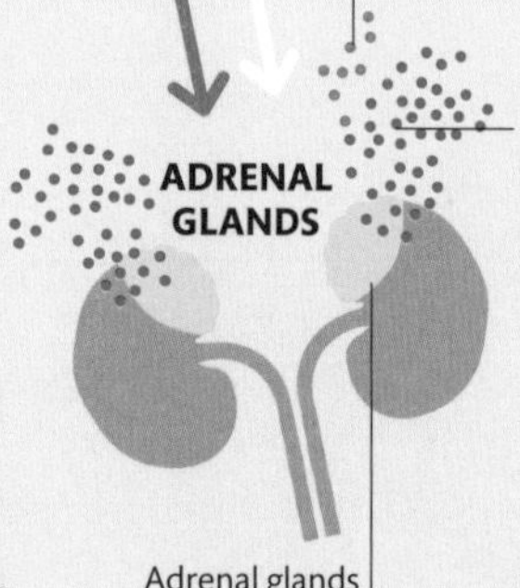
ADRENAL GLANDS

Nerve signals trigger the adrenal glands to secrete epinephrine

Adrenal glands are located one on each kidney

Bladder
Bladder relaxes; this may cause loss of control in extreme stress.

Blood vessels constrict
Blood flow is diverted to muscles and restricted in other parts of the body.

Sweat increases
In humans, sweating increases due to rise in body temperature.

Digestion slows
Bodily functions not essential to survival are put on hold.

Immune system
Changes occur to prepare the body for possible injury.

3 The body's response
The hormones from the pituitary trigger the adrenal glands to produce the hormones epinephrine (adrenaline) and cortisol, which cause changes in the body that prepare the animal to react to the threat.

4 The conscious perception
An image is formed in the cortex and analyzed to assess whether the threat is real; memories are consulted to determine if it has been faced before.

Possible responses to a threat

An animal or human facing a threat can react in various ways depending on the situation, the strength of the threat, and the possibility of self-defense. The most common reaction may be to freeze in order to remain unnoticed. Alternatively, the animal or person can take action—fighting back against an aggressor, or fleeing from trouble. These responses are activated by the sympathetic element of the autonomic system. In humans, they may occur in reaction to mental as well as physical threats—for example, in people with phobias.

Threat responses

An animal under threat has to respond well before its brain has consciously perceived the threat. A fast-track pathway from the brain to the body prepares the animal for action. This can make the difference between life and death.

The autonomic nervous system

The autonomic nervous system is part of the peripheral nervous system, which comprises all the nervous structures apart from the brain and spinal cord (the central nervous system). The autonomic system controls unconscious body functions such as heart rate, muscle contractions in the digestive tract, regulation of blood flow, and breathing. Within the autonomic nervous system, there are two networks: the sympathetic nervous system, which prepares the body for fight or flight (see opposite); and the parasympathetic nervous system, which prepares the body for rest and recovery. The two systems act to control the same organs and other parts of the body, but in exactly opposite ways.

DO INSECTS HAVE A FIGHT-OR-FLIGHT RESPONSE?

Insects do not have epinephrine, but they have a similar hormone, octopamine, which increases their heart rate and releases stored fat reserves ready for fight or flight.

Rest and digest
When an animal is not in fight-or-flight mode, the parasympathetic nervous system sends signals from the spinal cord to various organs, calming them and putting the body into rest-and-digest mode. Energy is directed toward carrying out maintenance functions such as absorbing nutrients and repairing the body (via the immune system).

Eyes
The pupils constrict, dilating only in low-light conditions.

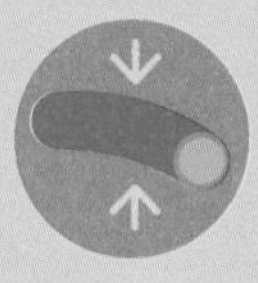

Blood vessels
Arteries return to normal diameter, ensuring even blood flow.

Liver
The liver builds energy reserves by storing sugars or turning them to fat.

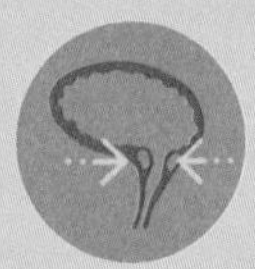

Bladder
The bladder neck tightens to prevent urine leakage.

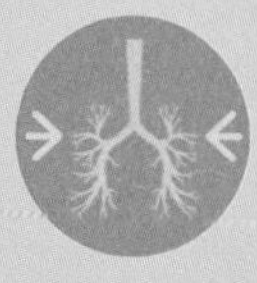

Lungs
The airways in the lungs return to their normal diameters.

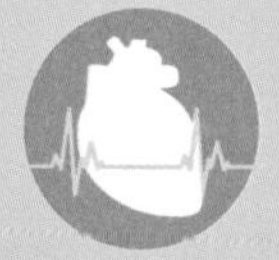

Heart
The heart beats at a normal resting heart rate.

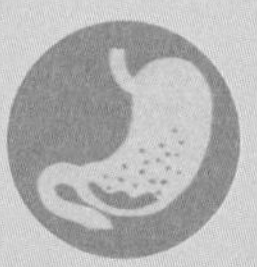

Stomach
Contractions in the stomach are stimulated to help digestion.

Intestines
Smooth muscles in the intestine walls contract to move waste.

PLAYING DEAD

Instead of fleeing or fighting, some animals play dead—a reaction called tonic immobility, or thanatosis—so as to appear dead or diseased and not worth eating. This reaction is stronger than just freezing. Heart rate and breathing rate slow. The body goes stiff, and the mouth may gape. The animal may release urine, feces, or foul-smelling fluid. The effect can be so powerful that some animals are "dead" for hours before they recover.

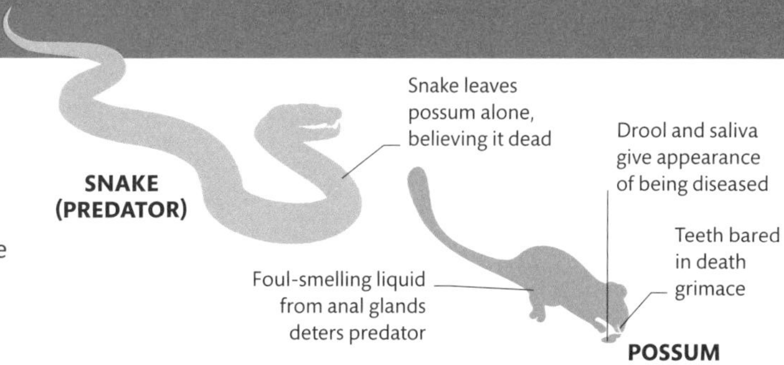
SNAKE (PREDATOR)
Snake leaves possum alone, believing it dead
Drool and saliva give appearance of being diseased
Teeth bared in death grimace
Foul-smelling liquid from anal glands deters predator
POSSUM

Defense against disease

Animals need defense mechanisms against pathogens, parasites, and pollutants that can cause disease if allowed to take hold.

Physical barriers

An animal's first line of defense against disease is its physical outer barrier, such as an exoskeleton (see p.157). Some animals have a thick layer of skin. For extra strength, skin also has special cells that produce mucus, which helps trap and sweep away invaders, much like a moat makes it tougher to reach a castle wall. However, the barrier also has to allow for exchange between an animal's internal and external environments, so invaders can sometimes get through.

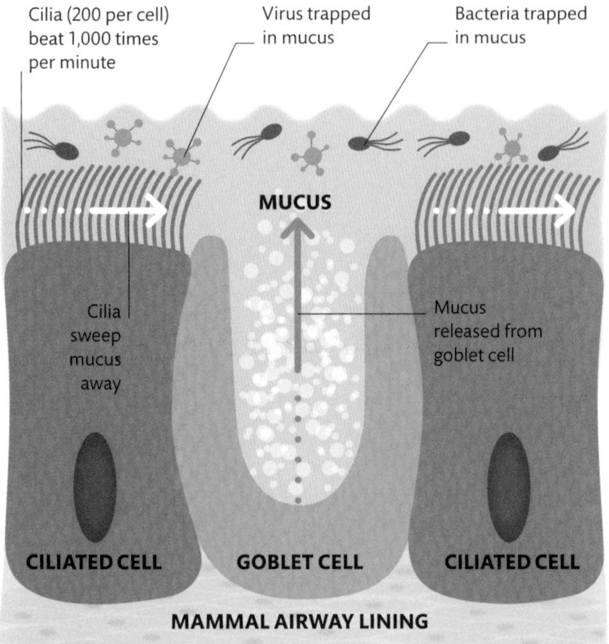

Cilia (200 per cell) beat 1,000 times per minute
Virus trapped in mucus
Bacteria trapped in mucus
MUCUS
Cilia sweep mucus away
Mucus released from goblet cell
CILIATED CELL
GOBLET CELL
CILIATED CELL

MAMMAL AIRWAY LINING

Mucus
Animals produce mucus to slow pathogens down. The mucus that coats the skin of amphibians and fish contains chemicals that kill microbes. Mammals also have mucus-secreting cells in their airways. The mucus, along with tiny hairs (cilia), catches invaders, which can then be blown out of the nose or swallowed.

BACTERIA FIND IT **HARD TO ATTACH** THEMSELVES TO **SHARK SKIN** DUE TO ITS UNIQUE **STRUCTURE**

How the immune system works

If invaders, or pathogens, do get past the body's physical barriers, a complex team of immune-system cells is waiting to attack them. The immune system has two lines of defense. The first part, the innate system, contains white blood cells (see p.74) that respond immediately to alarm signals coming from sick or damaged cells. These cells seek out the invader and engulf it. If this fails, a second adaptive immune system uses stored information from previous attacks, or infections, by pathogens to launch a targeted response.

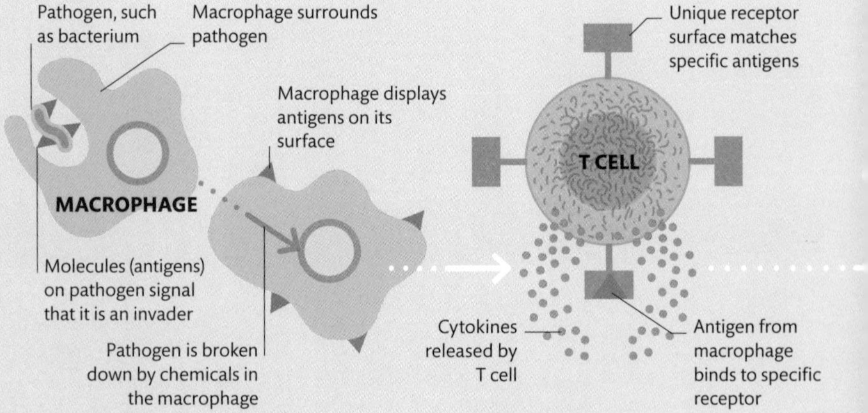

Pathogen, such as bacterium
Macrophage surrounds pathogen
Macrophage displays antigens on its surface
MACROPHAGE
Molecules (antigens) on pathogen signal that it is an invader
Pathogen is broken down by chemicals in the macrophage
Unique receptor surface matches specific antigens
T CELL
Cytokines released by T cell
Antigen from macrophage binds to specific receptor

1 Patrolling for pathogens
White blood cells called macrophages attack the pathogen by engulfing it. The macrophage then breaks down the pathogen and displays its antigen molecules on its own surface, alerting other cells to the problem.

2 T cells replicate
White blood cells called T cells recognize and bind to the antigens, and trigger responses that coordinate cell activity in the immune system. Cytokines tell "killer" T cells to replicate and activate B cells.

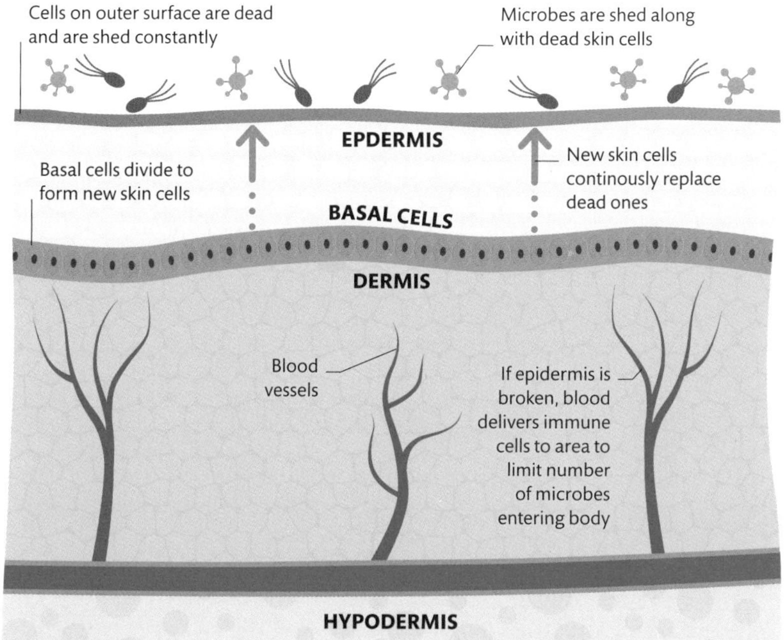
Cells on outer surface are dead and are shed constantly
Microbes are shed along with dead skin cells
EPDERMIS
Basal cells divide to form new skin cells
New skin cells continously replace dead ones
BASAL CELLS
DERMIS
Blood vessels
If epidermis is broken, blood delivers immune cells to area to limit number of microbes entering body
HYPODERMIS

Skin
The skin has three layers. The outer epidermis is constantly being worn away at the surface but is replenished by new cells formed at its base. The thicker dermis gives skin its strength and flexibility and has a rich blood supply. The hypodermis contains the fat that keeps the body warm and energized.

MUCUS BUBBLES

Fish produce a layer of mucus that not only slows down microscopic invaders but it also contains chemical compounds that kill microbes such as bacteria. Parrot fish build a large mucus cocoon around themselves at night to prevent parasites and microbes from attacking them as they sleep.

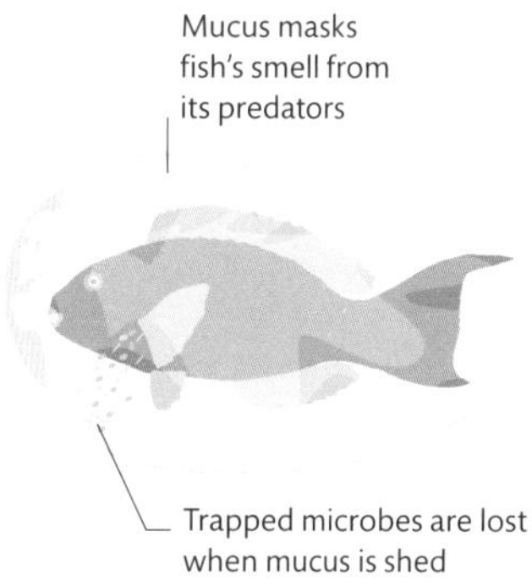
Mucus masks fish's smell from its predators
Trapped microbes are lost when mucus is shed

PARROT FISH

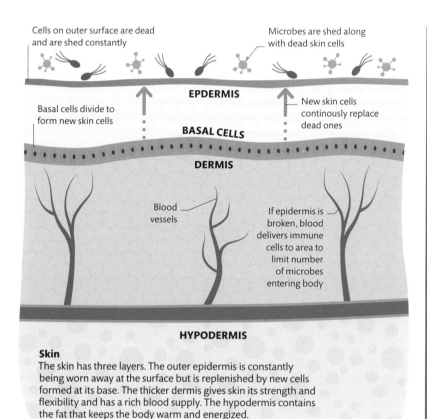

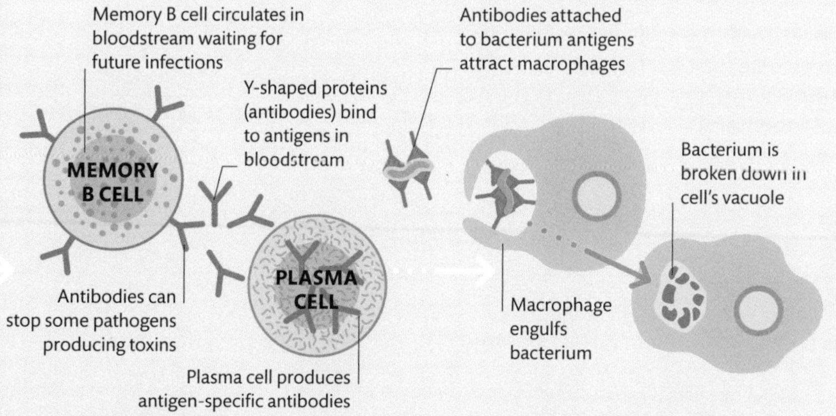
Memory B cell circulates in bloodstream waiting for future infections
Y-shaped proteins (antibodies) bind to antigens in bloodstream
MEMORY B CELL
PLASMA CELL
Antibodies can stop some pathogens producing toxins
Plasma cell produces antigen-specific antibodies
Antibodies attached to bacterium antigens attract macrophages
Bacterium is broken down in cell's vacuole
Macrophage engulfs bacterium

3 B cells attack
Activated B cells replicate and form both memory B cells and plasma cells. The memory cells hold information about an antigen; plasma cells release antibodies that stick to the surface of a specific pathogen.

4 Marked for destruction
Antibodies prevent the pathogen from binding to, and infecting, cells. They also help stick the pathogens together in a clump, making it easier for macrophages to seek out the invading pathogens and destroy them.

WHY DO BATS CARRY SO MANY DISEASES?

Bats appear to have a mutation in their DNA that enables them to live with more viruses than other mammals.

ECOLOGY

Ecosystems

Ecosystems are communities of plants, animals, and other organisms and their interactions with each other and the physical environment. They can be as small as a pond or as large as a desert.

1 Energy
Most of the energy for ecosystems comes from sunlight, which allows plants and algae to use photosynthesis to convert carbon dioxide and water into organic compounds. In environments where there is no light, such as hydrothermal vents (see p.25), some organisms derive energy from chemicals in a process called chemosynthesis.

2 Producer
Plants and algae are producers, or autotrophs. They make their own energy and provide food for other organisms through the processes of photosynthesis and chemosynthesis. Each level of a food chain is called a trophic level, and producers represent the first of these. When producers are eaten by consumers, energy is passed to the next trophic level.

A stable ecosystem
When an ecosystem is healthy, there is a balanced relationship between the different organisms and their environment. Food chains, such as this one for a woodland ecosystem, show how all living things rely on each other for food, how energy flows between them, and how waste is recycled.

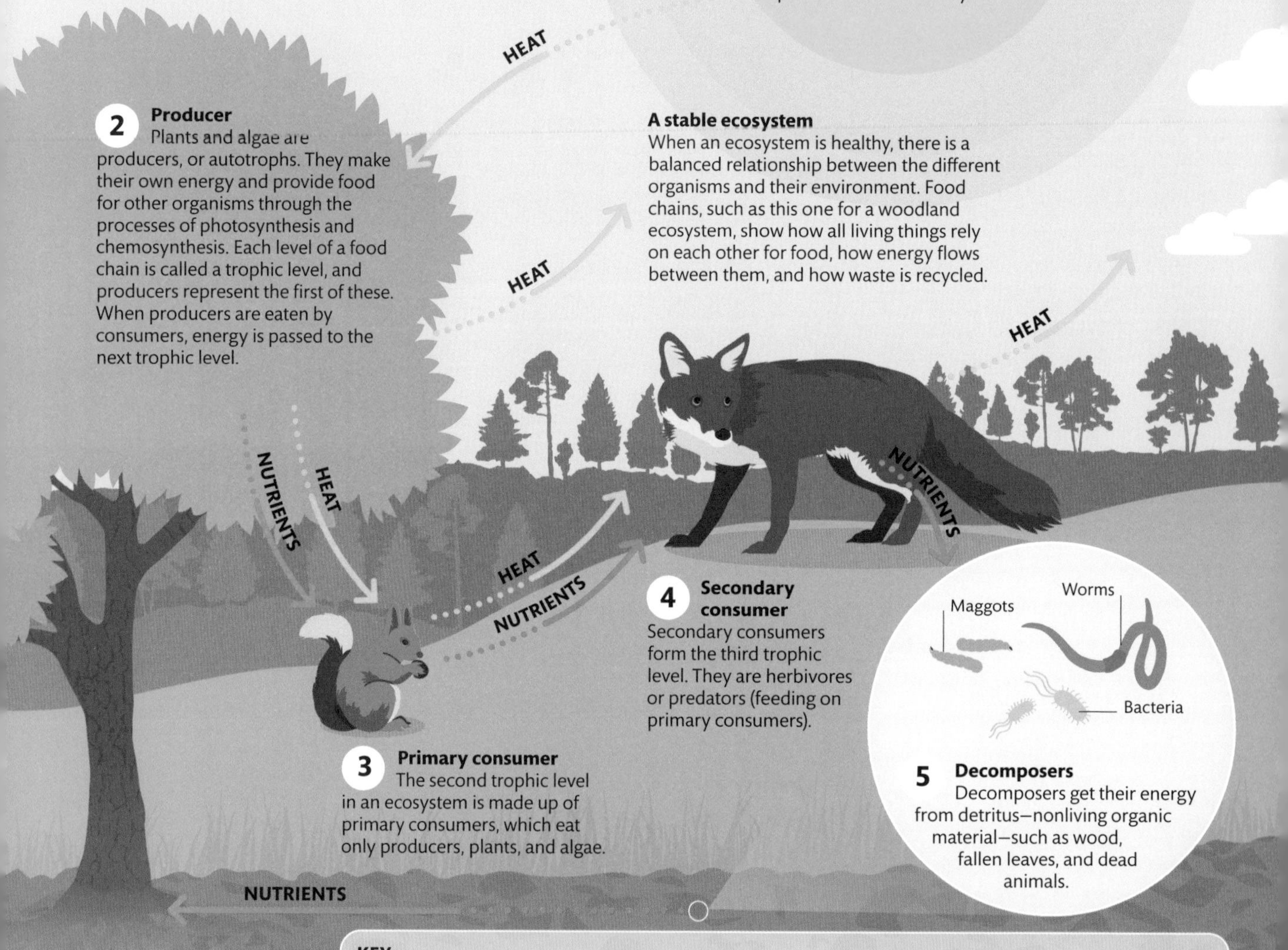

HEAT
HEAT
HEAT
NUTRIENTS
HEAT
HEAT
NUTRIENTS
NUTRIENTS
Maggots
Worms
Bacteria
NUTRIENTS

3 Primary consumer
The second trophic level in an ecosystem is made up of primary consumers, which eat only producers, plants, and algae.

4 Secondary consumer
Secondary consumers form the third trophic level. They are herbivores or predators (feeding on primary consumers).

5 Decomposers
Decomposers get their energy from detritus—nonliving organic material—such as wood, fallen leaves, and dead animals.

KEY

Energy Flow
From one trophic level to the next, energy flows through living things. Only about 10 percent passes between each trophic level because most of the energy is lost as heat (released during metabolic processes) and through movement.

Nutrient cycle
Plants take up vital nutrients such as carbon, oxygen, nitrogen, and calcium through their roots. They are either transferred to the consumers that eat them or are recycled into the soil when the plant dies.

Biological communities

A biological community is made up of all the organisms living in a geographic area. Environmental change, such as the increased supply of nutrients or a decrease in rainfall, will impact it. Changes in the population of any species will also have a knock-on effect on the whole community. For example, an increase in a predator will impact prey numbers (see p.187).

WHAT IS AN ECOLOGICAL NICHE?

The word niche comes from the French word nicher, which means "to nest." In ecology, it refers to the match of a species to a specific place in an ecosystem.

Biotic and abiotic factors

Two groups of factors mold an ecosystem: biotic and abiotic. The biotic factors are all the living organisms in an ecosystem because each one has a direct or indirect effect on all the others. The abiotic factors are the many nonliving variables that influence the variety and abundance of organisms. They include light, temperature, soil, water availability and acidity, nutrient supply, and pollution. Abiotic factors are determined by climate, geology, and topography.

BIOTIC	ABIOTIC
Food availability	Light
Predation	Wind
Diseases	Temperature
Competition	Rainfall

MICROHABITATS

A small part of an ecosystem that has its own conditions of, for example, temperature and light, and its own characteristic species is called a microhabitat. Some examples of microhabitats with their own flora and fauna include rock pools on sandy beaches, water-filled holes in trees, and patches of bare ground in grassland.

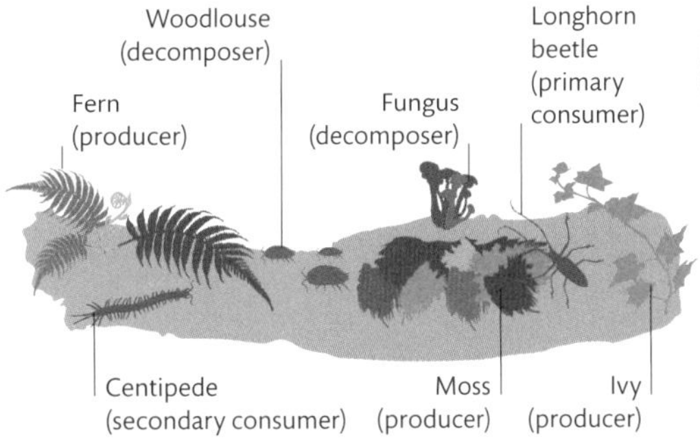
Woodlouse (decomposer)
Longhorn beetle (primary consumer)
Fern (producer)
Fungus (decomposer)
Centipede (secondary consumer)
Moss (producer)
Ivy (producer)

Decomposing log
Although a decomposing log may represent only a tiny part of the surrounding forest habitat, it has a distinct character, with its own populations of organisms.

Interdependence

All organisms in an ecosystem depend on each other to some extent. Some relationships are particularly strong, and this means that a change in the population of one species in an ecosystem can affect many other species in the same community. This is called interdependence.

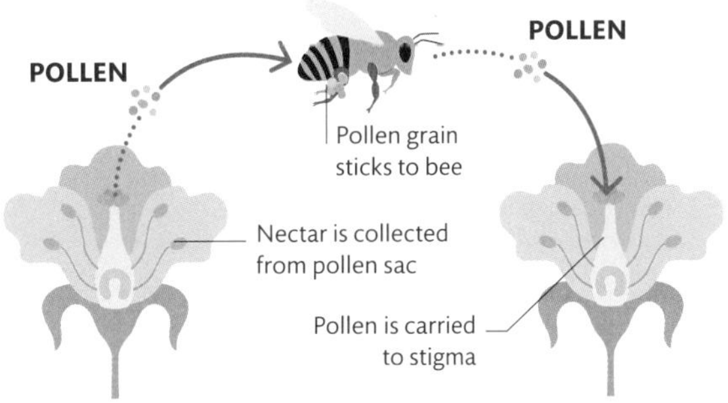
POLLEN
POLLEN
Pollen grain sticks to bee
Nectar is collected from pollen sac
Pollen is carried to stigma

Mutualistic relationship
Some plants depend upon insects to pollinate them, while the same insects need food that the plants provide. Bees fly from flower to flower, fertilizing plants in the process.

Biomes

Biomes are large geographic regions that share similar types of climate, soil, plant species, and animals. Because of their size, there is a lot of variation within each biome.

Types of biome

Earth's surface can be divided into about 10 major biomes, with their distribution defined mainly by climate. The same type of biome can be found across continents, such as savannas in Africa and Australia. While each biome has its own distinctive communities of organisms, with varying patterns of plants, animals, fungi, and many ecosystems, there are many similarities.

WORLD BIOMES MAP

Ecoregions

Biomes are made up of ecoregions, each of which has more closely aligned communities of species. For example, the island of Madagascar is dominated by tropical rainforests, deserts, and savanna biomes. These biomes are divided into ecoregions with many endemic plant species and animals due to their distinctive environmental conditions.

KEY
Dry deciduous forest
Heathery thicket
Lowland rainforest
Mangrove
Spiny thicket
Subhumid forest
Succulent woodland

Boreal forest
Freezing temperatures for six to eight months mean only very hardy plants and animals can survive. Pines and firs dominate the region. Many mammals hibernate, and most birds migrate south for the winter.

Marine
Life in Earth's largest biome ranges from blue whales, the planet's largest animals, to microscopic plankton. Most life is concentrated in shallow coastal waters and cool currents.

Savanna
This tropical and subtropical grassland and open-canopy woodland has a climate with marked dry and wet seasons. It is home to large mammals including zebras, giraffes, big cats, and elephants.

Polar
Despite the inhospitable climate of these regions, covered by ice for most of the year, they nevertheless support some hardy animals, including polar bears, penguins, seals, and narwhals.

Tropical rainforest
These hot and moist regions have rainfall in all seasons. The luxuriant forest supports the greatest diversity of trees, invertebrates, amphibians, birds, and mammals in the world.

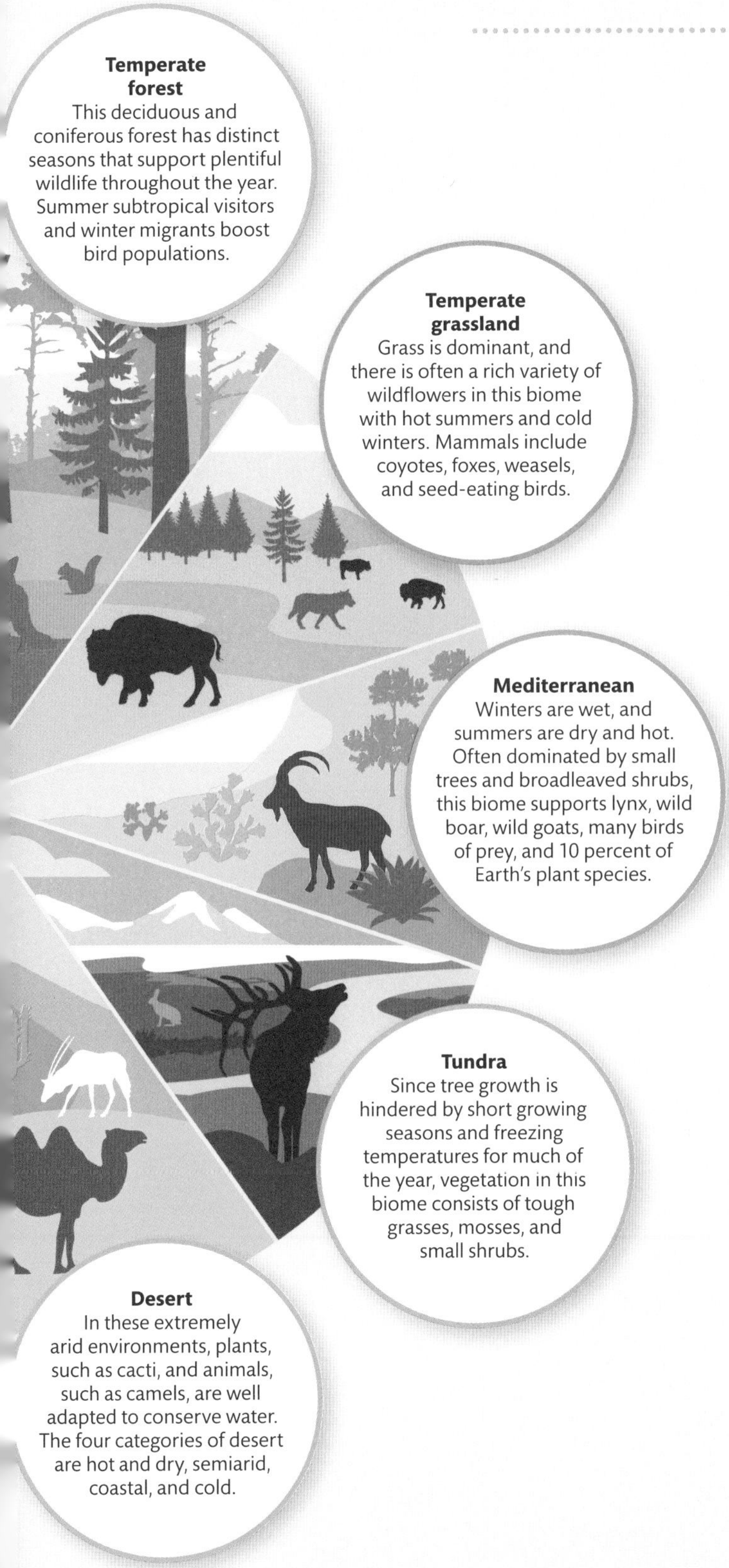

Temperate forest
This deciduous and coniferous forest has distinct seasons that support plentiful wildlife throughout the year. Summer subtropical visitors and winter migrants boost bird populations.

Temperate grassland
Grass is dominant, and there is often a rich variety of wildflowers in this biome with hot summers and cold winters. Mammals include coyotes, foxes, weasels, and seed-eating birds.

Mediterranean
Winters are wet, and summers are dry and hot. Often dominated by small trees and broadleaved shrubs, this biome supports lynx, wild boar, wild goats, many birds of prey, and 10 percent of Earth's plant species.

Tundra
Since tree growth is hindered by short growing seasons and freezing temperatures for much of the year, vegetation in this biome consists of tough grasses, mosses, and small shrubs.

Desert
In these extremely arid environments, plants, such as cacti, and animals, such as camels, are well adapted to conserve water. The four categories of desert are hot and dry, semiarid, coastal, and cold.

Biodiversity

Shorthand for biological diversity, biodiversity refers to the range of different organisms living in an area, be it large or small. Many factors contribute to this, including climate, geology, and stability over time. As a rule, biodiversity increases toward the equator, with tropical forests and warm coastal seas registering the greatest number of species. Biodiversity plays a vital role in sustaining healthy ecosystems.

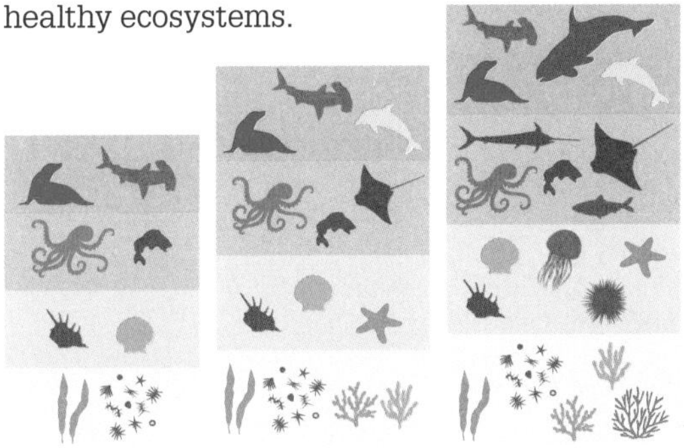
LOW DIVERSITY
MEDIUM DIVERSITY
HIGH DIVERSITY

9,000 SPECIES OF MARINE LIFE CAN BE FOUND IN THE BIODIVERSE GREAT BARRIER REEF

BIODIVERSITY HOTSPOTS

Conservation International has identified 36 global hotspots based on the richness of their biodiversity and the level of threat to this. All have more than 1,500 endemic plant species and have lost at least 70% of their vegetation.

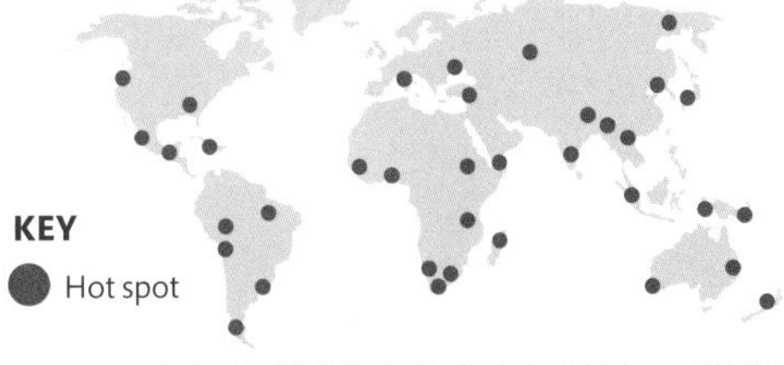
KEY
Hot spot

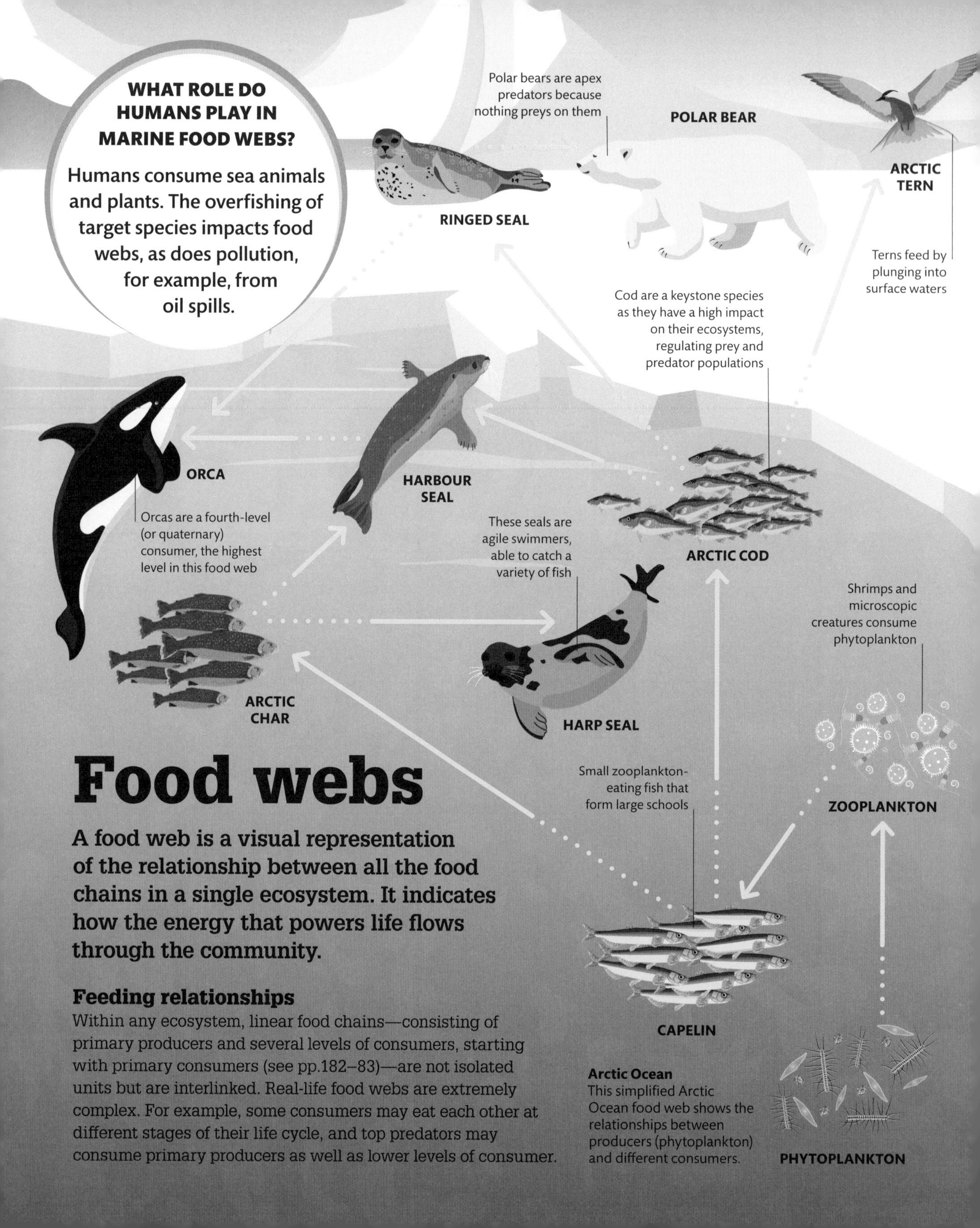

Polar bears are apex predators because nothing preys on them
POLAR BEAR
ARCTIC TERN
Terns feed by plunging into surface waters
RINGED SEAL
Cod are a keystone species as they have a high impact on their ecosystems, regulating prey and predator populations
ORCA
HARBOUR SEAL
Orcas are a fourth-level (or quaternary) consumer, the highest level in this food web
These seals are agile swimmers, able to catch a variety of fish
ARCTIC COD
Shrimps and microscopic creatures consume phytoplankton
ARCTIC CHAR
HARP SEAL
Small zooplankton-eating fish that form large schools
ZOOPLANKTON
CAPELIN
PHYTOPLANKTON

WHAT ROLE DO HUMANS PLAY IN MARINE FOOD WEBS?

Humans consume sea animals and plants. The overfishing of target species impacts food webs, as does pollution, for example, from oil spills.

Food webs

A food web is a visual representation of the relationship between all the food chains in a single ecosystem. It indicates how the energy that powers life flows through the community.

Feeding relationships

Within any ecosystem, linear food chains—consisting of primary producers and several levels of consumers, starting with primary consumers (see pp.182–83)—are not isolated units but are interlinked. Real-life food webs are extremely complex. For example, some consumers may eat each other at different stages of their life cycle, and top predators may consume primary producers as well as lower levels of consumer.

Arctic Ocean
This simplified Arctic Ocean food web shows the relationships between producers (phytoplankton) and different consumers.

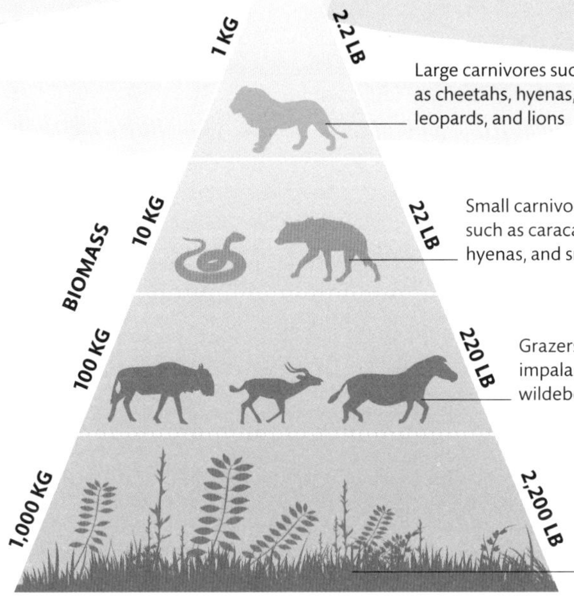
BIOMASS
1 KG
2.2 LB
10 KG
22 LB
100 KG
220 LB
1,000 KG
2,200 LB
Large carnivores such as cheetahs, hyenas, leopards, and lions
Small carnivores such as caracals, hyenas, and snakes
Grazers such as deer, impalas, zebras, and wildebeest
Variety of grasses and shrubs

Serengeti biomass pyramid
Plants are at the lowest trophic level of the biomass pyramid in the Serengeti grasslands, having the highest amounts of mass and energy.

Energy and biomass

The total mass of organisms at each stage of the food web, known as a trophic level, in an ecosystem can be represented by a biomass pyramid. There are two kinds of pyramid. In a terrestrial "upright" pyramid, producers (plants) at the bottom trophic level far outweigh consumers, and the highest level of consumer has the smallest biomass. Energy levels also decline from lower to higher trophic levels. An "inverted" pyramid is typical of marine ecosystems, with producers having less biomass than consumers.

Classifying feeding

There are several ways to classify how animals feed. A broad system groups animals according to whether they eat plants (herbivores), meat (carnivores), or both (omnivores). Some animals are highly specialized within these groups. For example, insectivores, such as anteaters, eat insects; and frugivores, such as fruit bats, eat fruit. An alternative system divides animals into predators, prey, and scavengers.

Predator
A secondary, tertiary, or quaternary consumer that obtains most of their food by killing and eating live animals.

Herbivore
A primary consumer that derives most of its food from plants. Examples include rabbits and sheep.

Prey
A live animal that is hunted and eaten by other animals. Any consumer, bar an apex predator, can be prey.

Carnivore
A meat-eating animal. Not all carnivores kill their prey; examples include vultures (which are also scavengers).

Scavenger
A carnivorous animal that consumes dead organisms that have been killed or have died from natural causes.

Omnivore
An animal with a mixed diet of plant and animal foods. Examples include bears, pigs, and many bird species.

PREDATOR-PREY CYCLES

The changing number of predators and prey in an ecosystem make up a predator-prey cycle. For example, hares are predated by lynx, so when hare numbers fall so do those of lynx after a delay of a few years.

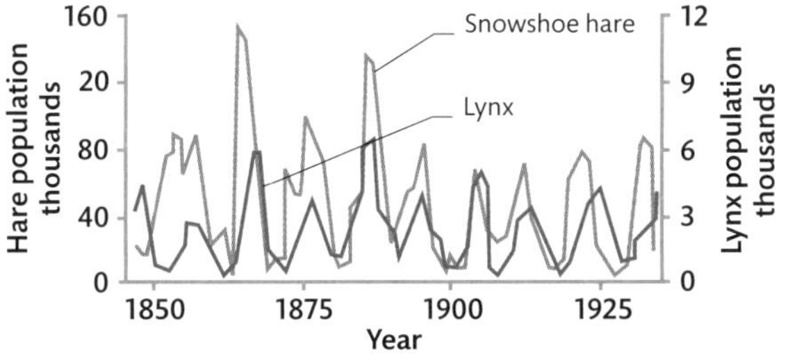
Hare population thousands
160
20
80
40
0
Snowshoe hare
Lynx
Lynx population thousands
12
9
6
3
0
1850
1875
1900
1925
Year

SOME BLUE WHALES EAT UP TO 16 TONS OF KRILL AND PHYTOPLANKTON EVERY DAY

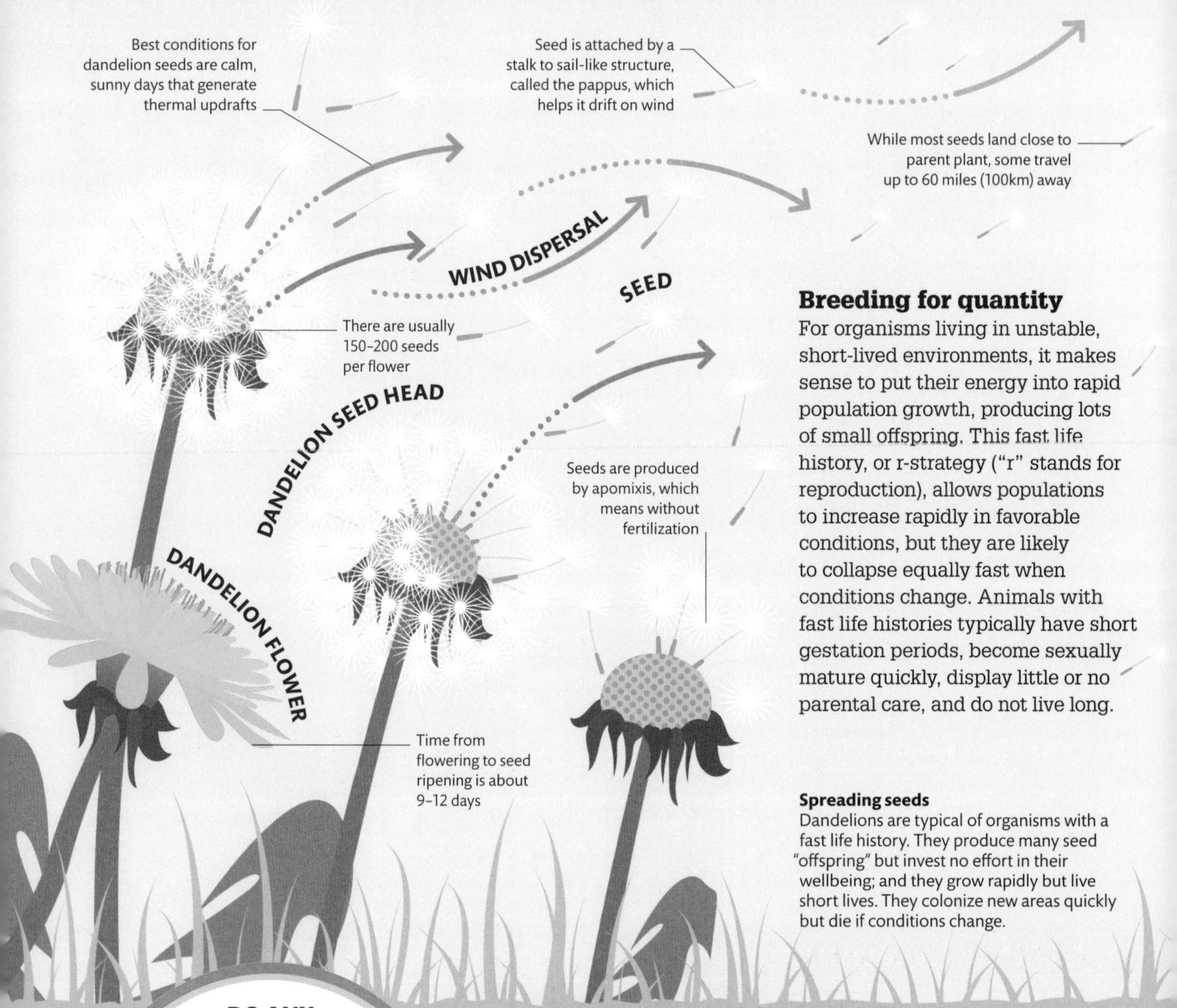
Best conditions for dandelion seeds are calm, sunny days that generate thermal updrafts
Seed is attached by a stalk to sail-like structure, called the pappus, which helps it drift on wind
While most seeds land close to parent plant, some travel up to 60 miles (100km) away
WIND DISPERSAL
SEED
There are usually 150–200 seeds per flower
DANDELION SEED HEAD
Seeds are produced by apomixis, which means without fertilization
DANDELION FLOWER
Time from flowering to seed ripening is about 9–12 days

Breeding for quantity

For organisms living in unstable, short-lived environments, it makes sense to put their energy into rapid population growth, producing lots of small offspring. This fast life history, or r-strategy ("r" stands for reproduction), allows populations to increase rapidly in favorable conditions, but they are likely to collapse equally fast when conditions change. Animals with fast life histories typically have short gestation periods, become sexually mature quickly, display little or no parental care, and do not live long.

Spreading seeds
Dandelions are typical of organisms with a fast life history. They produce many seed "offspring" but invest no effort in their wellbeing; and they grow rapidly but live short lives. They colonize new areas quickly but die if conditions change.

DO ANY ORGANISMS MIX ELEMENTS OF FAST AND SLOW LIFE HISTORIES?

Many species have traits of both strategies. For example, sea turtles and trees have long lifespans but produce large numbers of unnurtured offspring.

Breeding strategies

Some organisms produce only a small number of offspring over their lifetime but concentrate on protecting their "investment" well. Others produce huge numbers of offspring but put just a small amount of energy into nurturing them.

Breeding for quality

Some animals living in stable environments follow a slow life history, or K-strategy, model ("K" stands for carrying capacity), where a lot of effort is invested into "high-quality" offspring. Typical features are long gestation periods, offspring that are numerically few but large in size, maturity that is attained only slowly, generally stable populations, and long lifespans. Examples include many larger mammals—including humans—and large birds such as albatrosses.

2,000 EGGS CAN BE CARRIED BY MALE SEAHORSES BUT ONLY 0.5% SURVIVE

Calf is sometimes not fully weaned until it is five years old
Mother elephants stay with their calves until long after they are weaned
CALF
MOTHER

Elephants
Female African elephants have a gestation period of 22 months, and during their lifetime they give birth to just four or five young.

Growth and survivorship

The carrying capacity of a specific environment is the maximum population size of a species it can support. In a newly created environment, the population of a fast life history animal or plant will quickly exceed this figure, then collapse again since it will be unsustainable. A slow life history species will reach carrying capacity more slowly but will continue to remain around this figure.

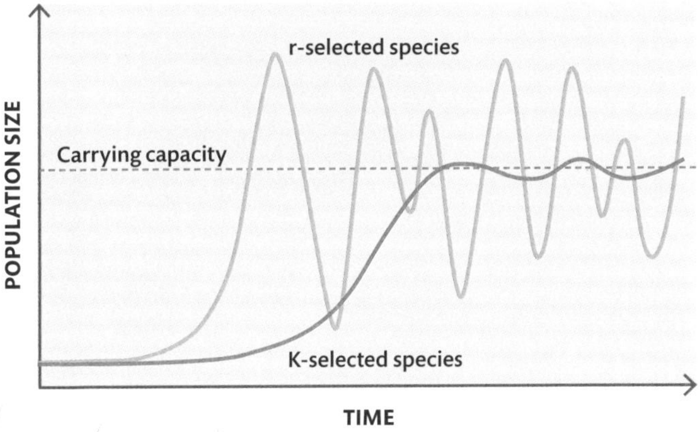
r-selected species
Carrying capacity
POPULATION SIZE
K-selected species
TIME

Growth curve
The populations of fast life history species can fluctuate dramatically with changes in the environment, whereas slow life history species' populations remain stable.

BROOD PARASITISM

Eurasian cuckoos are brood parasites—they rely on other species to raise their young, thereby dramatically reducing the effort they need to invest in raising the next generation. Females lay their eggs in the nests of other birds, tricking the host into looking after the chick when it hatches. The chick eventually grows bigger than its host.

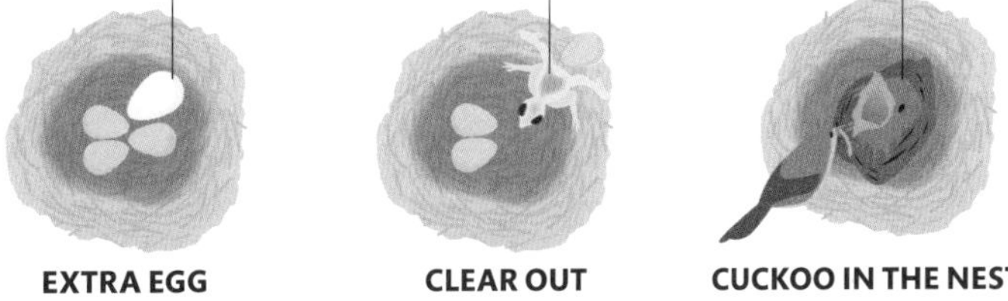
Cuckoo removes one of host's eggs and replaces it with her own
Cuckoo chick hatches first and removes other eggs
Growing cuckoo chick gets all food brought by its hosts
EXTRA EGG
CLEAR OUT
CUCKOO IN THE NEST

Social living

It is common for animals of the same species to live together in groups. Social groups range from simple gatherings that provide safety in numbers, to complex societies with designated roles.

Types of social behavior

In animals, social behavior involves interactions between unrelated and related individuals of the same species that benefit the whole group. Many animal species, such as lion prides, are social throughout the year. Some animals come together only during the breeding season but live solitary lives the rest of the time, in seabird colonies for example. Some of the most complex social behavior is seen in the insect world, where colonial societies work because of cooperation between many individuals.

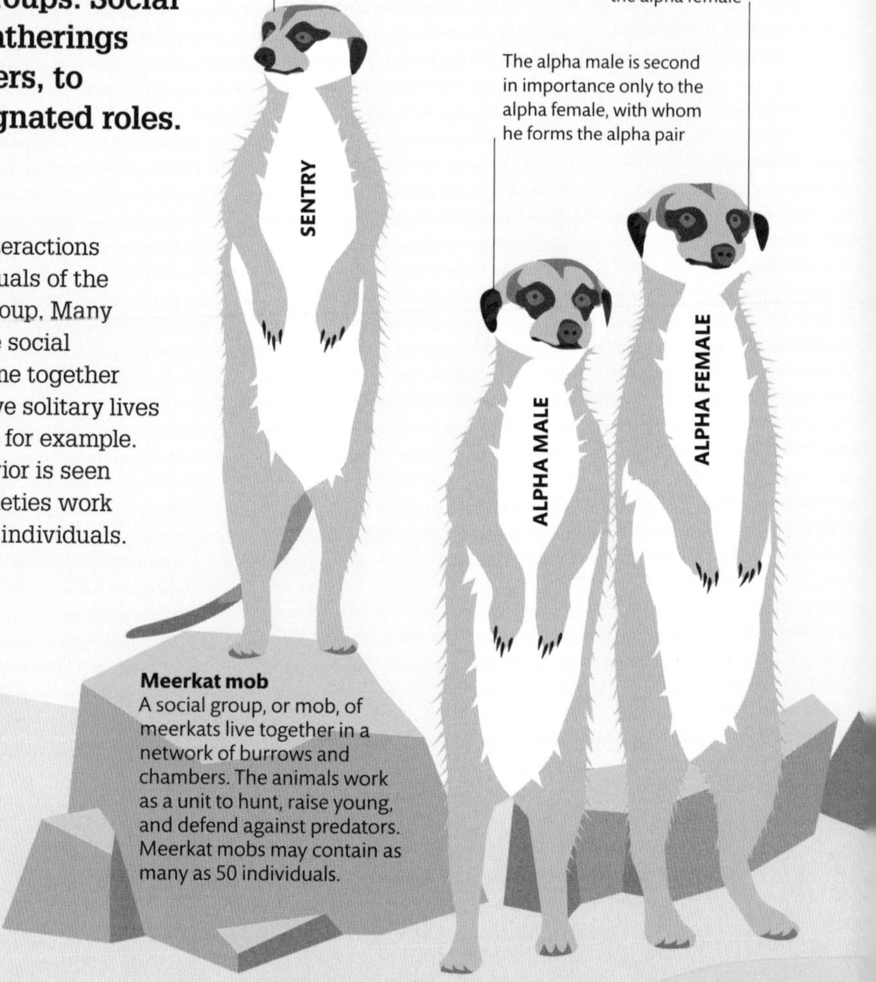
Group members take turns on sentry duty, watching for predators and barking a warning at the first sign of danger
The dominant meerkat in the group, most other mob members are offspring or siblings of the alpha female
The alpha male is second in importance only to the alpha female, with whom he forms the alpha pair
SENTRY
ALPHA MALE
ALPHA FEMALE

Meerkat mob
A social group, or mob, of meerkats live together in a network of burrows and chambers. The animals work as a unit to hunt, raise young, and defend against predators. Meerkat mobs may contain as many as 50 individuals.

CAN PLANTS ALSO FORM SOCIAL GROUPS?

The roots of aspen trees are interconnected so they can share nutrients and other resources to support each other.

Eusociality

An extreme form of social living, where there is a division of labor and individuals have specialized roles, is called eusociality. Most ants are eusocial—all the individuals in a colony work as part of a team to make it function effectively. Most of the ants in a nest of leafcutter ants are nonreproductive female workers, groups of which perform different functions. There are specialist foragers, protective soldiers, and even gardeners, all of whom work for the queen.

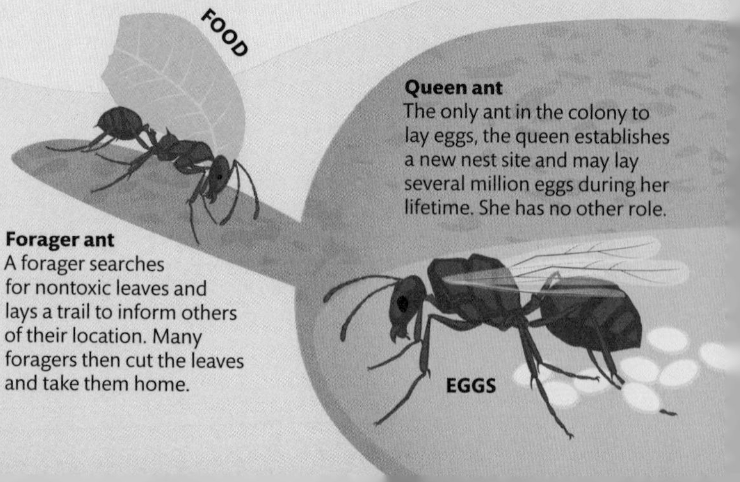
FOOD
EGGS

Forager ant
A forager searches for nontoxic leaves and lays a trail to inform others of their location. Many foragers then cut the leaves and take them home.

Queen ant
The only ant in the colony to lay eggs, the queen establishes a new nest site and may lay several million eggs during her lifetime. She has no other role.

FED VAMPIRE BATS OFTEN REGURGITATE BLOOD FOR HUNGRY BATS WITHIN THEIR SOCIAL ROOST

Group living

While living in a group has advantages over living a solitary life, it also has disadvantages. There is always a chance that members of the community will fight, and there are other potential problems, too.

Pros

Having more eyes to detect predators increases protection and makes it easier to fight them off, resulting in increased survival rates. There are more opportunities for cooperative hunting and feeding, as seen in lion prides and wolf packs. Duties can be divided, including taking care of the young, as is the case with giraffes and penguins.

Cons

Between densely associating individuals, there is increased competition for resources, especially for food and nest space. There is a greater chance of parasites and diseases being transmitted rapidly and the easy detection of the group by predators. It is also more likely that closely related individuals will have offspring with undesirable characteristics.

Young meerkats (pups) learn by observing and copying adult behavior, for example by watching how an older member of the mob removes a scorpion's stinger

BABYSITTER

PUP

BETA MALES AND FEMALES

Mob members who are neither pups nor part of the alpha pair are described as beta females and males

A baby sitter is an adult meerkat who watches over meerkat pups while other members forage away from the burrow

Male ant
Males are only present in a nest just before a new colony is formed. The males have wings and fly with fertile females, dying soon after mating.

Worker ant
The many duties of colony workers include caring for the eggs, larvae, and pupae, gardening, cleaning the nest, and defending it from attack.

LARVAE

COMMUNICATION

Effective communication reduces the need for physical conflict and is especially important for social animals. Communication can be through vocalization, body language, or facial expressions. Chimpanzees use a range of expressions to show how they are feeling.

Relaxed chimp has closed mouth with no teeth showing

RELAXED FACE

Mouth is open and lips are pulled back to expose teeth

FEAR GRIN

Upper and lower lips are pushed forward if chimp is uneasy

POUT

Top lips cover top teeth and bottom teeth are exposed

PLAY FACE

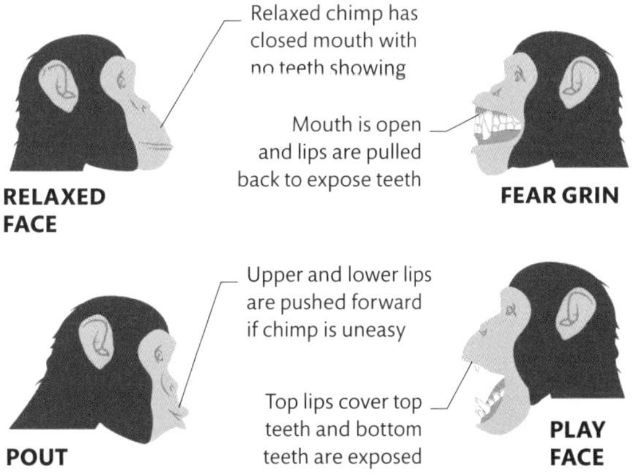

Ecological damage

Natural habitats have been damaged by nature and humans for millennia, but in recent centuries the pace has accelerated. This threatens the survival of many species of plants and animals.

WHAT IS OCEAN ACIDIFICATION?

Extra carbon dioxide dissolving in seawater makes an acid. This prevents sea animals making thick enough shells and disrupts the ocean's food chain.

Environmental threats

Natural processes—including severe weather, forest fires, volcanic eruptions, and the action of glaciers—have always damaged ecosystems. Huge areas of forest have also been cleared by people for agriculture, wetlands have been drained for development, overgrazing by animals has turned vast tracts of grassland into desert, and cities sprawl across large areas. This happened first in Europe and North America and now continues elsewhere.

Human impact

Scientists estimate that more than 75 percent of Earth's land area has been severely degraded by human activities. Our negative impacts on the environment can broadly be divided into pollution, the destruction of habitats, and the introduction of invasive species. These impacts have been driven by the relentless exploitation of natural resources—especially fossil fuels, minerals, trees, water, and soil—and the often toxic waste produced by industrial activity.

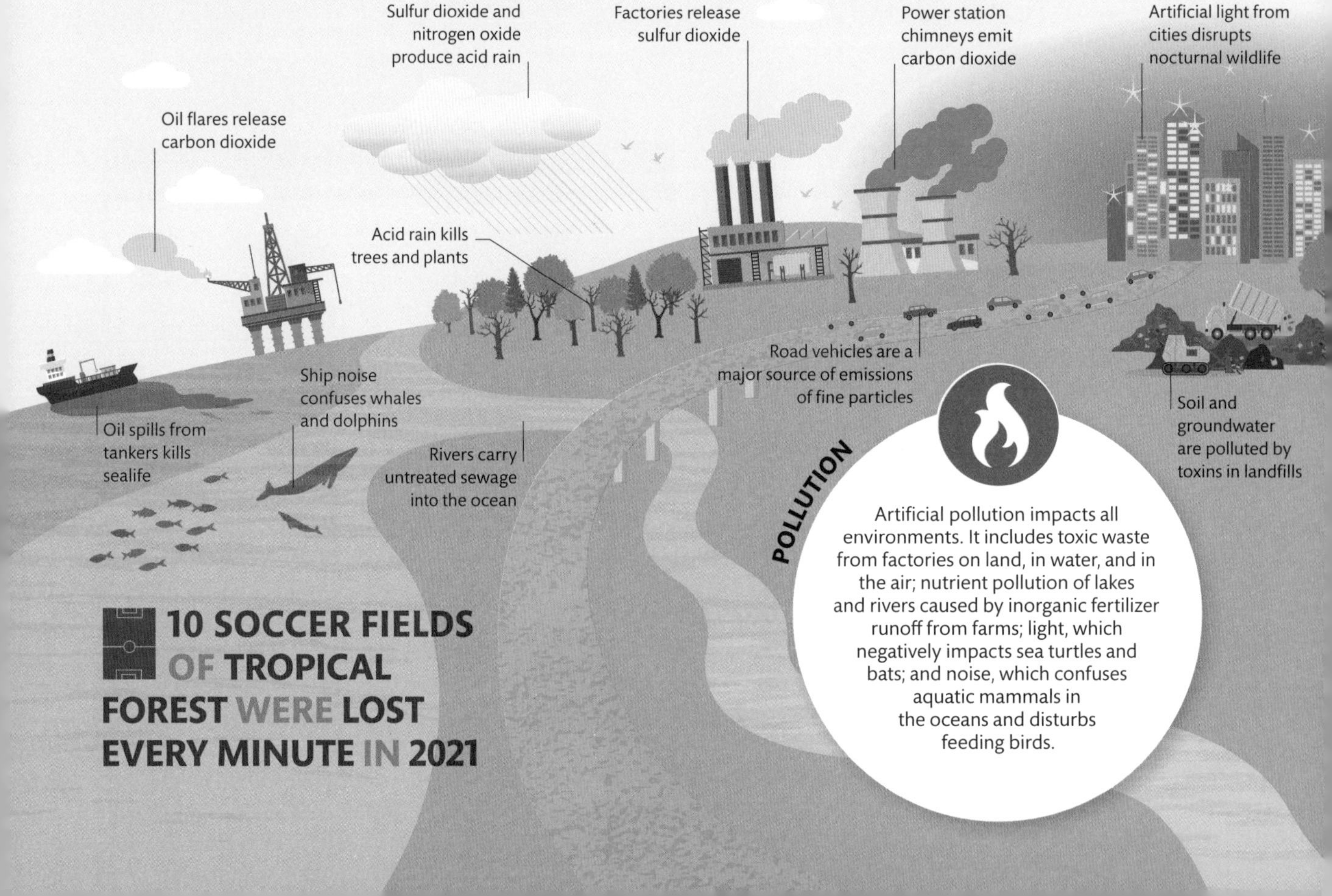

Sulfur dioxide and nitrogen oxide produce acid rain
Factories release sulfur dioxide
Power station chimneys emit carbon dioxide
Artificial light from cities disrupts nocturnal wildlife
Oil flares release carbon dioxide
Acid rain kills trees and plants
Road vehicles are a major source of emissions of fine particles
Ship noise confuses whales and dolphins
Oil spills from tankers kills sealife
Rivers carry untreated sewage into the ocean
Soil and groundwater are polluted by toxins in landfills

POLLUTION

Artificial pollution impacts all environments. It includes toxic waste from factories on land, in water, and in the air; nutrient pollution of lakes and rivers caused by inorganic fertilizer runoff from farms; light, which negatively impacts sea turtles and bats; and noise, which confuses aquatic mammals in the oceans and disturbs feeding birds.

10 SOCCER FIELDS OF TROPICAL FOREST WERE LOST EVERY MINUTE IN 2021

SUCCESSION

If new land opens up—due to a landslide, volcanic eruption, or human activity—the biological community develops from bare earth by a process called succession. Primary succession begins when an area of ground is colonized by living things for the very first time. Secondary succession takes place as an environment recovers after a major disturbance—for example, a devastating forest fire. In both cases, the plants and animals that inhabit the area undergo gradual changes.

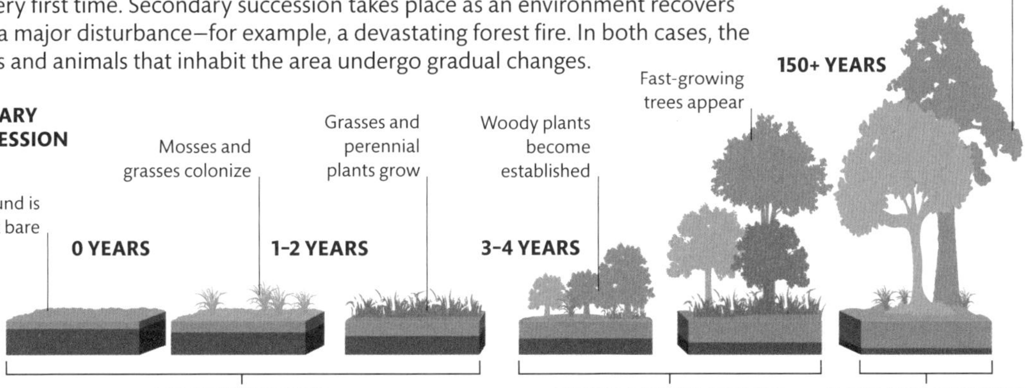
PRIMARY SUCCESSION
Climax community occurs when stable and complex forest is established
150+ YEARS
Fast-growing trees appear
Woody plants become established
Grasses and perennial plants grow
Mosses and grasses colonize
Ground is almost bare
0 YEARS
1–2 YEARS
3–4 YEARS
PIONEER SPECIES
INTERMEDIATE SPECIES
CLIMAX COMMUNITY

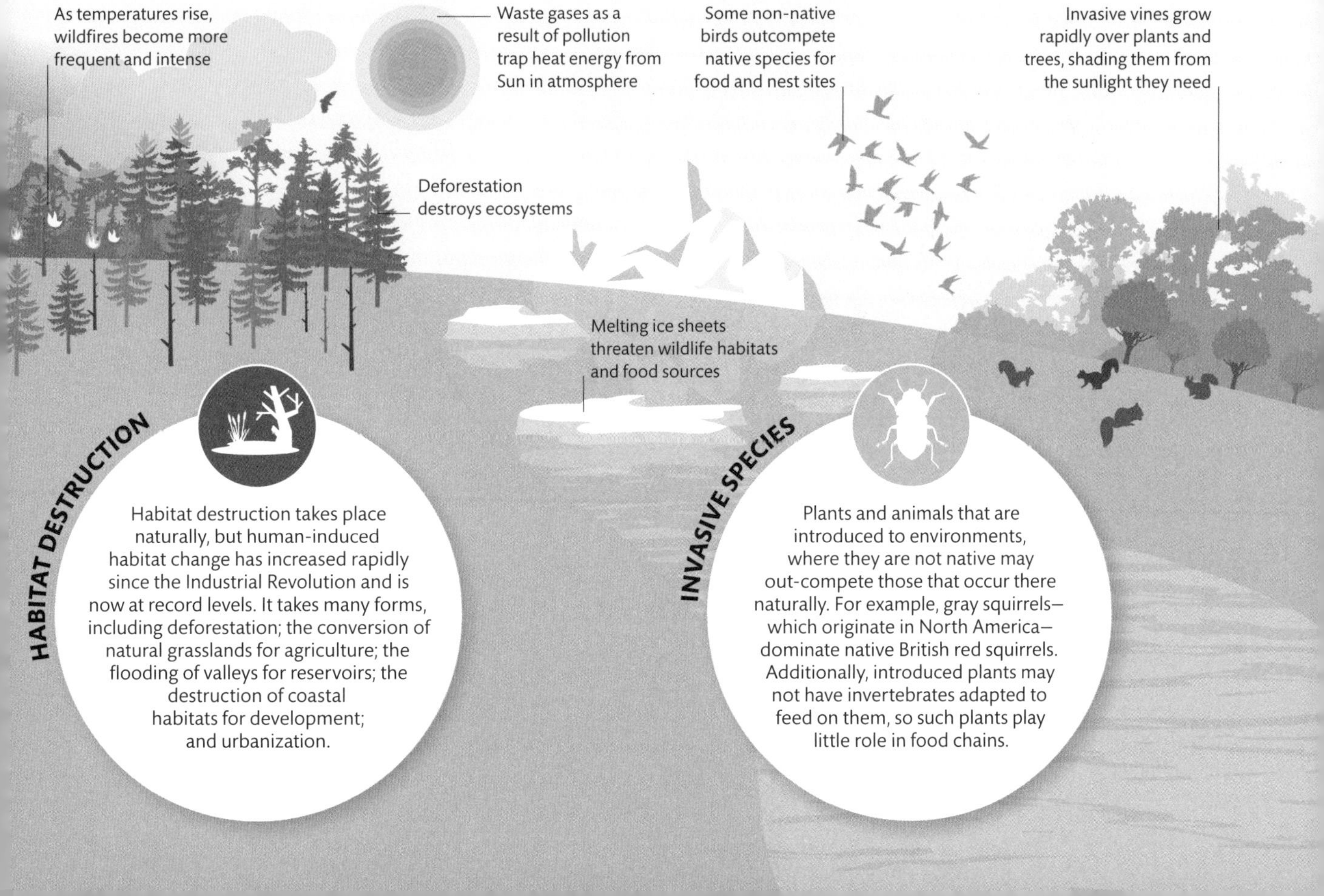
As temperatures rise, wildfires become more frequent and intense
Waste gases as a result of pollution trap heat energy from Sun in atmosphere
Some non-native birds outcompete native species for food and nest sites
Invasive vines grow rapidly over plants and trees, shading them from the sunlight they need
Deforestation destroys ecosystems
Melting ice sheets threaten wildlife habitats and food sources

HABITAT DESTRUCTION

Habitat destruction takes place naturally, but human-induced habitat change has increased rapidly since the Industrial Revolution and is now at record levels. It takes many forms, including deforestation; the conversion of natural grasslands for agriculture; the flooding of valleys for reservoirs; the destruction of coastal habitats for development; and urbanization.

INVASIVE SPECIES

Plants and animals that are introduced to environments, where they are not native may out-compete those that occur there naturally. For example, gray squirrels—which originate in North America—dominate native British red squirrels. Additionally, introduced plants may not have invertebrates adapted to feed on them, so such plants play little role in food chains.

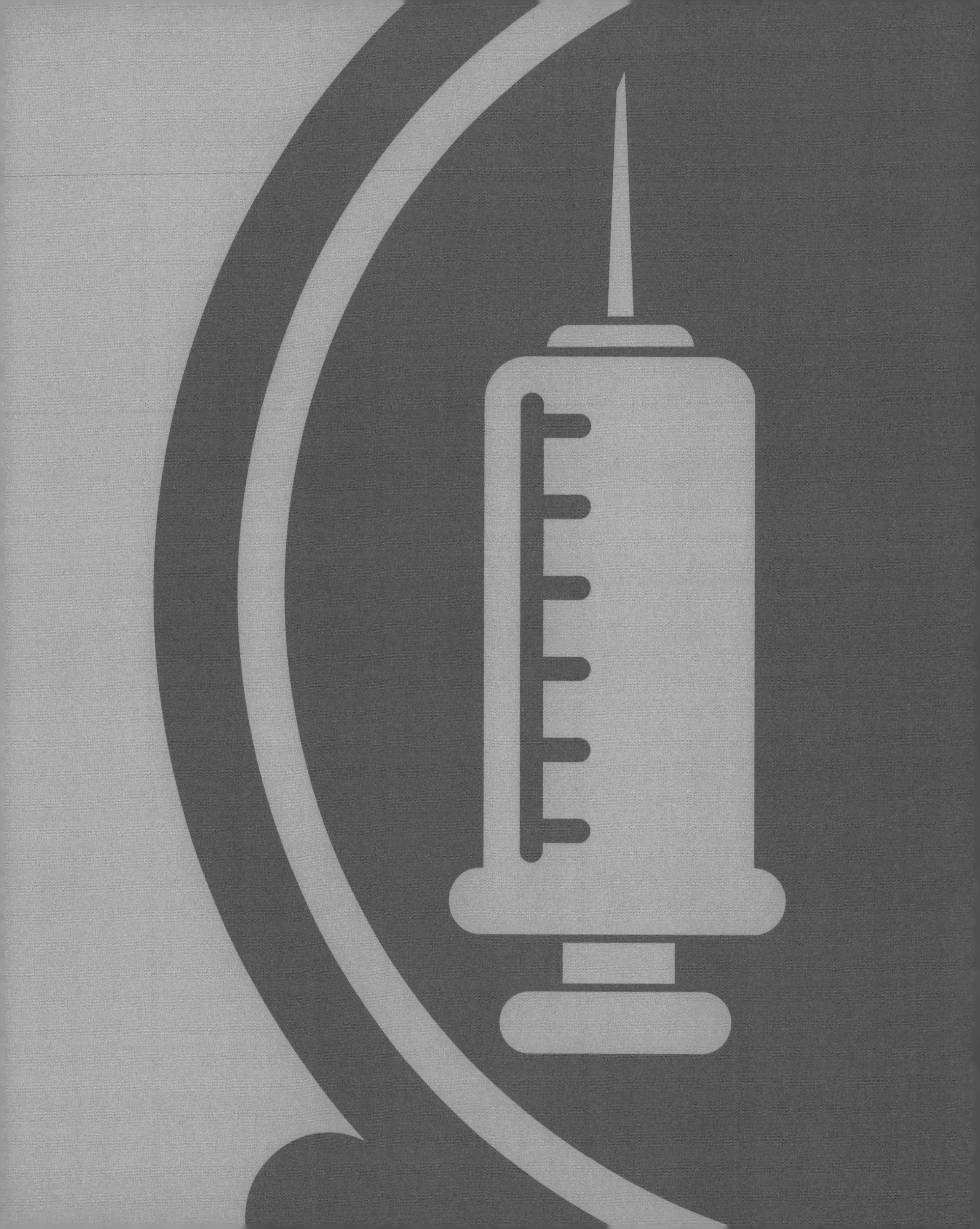

BIOTECHNOLOGY

Selective breeding

Humans have been practicing a form of biotechnology ever since they first formed settlements. By selective breeding, they have driven the development of animal breeds and plant varieties.

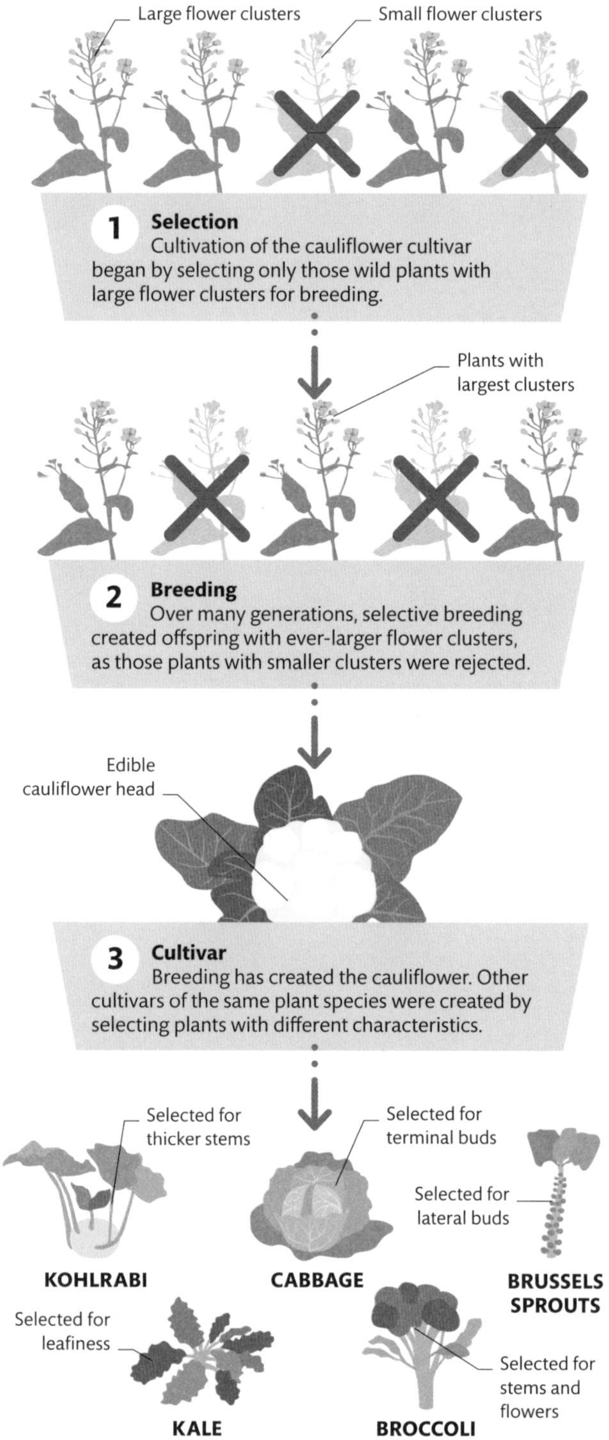

Large flower clusters
Small flower clusters
1 Selection
Cultivation of the cauliflower cultivar began by selecting only those wild plants with large flower clusters for breeding.
Plants with largest clusters
2 Breeding
Over many generations, selective breeding created offspring with ever-larger flower clusters, as those plants with smaller clusters were rejected.
Edible cauliflower head
3 Cultivar
Breeding has created the cauliflower. Other cultivars of the same plant species were created by selecting plants with different characteristics.
Selected for thicker stems
KOHLRABI
Selected for terminal buds
CABBAGE
Selected for lateral buds
BRUSSELS SPROUTS
Selected for leafiness
KALE
Selected for stems and flowers
BROCCOLI

Selecting traits
An example of the power of selective breeding is the development of cruciferous vegetables. Over hundreds of years, plant breeders have created several different varieties, or cultivars, of the same species, each with variations that give it a particular set of characteristics.

What is selective breeding?

In the wild, plants and animals evolve slowly, through genetic mutations and a random mixing of genes via sexual reproduction (see pp.84–85). The "fittest" survive and reproduce, passing on their genes. This process is driven by chance and by changes in the environment (see pp.104–105). In selective breeding, the influence of chance is greatly reduced, as the plants or animals given the chance to reproduce and pass on their genes are selected by farmers or breeders. The result is animals that produce more meat, eggs, or milk, and plants with higher yields of fruit, more protein, better flavor, or less waste.

THE GREEN REVOLUTION

The process of selective breeding can be accelerated and enhanced by crossbreeding many different species, and by genetic testing. An initiative that started in the 1950s, led by geneticist Norman Borlaug, used this approach to breed shorter, more drought-resistant strains of wheat, rice, and corn. This Green Revolution increased yields in many countries facing famine and saved millions of lives.

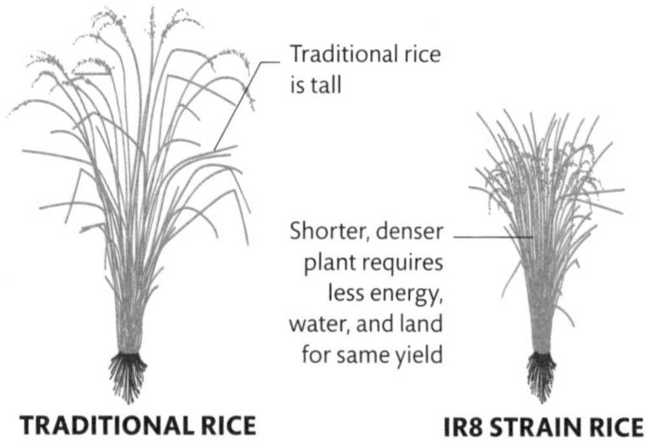

Traditional rice is tall
Shorter, denser plant requires less energy, water, and land for same yield
TRADITIONAL RICE
IR8 STRAIN RICE

Wild types

The crops and other cultivated plants and the livestock and pets we know today are descended from plants and animals that existed naturally. These wild types, as they are known, are very different from the domesticated versions. Corn, for example, originated from a wild type plant called teosinte, which grows in Mexico. The seeds were hard and the cobs small, but by selecting the plants with the softest seeds and largest cobs, the plant gradually developed into what we see today. The wild type of all dogs was a wolf that lived more than 10,000 years ago.

ARE WILD BANANAS EDIBLE?

Only the small amount of flesh around the many seeds of a wild banana is edible. It is much tougher and contains much less sugar than cultivated bananas.

GOATS WERE THE **FIRST ANIMALS** TO BE **DOMESTICATED**

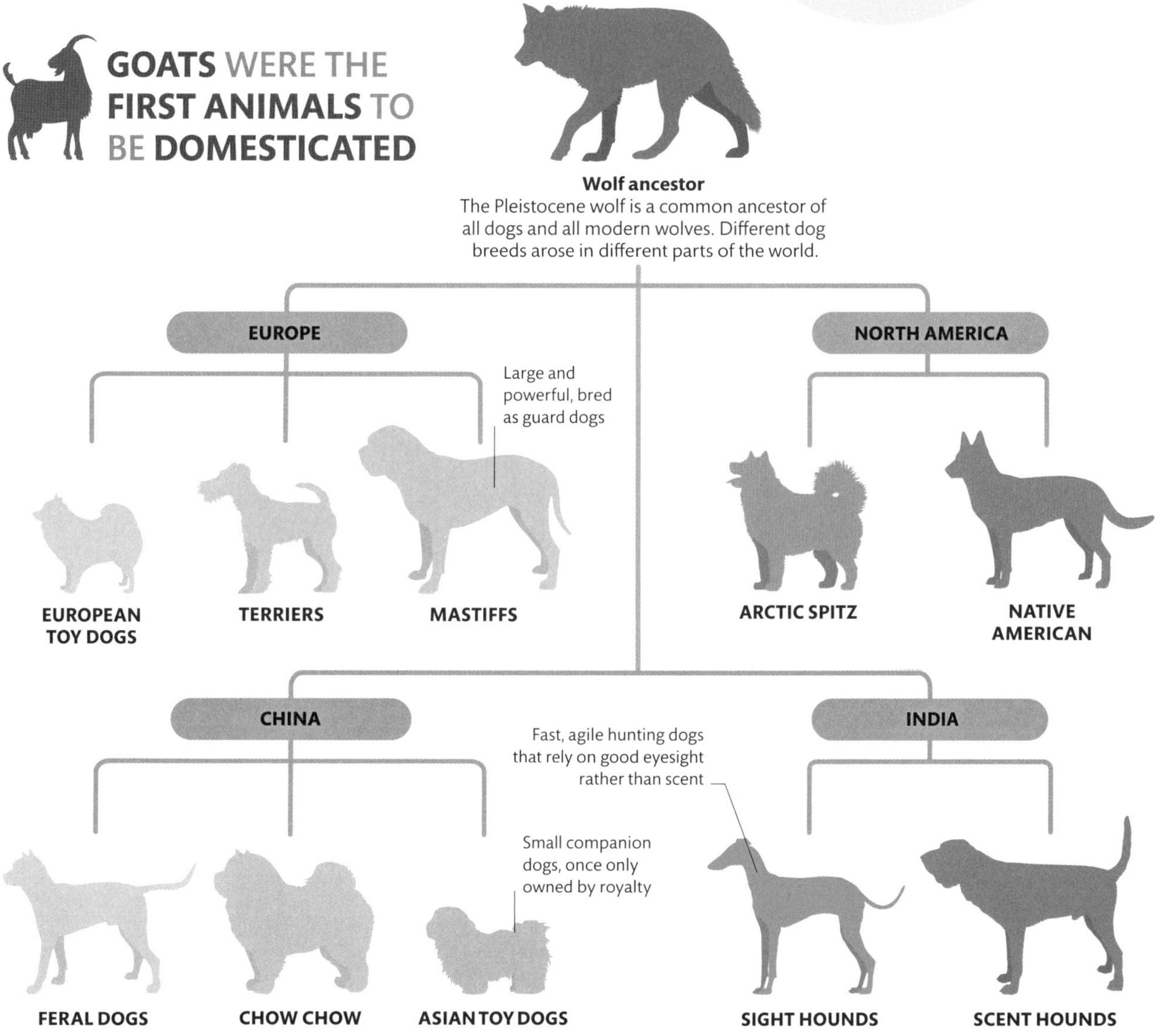
Wolf ancestor
The Pleistocene wolf is a common ancestor of all dogs and all modern wolves. Different dog breeds arose in different parts of the world.
EUROPE
NORTH AMERICA
Large and powerful, bred as guard dogs
EUROPEAN TOY DOGS
TERRIERS
MASTIFFS
ARCTIC SPITZ
NATIVE AMERICAN
CHINA
INDIA
Fast, agile hunting dogs that rely on good eyesight rather than scent
Small companion dogs, once only owned by royalty
FERAL DOGS
CHOW CHOW
ASIAN TOY DOGS
SIGHT HOUNDS
SCENT HOUNDS

Making food

One of the oldest examples of biotechnology is the exploitation of natural processes that take place inside living things, such as fermentation, in the preparation of food and drink.

DO ANIMALS GET DRUNK FROM FERMENTED FRUIT?

Naturally occurring yeasts cause fruits or flower nectar to ferment. As a result, many animals do ingest alcohol, and some even show signs of intoxication.

1 Glycolysis
The raw material for fermentation (and for respiration) is an organic compound called pyruvate (pyruvic acid). Two molecules of pyruvate are made from one molecule of the sugar glucose, which is found in foods and drinks. This process releases energy, as ATP (see p.48), for living organisms.

2 Fermenting process
During fermentation, pyruvate reacts further, releasing more ATP. The waste products are lactic acid in certain organisms and carbon dioxide and ethanol (alcohol) in others. Lactic acids and ethanol kill other microbes in the food, which helps preserve foods.

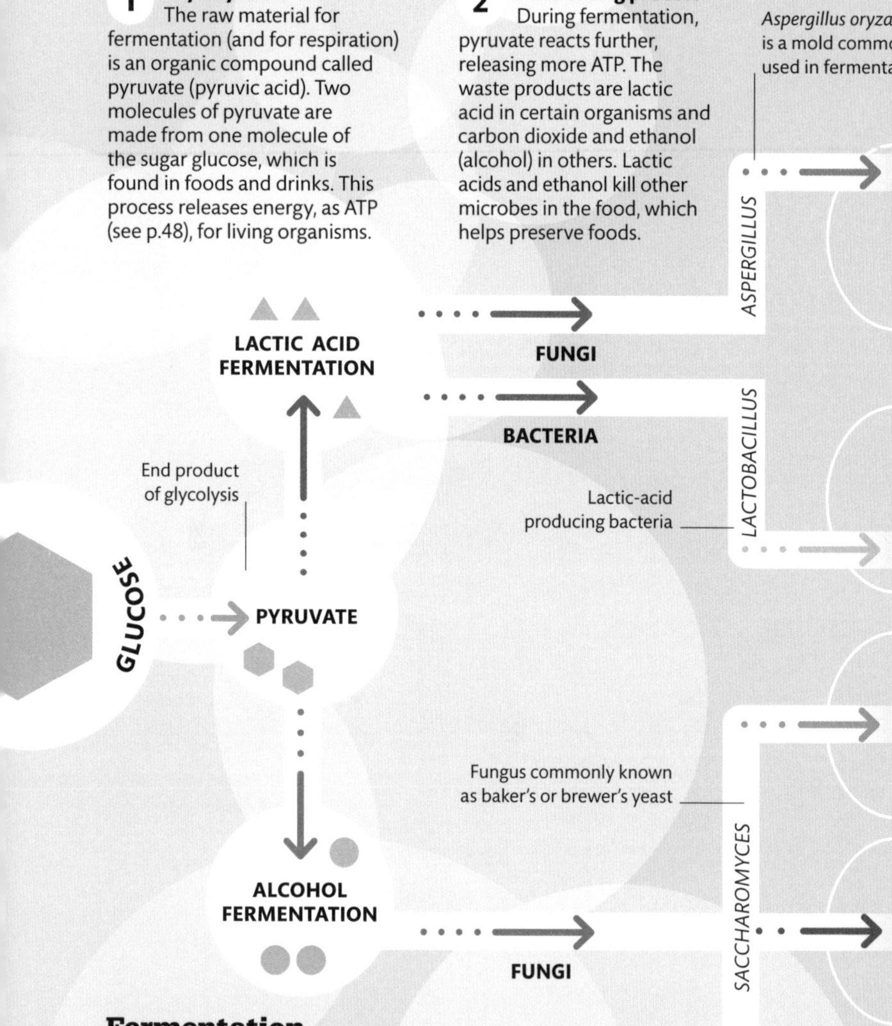

Aspergillus oryzae is a mold commonly used in fermentation
GLUCOSE
PYRUVATE
End product of glycolysis
LACTIC ACID FERMENTATION
FUNGI
BACTERIA
ASPERGILLUS
LACTOBACILLUS
Lactic-acid producing bacteria
Fungus commonly known as baker's or brewer's yeast
SACCHAROMYCES
ALCOHOL FERMENTATION
FUNGI

Adding flavor
Soy sauce, made by fermenting a mixture of soybeans, wheat, and brine (salty water), has a sharp taste because of the lactic acid produced by fermentation.

SOY SAUCE

Sticking together
Milk contains a protein called casein, which coagulates (sticks together) in acidic conditions. Lactic acid from the fermentation of milk produces those conditions in cheese-making.

CHEESE

Carbonated drink
Beer is made from grains that are malted (germinated then boiled, to release sugars). Fermentation produces carbon dioxide, which makes beer fizzy, and alcohol.

BEER

Smooth drink
Making wine uses the same kind of fermentation as beer brewing, but the carbon dioxide is released, leaving a flat drink rather than a fizzy one—apart from in sparkling wine.

WINE

Leavened bread
Most bread rises because bubbles of carbon dioxide gas produced by fermentation are trapped in the dough. Alcohol is also produced, but evaporates away during baking.

BREAD

Fermentation

A biochemical reaction called fermentation, which is caused by bacteria or fungi, is used to help preserve certain foods and drinks, or to improve their flavor, texture, or nutritional value. It is the waste products of fermentation, such as lactic acid and carbon dioxide, that are important in making foods.

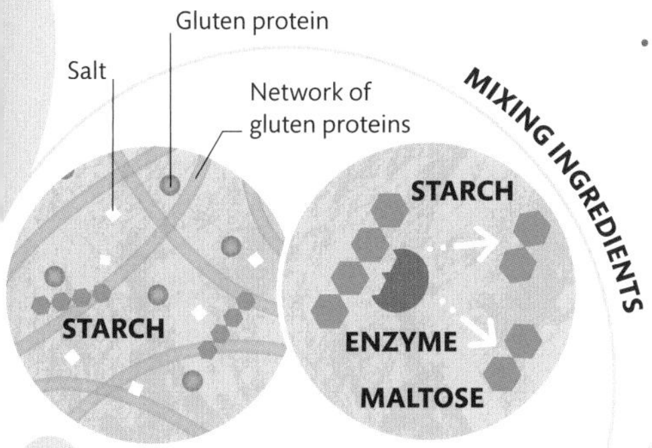
MIXING INGREDIENTS
Gluten protein
Salt
Network of gluten proteins
STARCH
STARCH
ENZYME
MALTOSE

1 Flour, water, salt, and yeast are mixed to form dough. Inside the dough, gluten proteins form a network, and enzymes in the flour break down the flour's starch, forming maltose (a sugar).

Making bread
Bread is an old kind of prepared food, and fermentation plays an important role in baking some varieties of leavened bread.

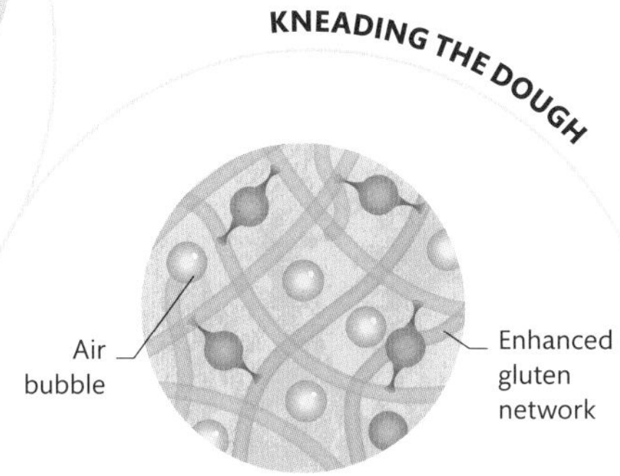
KNEADING THE DOUGH
Air bubble
Enhanced gluten network

2 Manipulating, or kneading, the dough encourages more bonds to form between gluten proteins, and traps air within the network that the bonds create inside the dough.

Controlling conditions

When using living processes or biological compounds such as enzymes (see pp.40–41) in food preparation, it is important to provide the best conditions, by controlling variables, such as temperature and acidity. To safely preserve food until it is consumed, it is necessary to understand how natural processes—including microbial growth, temperature, and oxidation—can contaminate or degrade food.

THE **FIRST KOREAN ASTRONAUT** TO VISIT THE INTERNATIONAL **SPACE STATION** TOOK **KIMCHI**, A FERMENTED PICKLE

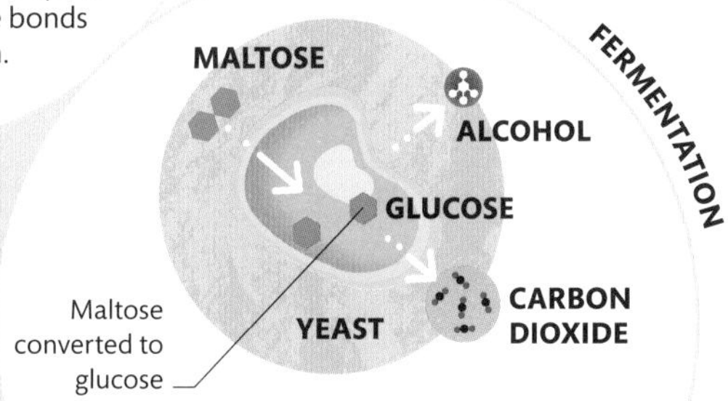
FERMENTATION
MALTOSE
ALCOHOL
GLUCOSE
CARBON DIOXIDE
YEAST
Maltose converted to glucose

3 Inside the yeast cells, enzymes convert the maltose into glucose, and ferment the glucose, producing alcohol and carbon dioxide gas, which enlarges the air bubbles and causes the dough to rise.

PASTEURIZATION

Pasteurization aims to stop, rather than use, biological processes to make a range of foods safer by using heat to kill most of the microorganisms naturally present in them. Originally designed to prevent wine from spoiling, and turning to vinegar, it is most associated with milk.

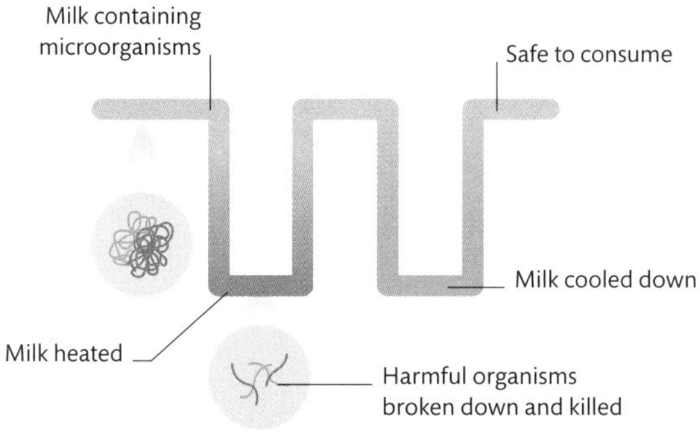
Milk containing microorganisms
Safe to consume
Milk cooled down
Milk heated
Harmful organisms broken down and killed

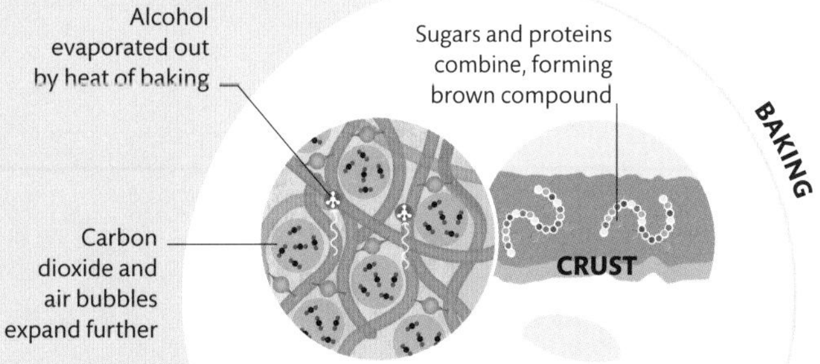
BAKING
Alcohol evaporated out by heat of baking
Sugars and proteins combine, forming brown compound
Carbon dioxide and air bubbles expand further
CRUST

4 The bubbles of trapped air and carbon dioxide expand as the dough cooks, giving the bread a light, spongy texture. Reactions between sugars and proteins at the surface form the brown crust.

Making medicine

Medicines are drugs used to treat, cure, or prevent diseases. Most modern medicines are manufactured in laboratories or factories and must be rigorously tested to make sure they are safe as well as effective.

WHAT ARE DRUGS?

A drug is any chemical compound or a mixture of compounds that has an effect on the body. Medicines are drugs.

Natural sources

People have used compounds found in nature to cure or prevent diseases for hundreds, or even thousands, of years. Many of these traditional medicines are effective—including most of the examples below—while some have little or no benefit. Modern pharmaceutical companies study traditional medicines and synthesize the compounds with the desired effects in large quantities. In some cases, where the active compounds are difficult to synthesize, they extract and use natural compounds.

IT TAKES ABOUT 12 YEARS TO DEVELOP AND APPROVE A NEW DRUG

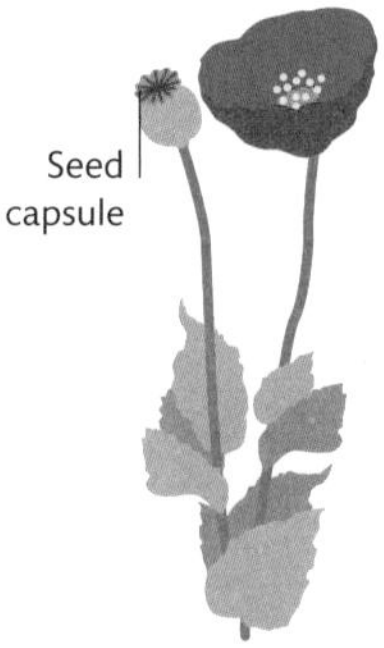
Seed capsule

Opium poppies
The potent opioid painkillers morphine and codeine are found in opium, which is extracted from the seed capsules of poppies.

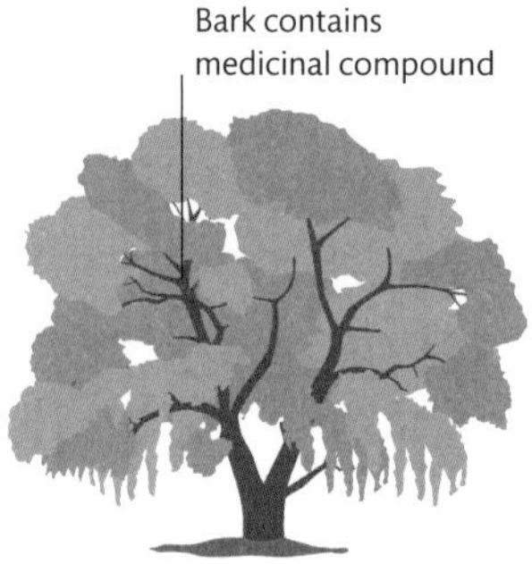

Willow
Has been used for thousands of years to treat fever and pain. Aspirin is a synthesized version of the active compound.

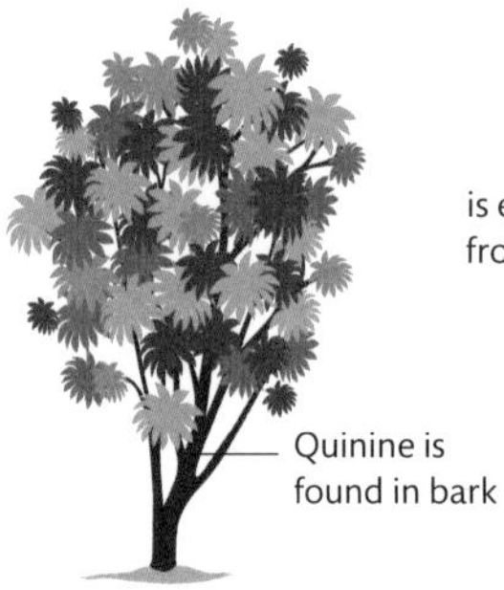
Quinine is found in bark

Cinchona tree
Used for centuries to treat malaria, the active ingredient, quinine, is now synthesized in large quantities.

Digitalis is extracted from leaves

Foxglove
Contains the drug digoxin, or digitalis, used to treat arrhythmias (irregular heartbeat) and heart failure.

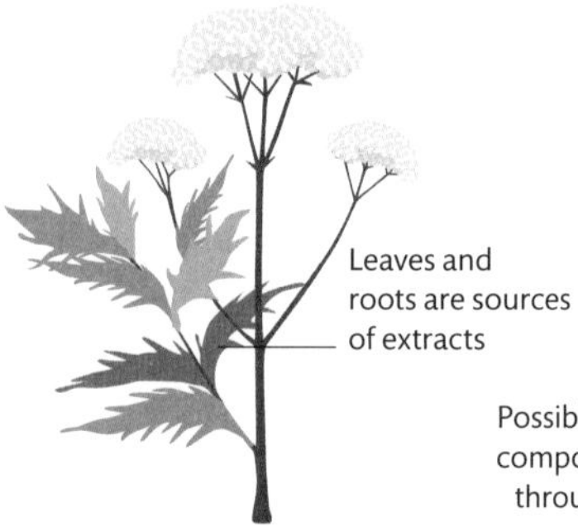

Leaves and roots are sources of extracts

Valerian
Used for at least two thousand years, valerian extracts contain oils that may help reduce anxiety and improve sleep.

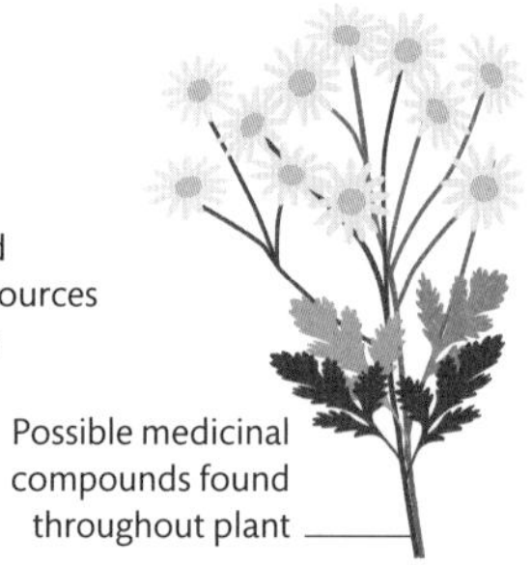
Possible medicinal compounds found throughout plant

Feverfew
Traditionally used to treat fever and headaches but there is little scientific evidence that extracts of this plant are effective.

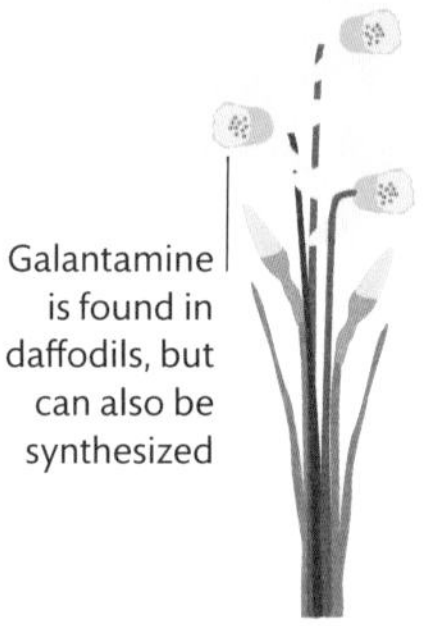
Galantamine is found in daffodils, but can also be synthesized

Daffodil
Traditionally used to treat a wide range of illnesses, daffodils also contain a compound, galantamine, used to ease Alzheimer's disease.

Paclitaxel occurs in bark

Pacific yew tree
After an extensive search for anticancer compounds in plants, researchers identified a potent drug, paclitaxel, in 1971.

DRUG TRIALS

Before a medicine can be tested on people, it has to pass rigorous preclinical tests in the lab and then, normally, on animals such as mice, to make sure it is safe for use in humans. Then, there are three main phases of clinical research, each one going ahead only if the drug passes the previous phase.

Phase I
A Phase I trial involves giving the drug to a few tens of healthy volunteers, to check safety and dosage.

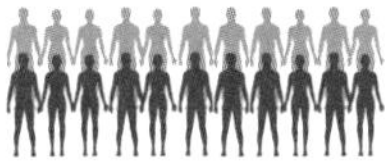

Phase II
Next, the drug is given to people who have the disease (or are at risk), to test for side effects.

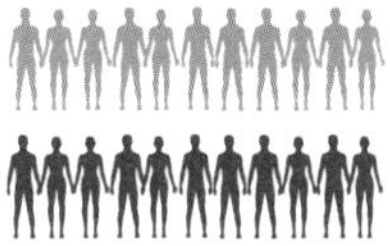

Phase III
This stage involves a few thousand volunteers in a "double-blind trial": half have the drug and half, a placebo.

Developing new drugs

The process to develop a new medicine begins by searching for compounds that interrupt disease pathways, which are sets of chemical reactions that cause diseases or their symptoms. Some of these candidate compounds are found in nature, while others are designed on computer and then synthesized in the laboratory. Any compounds that show promising results are tested to make sure they are safe and effective. After successful trials, the new medicine is registered with a drug agency and manufactured in large quantities.

1 Identify pathways
Researchers begin by studying a disease. In most cases, this involves searching for a set of proteins produced by pathogens (disease-causing microorganisms) or by the body itself. It is the interactions between these proteins, called a pathway, that cause the disease or its symptoms.

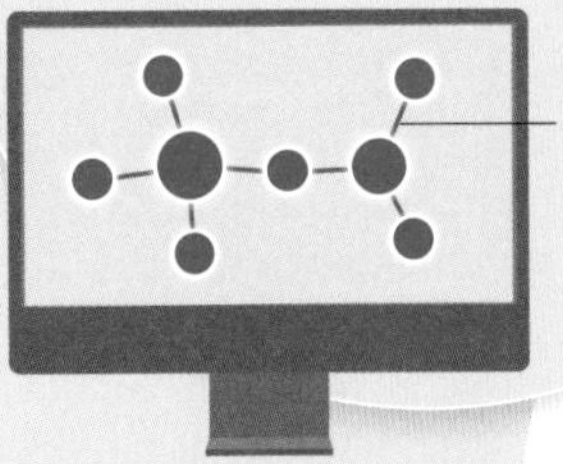
Networks of proteins disrupt cellular activity

2 Search for compounds
Next, researchers look for compounds that interrupt the disease pathways by binding directly to the target proteins. This can be done by trying thousands of possible compounds in the laboratory, or by designing molecules that will bind, and then synthesizing them.

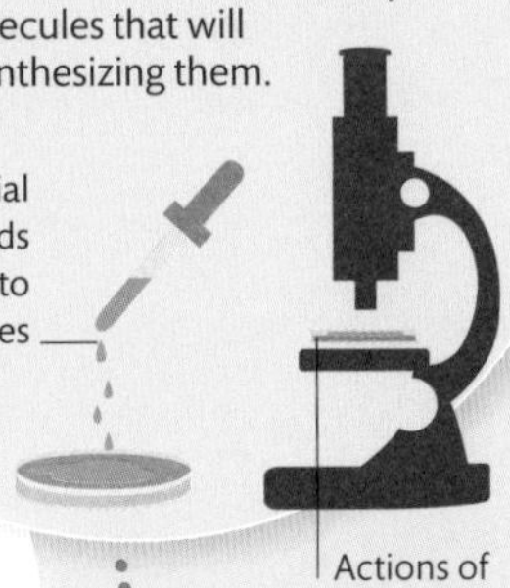
Potential compounds administered to cell cultures
Actions of potential medicines monitored under a microscope

3 Testing new drugs
Once a promising compound has been identified in the laboratory, it must be trialled in living organisms—first preclinically, typically in animals such as mice, and then in clinical trials involving humans. Clinical trials are carried out in several stages (see above).

4 Manufacture
Once proven safe and effective, a new drug is registered with a government drug agency. Then it can be manufactured in large quantities and made available to doctors to administer to their patients. The drug is still monitored, in a Phase IV trial, for long-term side effects.

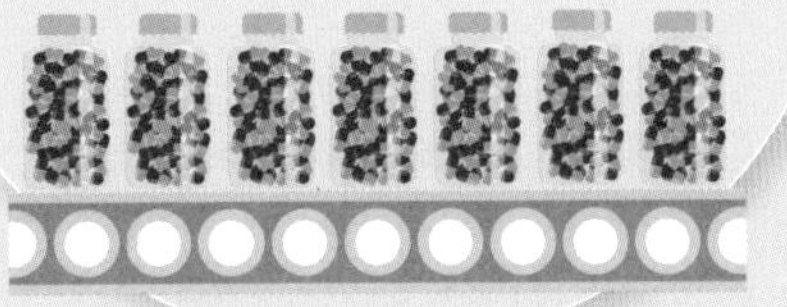

Vaccines

A vaccine is a medicine that teaches a person's immune system to be ready to fight an infectious disease, so that the disease will be prevented or reduced in that person and its spread restricted.

How vaccines work

A vaccine equips the body with the necessary antibodies to fight an infectious disease. The same process happens when the body builds up natural immunity after suffering a disease—but in the case of a vaccine, the person achieves the immunity without ever having the disease.

HOW LONG DOES A VACCINATION LAST?

As with natural immunity, vaccine-mediated immunity for some diseases lasts a lifetime, while for some, a booster is needed within a few years or even months.

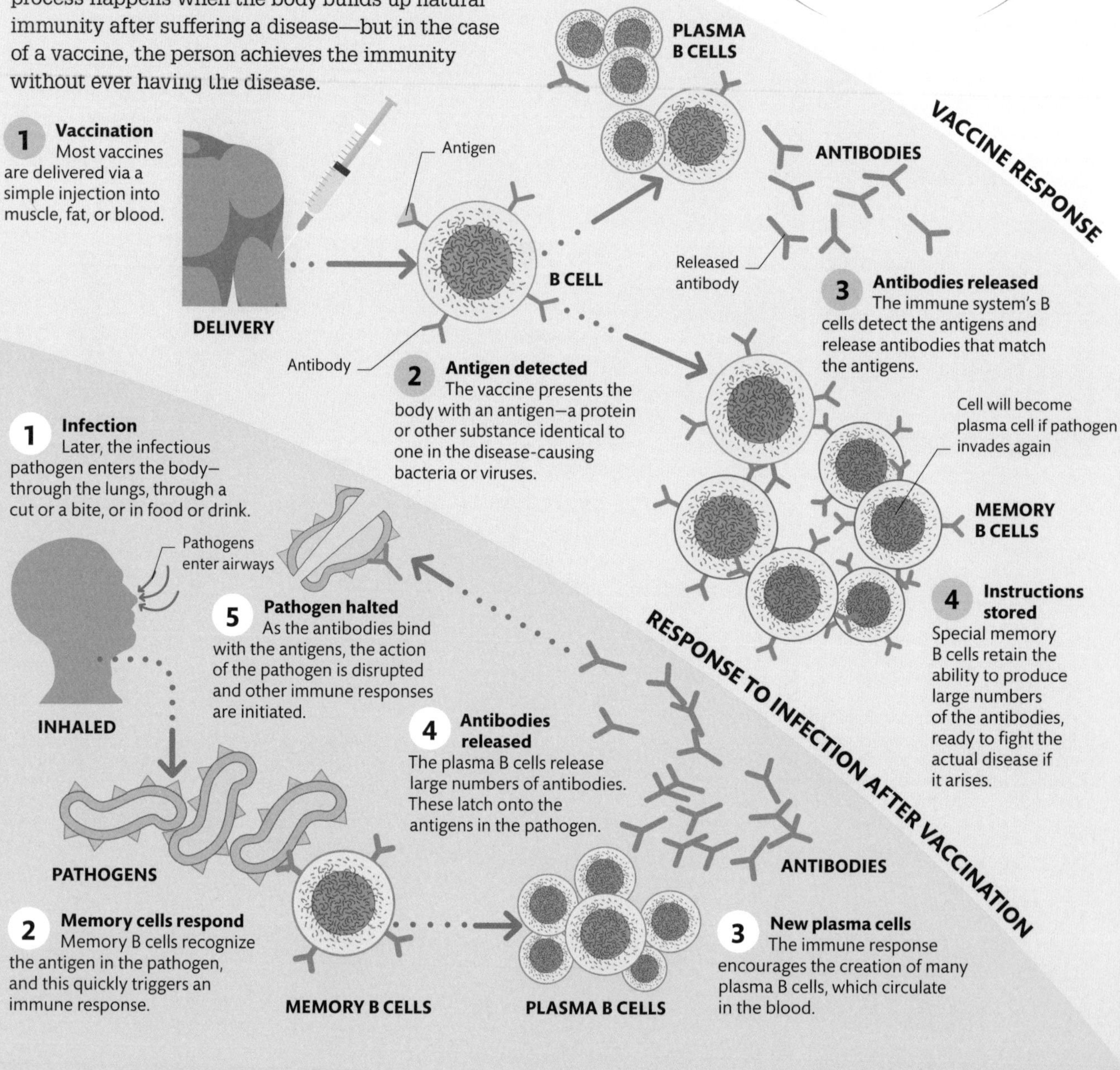

VACCINE RESPONSE
DELIVERY
Antigen
Antibody
B CELL
PLASMA B CELLS
ANTIBODIES
Released antibody
Cell will become plasma cell if pathogen invades again
MEMORY B CELLS
RESPONSE TO INFECTION AFTER VACCINATION
Pathogens enter airways
INHALED
PATHOGENS
MEMORY B CELLS
PLASMA B CELLS
ANTIBODIES

Vaccine response

1 Vaccination Most vaccines are delivered via a simple injection into muscle, fat, or blood.

2 Antigen detected The vaccine presents the body with an antigen—a protein or other substance identical to one in the disease-causing bacteria or viruses.

3 Antibodies released The immune system's B cells detect the antigens and release antibodies that match the antigens.

4 Instructions stored Special memory B cells retain the ability to produce large numbers of the antibodies, ready to fight the actual disease if it arises.

Response to infection after vaccination

1 Infection Later, the infectious pathogen enters the body—through the lungs, through a cut or a bite, or in food or drink.

2 Memory cells respond Memory B cells recognize the antigen in the pathogen, and this quickly triggers an immune response.

3 New plasma cells The immune response encourages the creation of many plasma B cells, which circulate in the blood.

4 Antibodies released The plasma B cells release large numbers of antibodies. These latch onto the antigens in the pathogen.

5 Pathogen halted As the antibodies bind with the antigens, the action of the pathogen is disrupted and other immune responses are initiated.

TYPES OF VACCINE

The first vaccines, produced more than 200 years ago, were pathogens of a related but less serious disease. Today, there are several ways in which a vaccine can deliver the antigens needed to produce the necessary immune response.

TYPE	HOW IT WORKS	EXAMPLES
Inactivated	A pathogen is killed or inactivated using heat, radiation, or chemicals, so that it cannot cause the disease.	Vaccines against flu, polio, hepatitis A, cholera, and bubonic plague
Related microbe	Typically, this kind of vaccine contains a pathogen that causes a related disease in another species.	Vaccines against smallpox (now eradicated) and tuberculosis
DNA	These contain small pieces of DNA, which the body's cells use to code for the disease antigen.	COVID-19
Tame toxins	These vaccines contain safe versions of disease-causing toxins produced by some pathogens.	Vaccines against diphtheria, pertussis (whooping cough), and tetanus
Pieces of pathogen	These contain proteins or fragments of proteins found on the pathogen's surface to create immune response.	Vaccines against hepatitis B and human papilloma virus (HPV)
Live attenuated vaccine	These vaccines contain a viable, or "live," version of the pathogen that has been bred to be harmless.	Vaccines against measles, mumps, rubella, and yellow fever
mRNA	mRNA in a vaccine instructs the body's cells to make proteins identical to those on the target pathogen.	COVID-19

New vaccines

As with all new medicines (see p.201), new vaccines are developed in laboratories and must then go through stringent preclinical and clinical trials before they are approved. The first messenger RNA (mRNA, see p.36) vaccine was developed and approved more quickly than most, in the early stages of the COVID-19 pandemic, to create immunity against the pathogen that causes that disease, the coronavirus SARS-CoV-2.

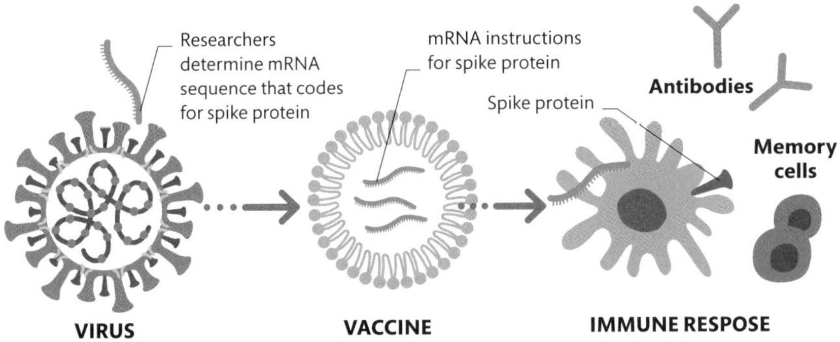
Researchers determine mRNA sequence that codes for spike protein
mRNA instructions for spike protein
Spike protein
Antibodies
Memory cells
VIRUS
VACCINE
IMMUNE RESPOSE

1 Identify target The target of the COVID-19 vaccine was the spike proteins on the virus's coat that help the virus gain entry into cells.

2 Create vaccine The vaccine contains millions of copies of the mRNA that codes for the spike protein, encapsulated in tiny fat globules.

3 Vaccination Once injected into the body, the body's cells use the mRNA to make spike proteins that create an immune response.

VACCINES PREVENT ABOUT 4–5 MILLION DEATHS A YEAR

ANTIBODIES

Antibodies are Y-shaped protein molecules that bind to antigens. There is an antibody for almost every possible antigen; those matching a disease to which the body is immune are made quickly, in large numbers.

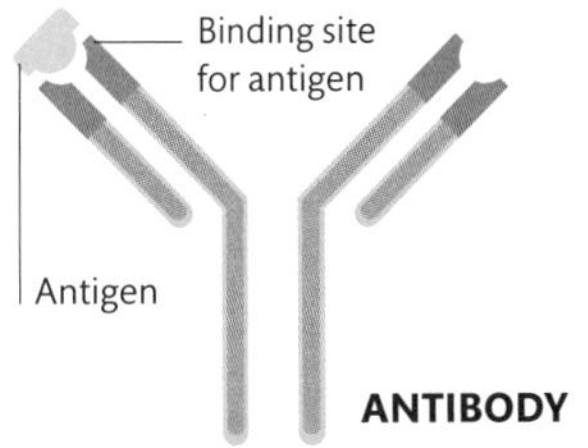
Binding site for antigen
Antigen
ANTIBODY

DNA microarrays

Many kinds of DNA test involve a DNA microarray: a small glass or plastic surface coated with hundreds or thousands of dots, each with different DNA probes attached. Each probe is a small piece of DNA. Specific lengths of the test DNA bind to these probes. The probes have fluorescent markers attached, which glow when the test DNA binds. A computer then analyses a high-resolution photograph of the fluorescence to determine which versions of certain genes are present or which genes are active.

Comparative test
DNA microarrays can find mutations in certain genes, or determine which genes are expressed in different cells. This can be used to compare different tissues or cells from pathological (diseased) and healthy tissues.

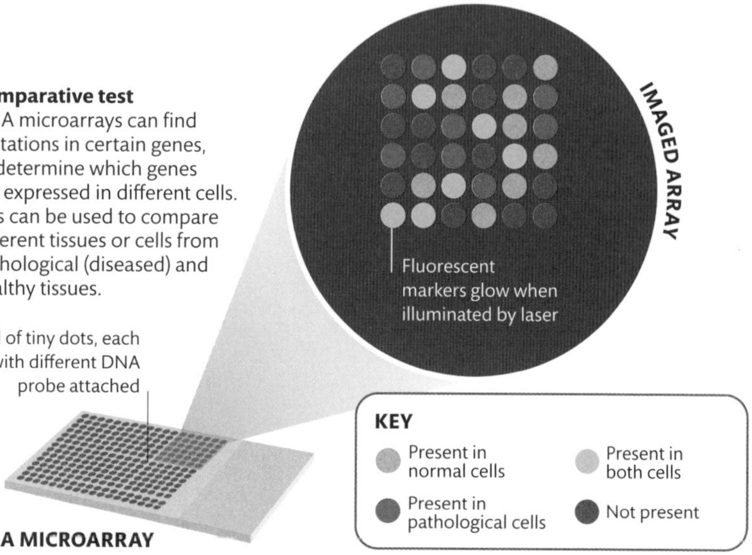

IMAGED ARRAY
Fluorescent markers glow when illuminated by laser
Grid of tiny dots, each with different DNA probe attached
DNA MICROARRAY
KEY
Present in normal cells
Present in both cells
Present in pathological cells
Not present

DNA testing

DNA testing can be used to find inheritable diseases, to identify people in paternity tests or through samples left at a crime scene, to determine a person's ancestry, or to sequence entire genomes.

WHAT IS ENVIRONMENTAL DNA?

All organisms shed DNA. By studying this environmental DNA, collected from the soil and water, biologists can determine which organisms live in a particular habitat.

Polymerase chain reaction

Before most DNA tests, the long strands of DNA to be tested are cut into smaller sections, typically using restriction enzymes, which cut the DNA at specific sequences along its length. Then, in order to achieve reliable results, these smaller sections are replicated in large numbers by repeating a technique called polymerase chain reaction (PCR) several times. In every run, the number of each segment present doubles.

Replicating DNA
In PCR, the stages of denaturing, annealing and extension, or synthesis, are repeated several hundred times, making millions of copies of each DNA segment.

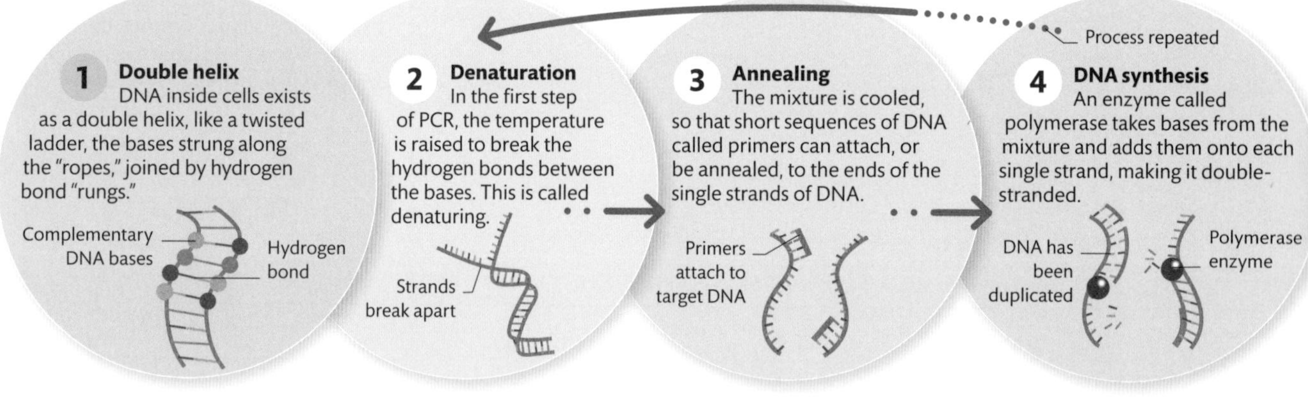

Process repeated
1 Double helix
DNA inside cells exists as a double helix, like a twisted ladder, the bases strung along the "ropes," joined by hydrogen bond "rungs."
Complementary DNA bases
Hydrogen bond
2 Denaturation
In the first step of PCR, the temperature is raised to break the hydrogen bonds between the bases. This is called denaturing.
Strands break apart
3 Annealing
The mixture is cooled, so that short sequences of DNA called primers can attach, or be annealed, to the ends of the single strands of DNA.
Primers attach to target DNA
4 DNA synthesis
An enzyme called polymerase takes bases from the mixture and adds them onto each single strand, making it double-stranded.
DNA has been duplicated
Polymerase enzyme

Sequencing DNA

DNA sequencing is achieved using sophisticated machines that can read the sequence of bases along specific sections of DNA or an entire genome. Direct comparisons to healthy versions of the same section or gene can reveal diseases such as cancers, or a propensity to developing or passing on diseases.

IT TOOK 13 YEARS TO COMPLETE THE FIRST HUMAN GENOME SEQUENCE

Reading DNA

There are several different kinds of DNA-sequencing technology. The most sophisticated can read an entire human genome in less than a day. One kind, known as the Sanger method, is shown below.

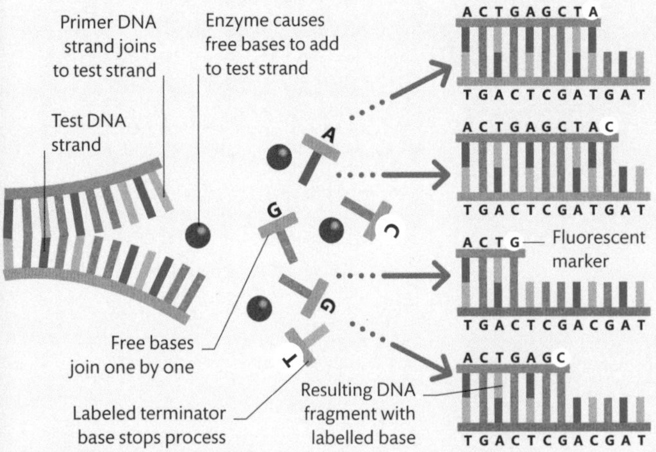
Primer DNA strand joins to test strand
Enzyme causes free bases to add to test strand
Test DNA strand
Free bases join one by one
Labeled terminator base stops process
Resulting DNA fragment with labelled base
Fluorescent marker
A C T G A G C T A
T G A C T C G A T G A T
A C T G A G C T A C
T G A C T C G A T G A T
A C T G
T G A C T C G A C G A T
A C T G A G C
T G A C T C G A C G A T

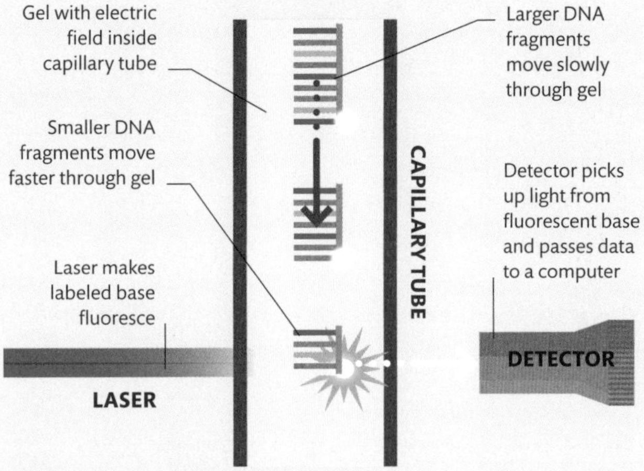
Gel with electric field inside capillary tube
Smaller DNA fragments move faster through gel
Laser makes labeled base fluoresce
LASER
CAPILLARY TUBE
Larger DNA fragments move slowly through gel
Detector picks up light from fluorescent base and passes data to a computer
DETECTOR

1 DNA bases labeled

First, a primer attaches at a specific location on a single-stranded length of test DNA. Polymerase enzymes attach DNA bases onto the strand one at a time. In solution with the normal bases are terminator bases. These stop the process and are labeled with fluorescent markers.

2 Bases read by laser

The newly formed complementary lengths of DNA break away from the original strand. The single-stranded lengths pass along a capillary tube, pulled by an electric field. A laser at the far end of the tube causes the terminator bases to glow near a detector. The fragments arrive at the laser in length order, from shortest to longest.

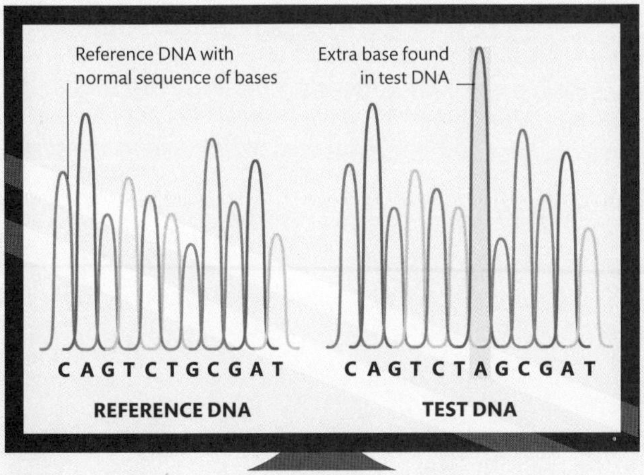
Reference DNA with normal sequence of bases
Extra base found in test DNA
C A G T C T G C G A T
REFERENCE DNA
C A G T C T A G C G A T
TEST DNA

3 Analysis

A computer records the colored flashes, and pieces together the exact sequence of bases. By comparing the sequence of a particular gene in one patient with a known healthy version of a gene, additions or deletions that may cause disease can be identified.

CHROMOSOME TESTING

A human body cell has 23 pairs of chromosomes (see p.58). Scientists may study the full set of a person's chromosomes, called a karyotype, to see if there are any extra, missing, or abnormal chromosomes.

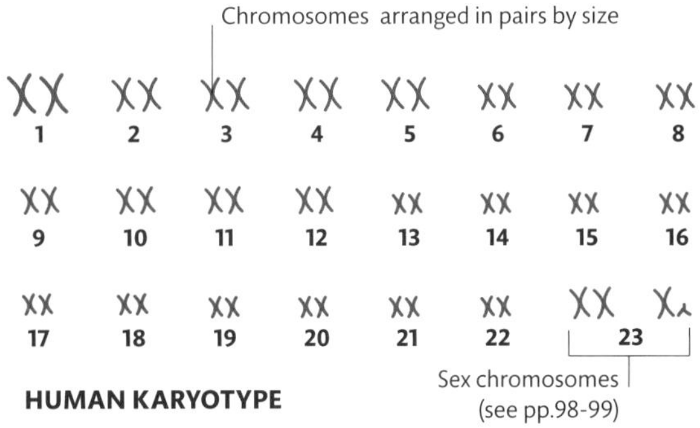
Chromosomes arranged in pairs by size
1 2 3 4 5 6 7 8
9 10 11 12 13 14 15 16
17 18 19 20 21 22 23
Sex chromosomes (see pp.98-99)
HUMAN KARYOTYPE

Genetic engineering

Genetic engineering (or genetic modification) alters the genome of an organism to enhance that organism's capabilities. It typically involves transferring a gene from one species to another.

Genetic modification

Humans have been modifying the genomes of plants and animals for thousands of years, by domestication and selective breeding (see p.196). Genetic engineering enables researchers to make more targeted, and much quicker, modifications. All cells share the same genetic language, DNA, and that makes it possible to transfer genes between species—even from humans to bacteria. Applications of genetic engineering include the production of vaccines and other medicines and genetically modified plants and animals.

CAN MODIFIED GENES SPREAD?

Horizontal gene transfer does happen in nature, and so it is possible for modified genes in crops to spread—but it is extremely rare and unlikely to cause any problems.

Disease resistance
Certain bacteria found in soil have genes that produce proteins that are toxic to insects. Inserting these genes into corn-plant genomes protects the plants from pests, reducing the need for pesticides.

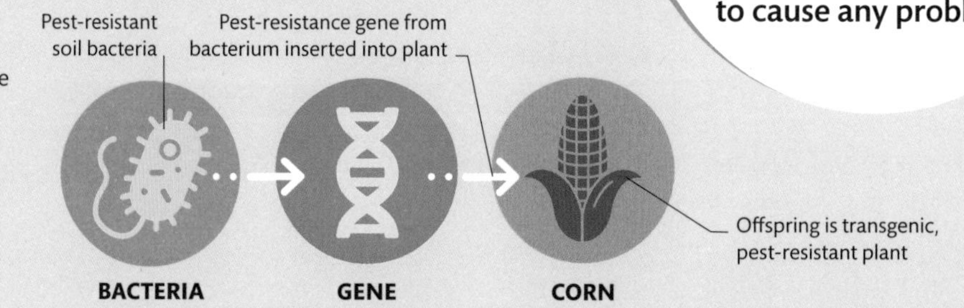
Pest-resistant soil bacteria
Pest-resistance gene from bacterium inserted into plant
Offspring is transgenic, pest-resistant plant
BACTERIA
GENE
CORN

Altering output
Some people's bodies do not produce enough growth hormone, an important compound for healthy development. One convenient way of producing it is to transfer the gene from the human genome into the genome of a goat egg or embryo.

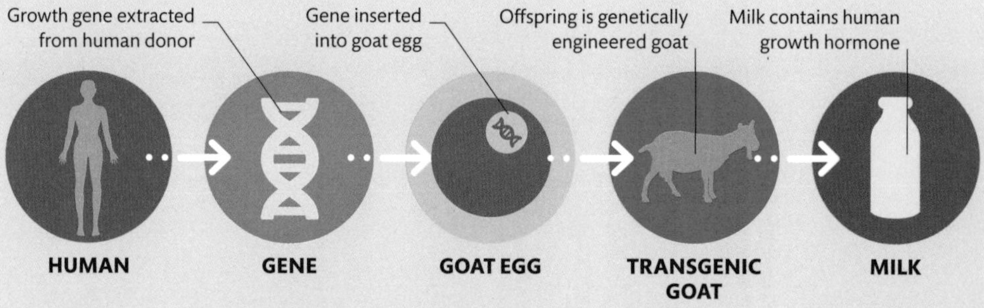
Growth gene extracted from human donor
Gene inserted into goat egg
Offspring is genetically engineered goat
Milk contains human growth hormone
HUMAN
GENE
GOAT EGG
TRANSGENIC GOAT
MILK

Living factories
Millions of people with diabetes depend upon insulin produced by transgenic (genetically engineered) bacteria that carry the gene for producing human insulin.

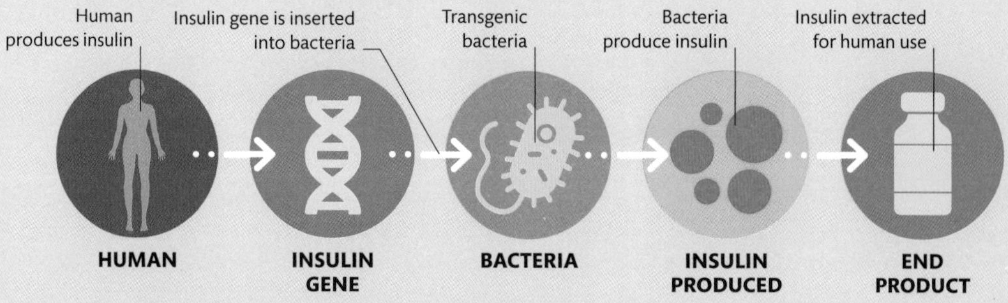
Human produces insulin
Insulin gene is inserted into bacteria
Transgenic bacteria
Bacteria produce insulin
Insulin extracted for human use
HUMAN
INSULIN GENE
BACTERIA
INSULIN PRODUCED
END PRODUCT

Recombination

The mixing of genes from different organisms is called recombination. It happens naturally during meiosis (see pp.82–83)—when genes from two parents are included in the genome of their offspring—and in horizontal gene transfer, typically when single-celled organisms take in genes from other species. There are several methods to artificially create recombinant DNA, including by physically shooting the gene into living cells and by using enzymes to insert the gene into a bacterial cell.

UP TO 90% OF ALL **SOYBEANS** ON THE MARKET HAVE BEEN **GENETICALLY MODIFIED**

GM ORGANISMS

Crops can be engineered to make them more resistant to pests or diseases or to stay fresh for longer. Although many genetically modified organisms have been approved—including salmon that grow faster, tomatoes that stay firm for longer, and pest-resistant cotton—the use of genetic engineering in food remains a controversial issue.

SALMON

TOMATO

COTTON

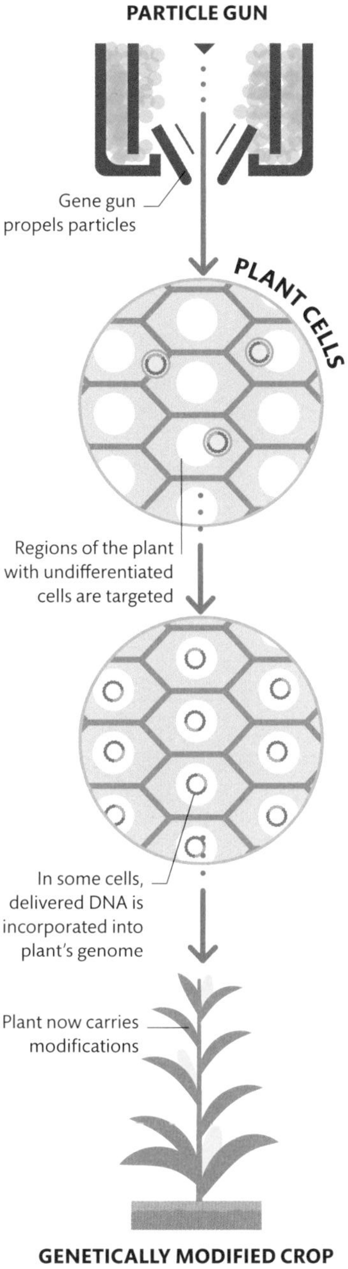
PARTICLE GUN
Gene gun propels particles
PLANT CELLS
Regions of the plant with undifferentiated cells are targeted
In some cells, delivered DNA is incorporated into plant's genome
Plant now carries modifications
GENETICALLY MODIFIED CROP

Biolistics

Biolistics, short for biological ballistics, is a method that is often used for creating transgenic plants. Tiny metal particles coated with the gene are fired at high speed into the target cells. If the delivered DNA is incorporated into the plant's genome, it is passed down to future generations of that plant.

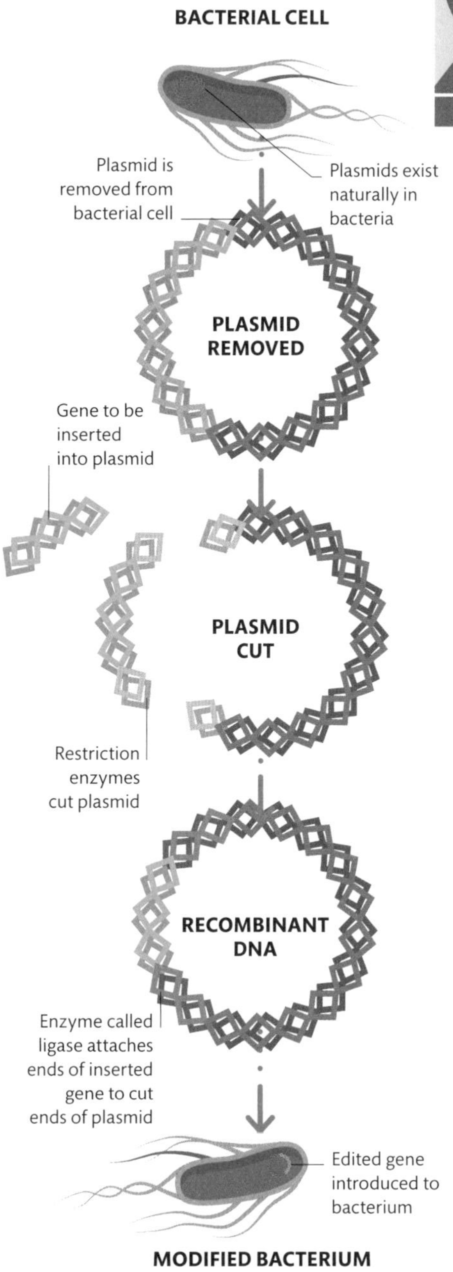
BACTERIAL CELL
Plasmid is removed from bacterial cell
Plasmids exist naturally in bacteria
PLASMID REMOVED
Gene to be inserted into plasmid
PLASMID CUT
Restriction enzymes cut plasmid
RECOMBINANT DNA
Enzyme called ligase attaches ends of inserted gene to cut ends of plasmid
Edited gene introduced to bacterium
MODIFIED BACTERIUM

Bacteria and enzymes

In addition to their main genome, bacteria have small loops of DNA, called plasmids. Geneticists use restriction enzymes to cut the plasmids so that genes from other species can be inserted. Once the plasmid is modified, heat or electric shock is used to encourage the bacterium to take in the altered plasmid.

Gene therapy

Some diseases are the result of a faulty allele (version of a gene), with errors in the sequence of bases in its DNA (see p.37). For many of these genetic diseases, treatment is available via a technology called gene therapy.

IS GENE THERAPY SAFE?

Gene therapy comes with some risks—but as with all emerging medical technologies, gene therapy is subject to strict testing and regulation.

1 Cells harvested Carrying out gene therapy on cells outside the body (ex vivo) reduces the risk that the body's immune system will produce inflammation.

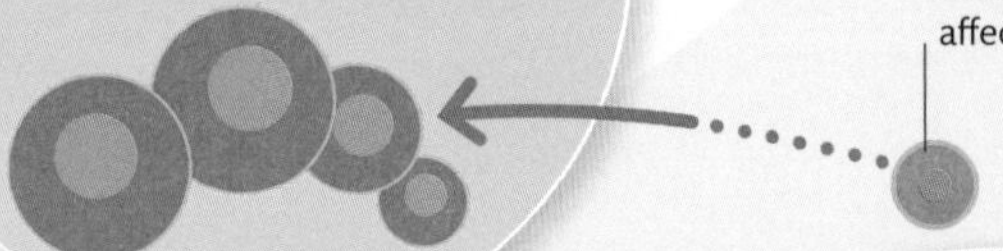

Some of the patient's own cells are removed from area of body affected by the disease

2 Virus disabled Viruses are used because they are good at entering cells. First, a virus's own genetic material is removed or deactivated.

DNA or RNA is removed or deactivated

PATIENT'S CELLS

3 Gene inserted Next, the healthy allele is inserted into the virus's protein coat (see p.16).

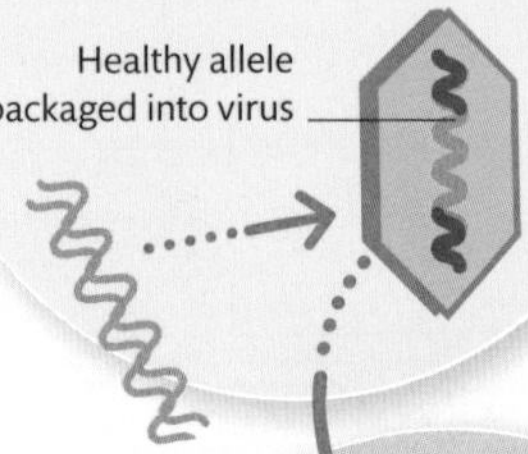

Healthy allele packaged into virus

Delivery methods
There are several ways of delivering healthy alleles to a patient. Most commonly, viruses carry the DNA into the patient's own cells. This process can happen inside the body (see opposite page) or outside (shown here).

4 Virus added to cells The virus is introduced into the cells taken from the patient. Once inside, the kind of virus used in gene therapy will insert the DNA directly into a cell's genome.

5 Healthy gene inserted into cells This kind of treatment is often used for dividing cells, such as skin and blood cells. The gene is then copied to daughter cells before being introduced into the body.

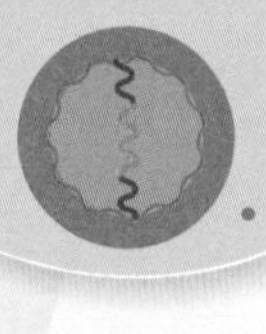

6 Cells inserted into body The cells, now containing a healthy allele, are inserted into the patient, typically as a single injection into the tissue affected by the disorder.

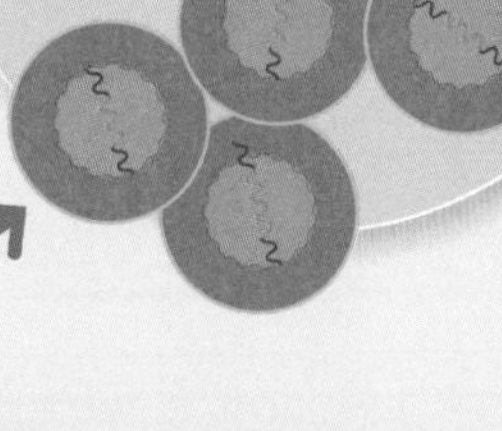

Modified cells introduced into patient's body

7 Protein production Now, the correct protein can be made, and the disease diminished or cured. In some therapies, only one treatment is required.

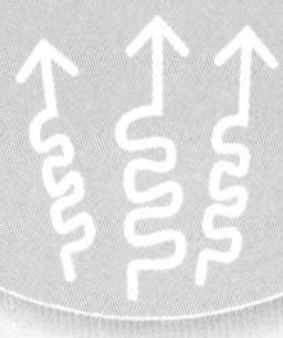

Delivering healthy alleles

Errors in the DNA of a defective allele might be inherited or random mutations. Either way, that allele might produce a defective version of the protein for which that gene codes, or no protein at all. Gene therapy delivers a working allele of that gene, or in some cases removes or modifies genetic material. It is most effective for treating diseases that are caused by a single defective gene or that affect a single type of tissue, such as blood or lung cells.

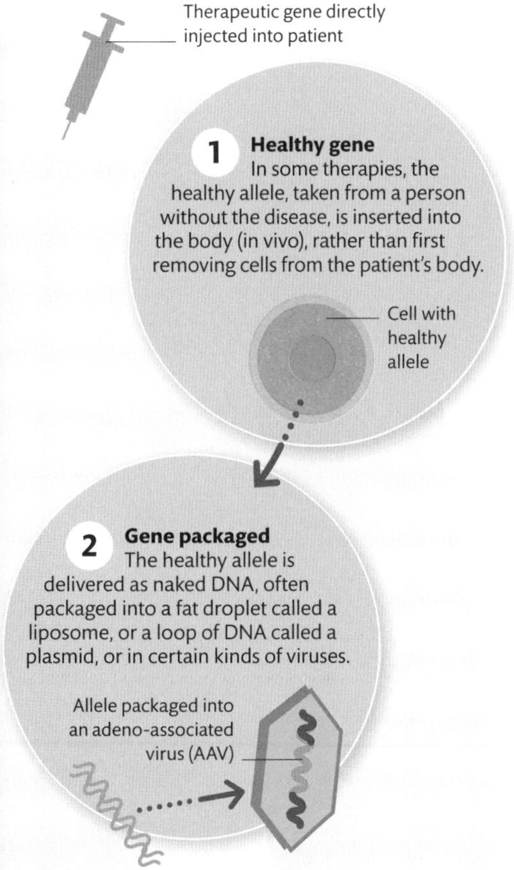
Therapeutic gene directly injected into patient
1 Healthy gene
In some therapies, the healthy allele, taken from a person without the disease, is inserted into the body (in vivo), rather than first removing cells from the patient's body.
Cell with healthy allele
2 Gene packaged
The healthy allele is delivered as naked DNA, often packaged into a fat droplet called a liposome, or a loop of DNA called a plasmid, or in certain kinds of viruses.
Allele packaged into an adeno-associated virus (AAV)

GENE-THERAPY TRIALS HAVE SHOWN SUCCESS IN TREATING SOME CANCERS

Gene editing

With traditional gene therapy, there is no guarantee that a gene carried into a cell will be inserted into the correct place on the genome. A more accurate approach is to edit the gene directly (either inside or outside the body), using a gene-editing technique called CRISPR (clustered regularly interspaced short palindrome repeats). The faulty allele can be removed or repaired, or a healthy one inserted in its place.

1 Preparing to edit A DNA-cutting enzyme called Cas9 forms a complex with a short piece of RNA that has been designed to match a specific DNA sequence.

2 Search The complex moves along the DNA until it finds the matching sequence on the patient's genome, to which it attaches itself.

3 Cut The Cas9 enzyme cuts the DNA at exactly the desired point in the patient's genome. Once the cut is made, the guide RNA and Cas9 enzyme depart.

4 Paste A copy of the replacement DNA is also present, and the cell uses its own repair capabilities to insert the DNA in place.

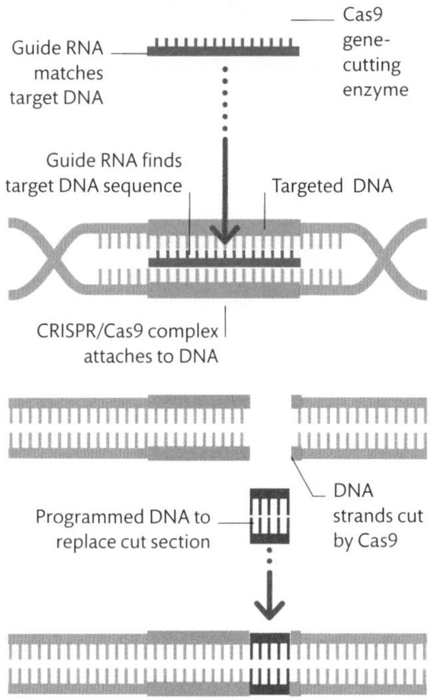
Guide RNA matches target DNA
Cas9 gene-cutting enzyme
Guide RNA finds target DNA sequence
Targeted DNA
CRISPR/Cas9 complex attaches to DNA
Programmed DNA to replace cut section
DNA strands cut by Cas9

SOMATIC VERSUS GERMLINE THERAPY

Somatic gene therapy affects genes in cells other than egg and sperm cells, so only affects the individual who receives the treatment. Germline gene therapy affects the genes in egg or sperm cells, and so will be passed on if that person has children. Germline gene therapy is banned in most countries.

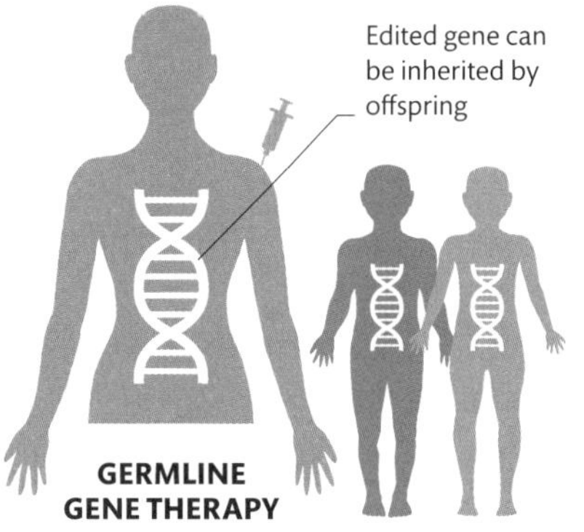
Edited gene can be inherited by offspring
GERMLINE GENE THERAPY

Cloning animals

There are two main ways of cloning entire animals. The first is embryo splitting, in which a scientist divides an embryo in two: each half can develop into an adult animal. This procedure can only make clones of developing embryos, not of adult animals. An adult mammal can be cloned by a process known as somatic cell nuclear transfer (SCNT). In this procedure, the nucleus is removed from a somatic cell (a cell other than an egg or sperm) and implanted into an egg cell whose nucleus has been removed.

THE **FIRST ANIMAL CELL** TO BE **CLONED** WAS THAT OF A **TADPOLE, IN 1952**

EMBRYO SPLITTING

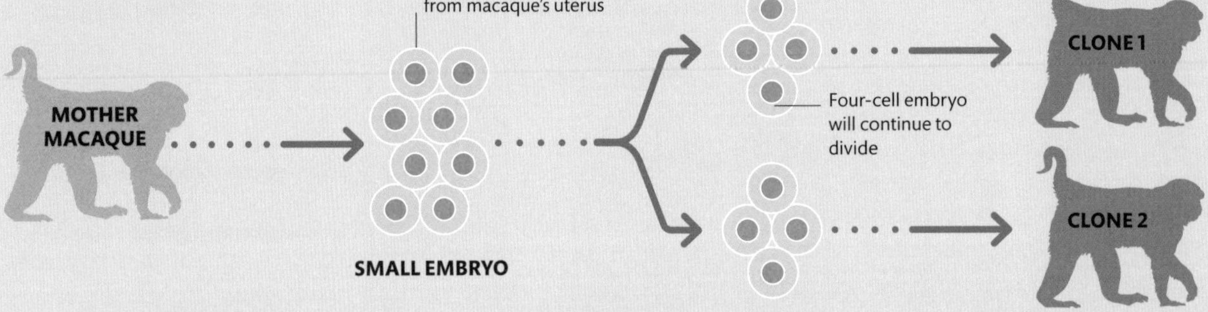
Eight-cell embryo extracted from macaque's uterus
MOTHER MACAQUE
SMALL EMBRYO
Four-cell embryo will continue to divide
CLONE 1
CLONE 2

1 Embryo extraction
Embryo splitting, also called twinning, has been carried out with many species. The first step of the procedure is to remove an embryo, typically consisting of about eight cells, from the womb of a pregnant female.

2 Dividing the embryo
A technician divides the embryo into two under a microscope. Each half consists of identical cells that are pluripotent: they can become any kind of cell, and as they divide they can produce any kind of tissue.

3 Implantation and development
The new embryos are the same as an embryo at an earlier stage of development. When implanted into the wombs of adult females of the same species, both can develop normally, resulting in identical offspring.

SOMATIC CELL NUCLEAR TRANSFER

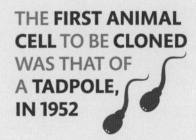

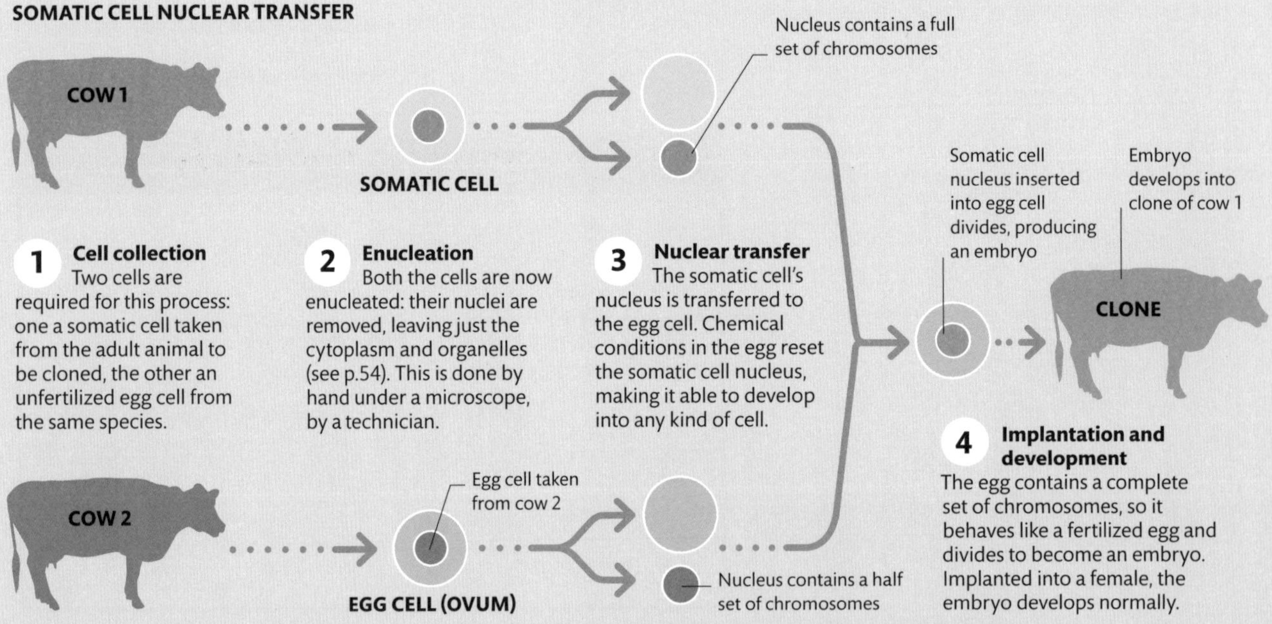
COW 1
SOMATIC CELL
Nucleus contains a full set of chromosomes
Somatic cell nucleus inserted into egg cell divides, producing an embryo
Embryo develops into clone of cow 1
CLONE
COW 2
Egg cell taken from cow 2
EGG CELL (OVUM)
Nucleus contains a half set of chromosomes

1 Cell collection
Two cells are required for this process: one a somatic cell taken from the adult animal to be cloned, the other an unfertilized egg cell from the same species.

2 Enucleation
Both the cells are now enucleated: their nuclei are removed, leaving just the cytoplasm and organelles (see p.54). This is done by hand under a microscope, by a technician.

3 Nuclear transfer
The somatic cell's nucleus is transferred to the egg cell. Chemical conditions in the egg reset the somatic cell nucleus, making it able to develop into any kind of cell.

4 Implantation and development
The egg contains a complete set of chromosomes, so it behaves like a fertilized egg and divides to become an embryo. Implanted into a female, the embryo develops normally.

Cloning

Clones are cells, tissues or entire organisms that have the same genome. Cloning is common in nature: any organism that reproduces asexually produces clones of itself. Clones can also be made artificially, in the laboratory.

Cuttings and cultures

All plants reproduce sexually, but most can also reproduce asexually, in which case the offspring are clones of the parent plant. This is called vegetative propagation. For example, new strawberry plants grow from nodes on elongated stems called runners, while new potato plants grow from swollen roots called tubers. Vegetative propagation can be achieved artificially, too, typically by taking cuttings. A biotechnology called micropropagation is used to make large numbers of identical plants, for research or for selling at garden centres.

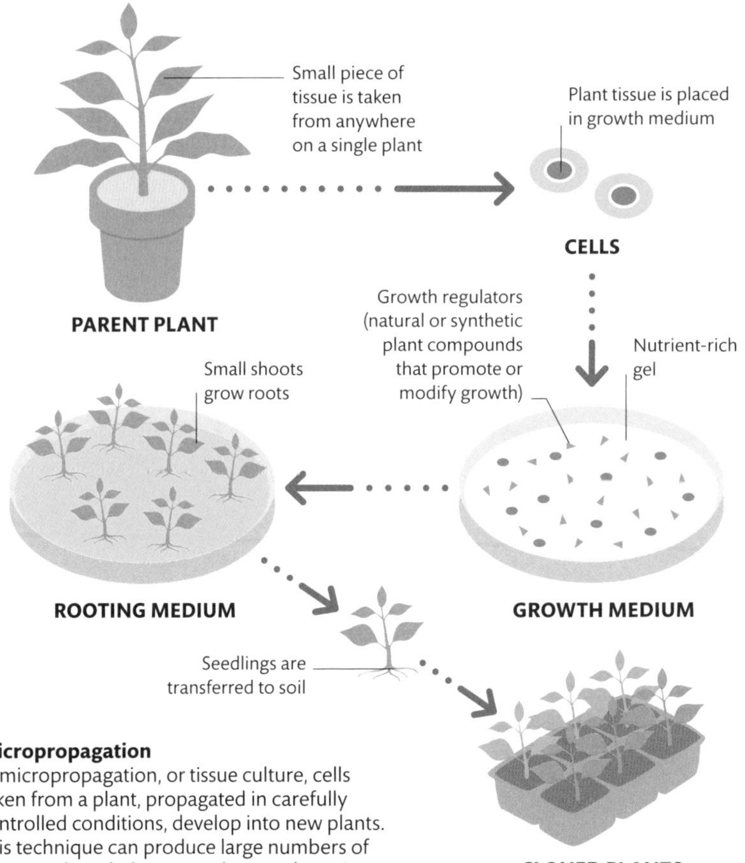
Small piece of tissue is taken from anywhere on a single plant
PARENT PLANT
Plant tissue is placed in growth medium
CELLS
Growth regulators (natural or synthetic plant compounds that promote or modify growth)
Nutrient-rich gel
GROWTH MEDIUM
Small shoots grow roots
ROOTING MEDIUM
Seedlings are transferred to soil
CLONED PLANTS

Micropropagation
In micropropagation, or tissue culture, cells taken from a plant, propagated in carefully controlled conditions, develop into new plants. This technique can produce large numbers of plants and can help save endangered species.

CAN DINOSAURS BE CLONED?

Although some biological material can be preserved over millions of years, DNA cannot. Since cloning requires an entire genome's worth of DNA, there will never be dinosaur clones.

NATURAL CLONES

About four of every thousand human births result in monozygotic twins: two babies that developed from a single zygote (fertilized egg). In the first few days after fertilization, the developing embryo splits in two. Since all the cells that make up the embryo share the same genome, so too do identical twins.

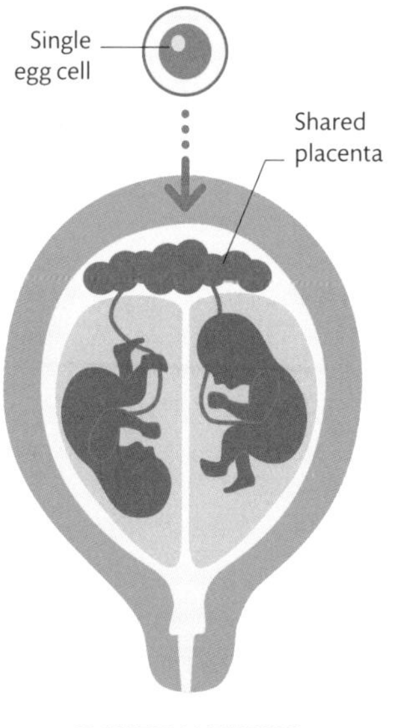
Single egg cell
Shared placenta

IDENTICAL TWINS

Anti-aging

Anti-aging technologies seek to slow down the aging process. They are not necessarily aimed at lengthening our lifespan, but rather at making us less prone to age-related diseases such as cancers and Alzheimer's disease.

What is aging?

Aging is the gradual deterioration of the adult body over time. As the human body grows older, it is subject to all kinds of damage—importantly, damage at the molecular level. Chemicals called reactive oxygen species, created by the normal processes inside cells, cause damage to DNA. Aging is also related to cell division (see pp.68–69). Individual cells grow older each time they divide and eventually stop dividing—a phenomenon known as cellular senescence.

CELLS BEGIN TO **DETERIORATE** AFTER ABOUT **50 ROUNDS OF DIVISION**

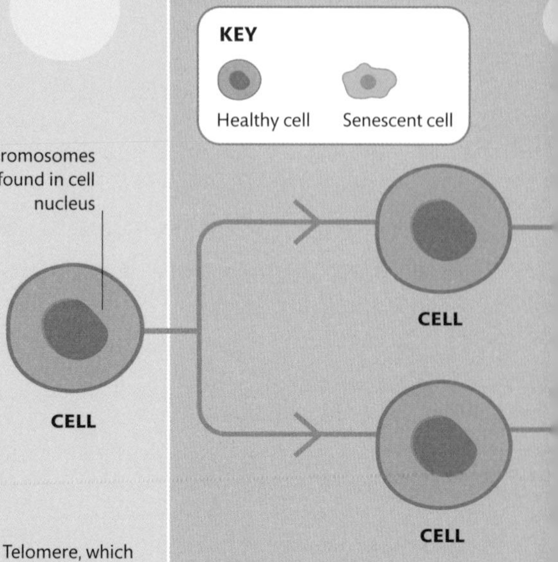

KEY
Healthy cell
Senescent cell
Chromosomes are found in cell nucleus
CELL
CELL
CELL

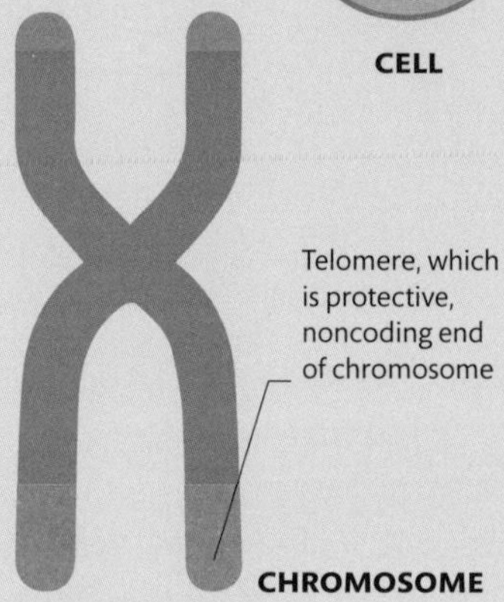

Telomere, which is protective, noncoding end of chromosome
CHROMOSOME

Senescence and chromosomes
Cellular senescence is triggered by damage but also by the shrinking of regions at the ends of chromosomes called telomeres.

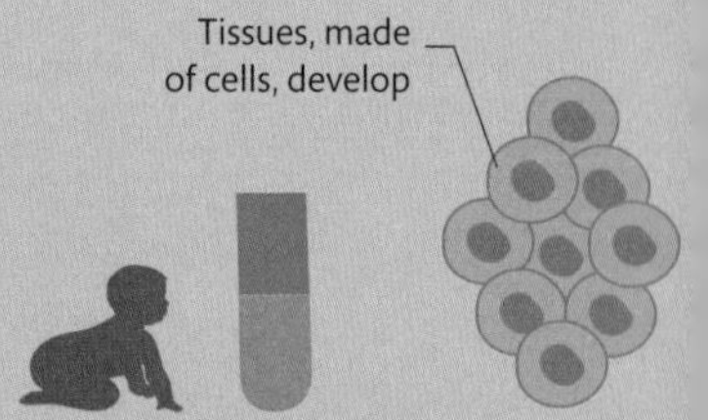

Tissues, made of cells, develop

Infancy
Growth is achieved by the division of cells. With each division, the telomeres are damaged but, as they comprise non-coding DNA, the cell function is not affected.

Rejuvenation therapies

There are several approaches to creating anti-aging technologies aimed at cellular rejuvenation. One promising approach is cellular reprogramming. As cells age, chemicals called methyl groups attach at particular points along DNA strands. Certain compounds can remove this methylation, resetting the cells to a younger state. This has to be done carefully, as removing too much methylation turns a cell back into a pluripotent stem cell, which can become any type of cell. This can lead to uncontrolled cell growth and cancer.

1 Stem cells
All cells begin as stem cells (see pp.88–89) that have little methylation and are young.

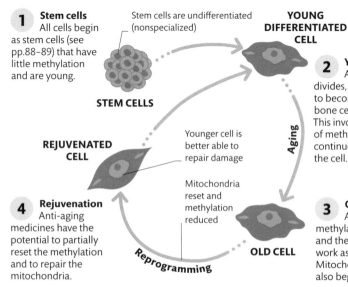

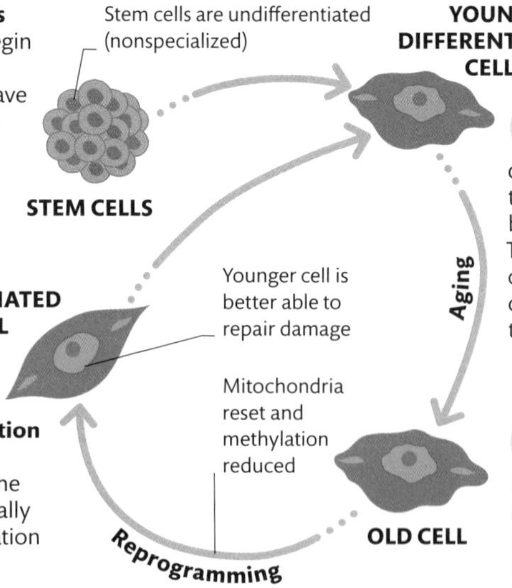

Stem cells are undifferentiated (nonspecialized)
STEM CELLS
YOUNG DIFFERENTIATED CELL
Aging
OLD CELL
Mitochondria reset and methylation reduced
Reprogramming
Younger cell is better able to repair damage
REJUVENATED CELL

2 Young cell
As a stem cell divides, it differentiates to become a skin or bone cell, for example. This involves the process of methylation, which continues, and ages the cell.

3 Old cell
As time passes, methylation increases, and the cell does not work as well as it did. Mitochondria (see p.60) also begin to degrade.

4 Rejuvenation
Anti-aging medicines have the potential to partially reset the methylation and to repair the mitochondria.

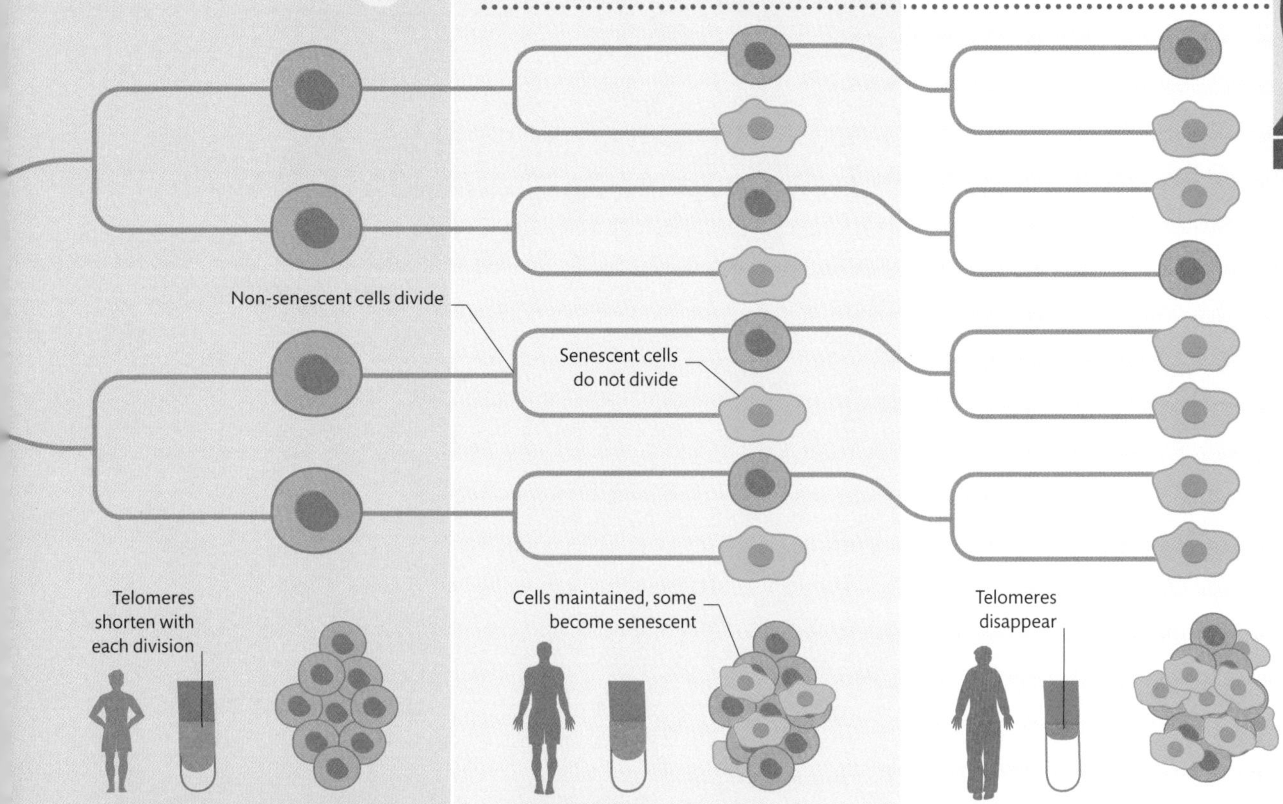
Non-senescent cells divide
Senescent cells do not divide
Telomeres shorten with each division
Cells maintained, some become senescent
Telomeres disappear

Childhood and adolescence
Growth continues during childhood and adolescence, as the body develops. The telomeres continue to shorten with every round of cell division.

Adulthood
In adulthood, the body stops growing and developing, and is instead maintained. The telomeres continue to shorten and some cells stop dividing altogether.

Old age
The body declines in old age. This is largely the result of increased cellular senescence—more cells stop dividing as the telomeres continue to shorten and disappear.

SIGNS OF AGING

As more and more cells reach senescence and stop dividing, the tissues of the body are no longer replenished, and damage is no longer repaired. These give rise to familiar signs of aging. For example, collagen in the skin breaks down, and when it is no longer replenished at the same rate, the skin begins to wrinkle and sag. In the eye, cellular debris builds up, leading to a common condition that affects eyesight, called macular degeneration.

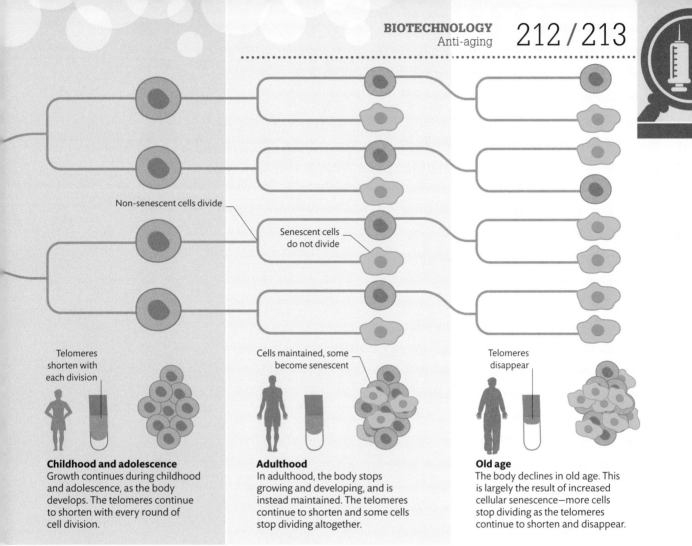

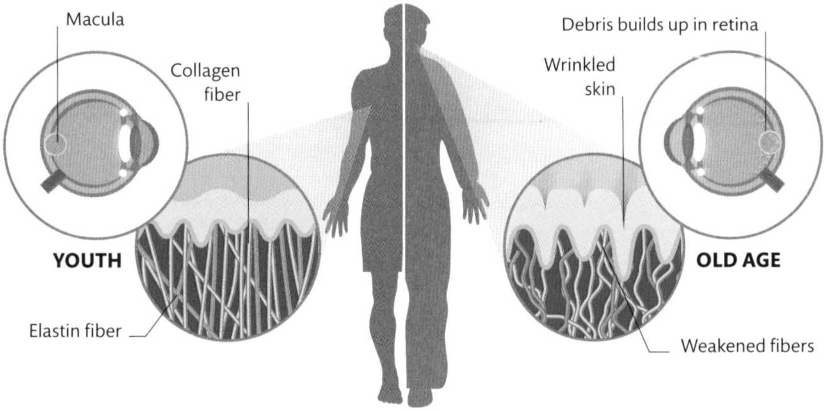
Macula
Collagen fiber
YOUTH
Elastin fiber
Debris builds up in retina
Wrinkled skin
OLD AGE
Weakened fibers

Body repair

Technologies that can replace body parts or capabilities lost through disease or injury include plastic surgery, prosthetic limbs, and devices such as heart pacemakers and cochlear implants.

Replacement limbs

Doctors have been making prosthetic limbs for hundreds of years. Some modern prosthetic limbs combine electronic and mechanical technology with detailed understanding of the nervous system. Although still largely experimental, some of these robotic prosthetics can return much of the function of the limb, including the sensation of touch.

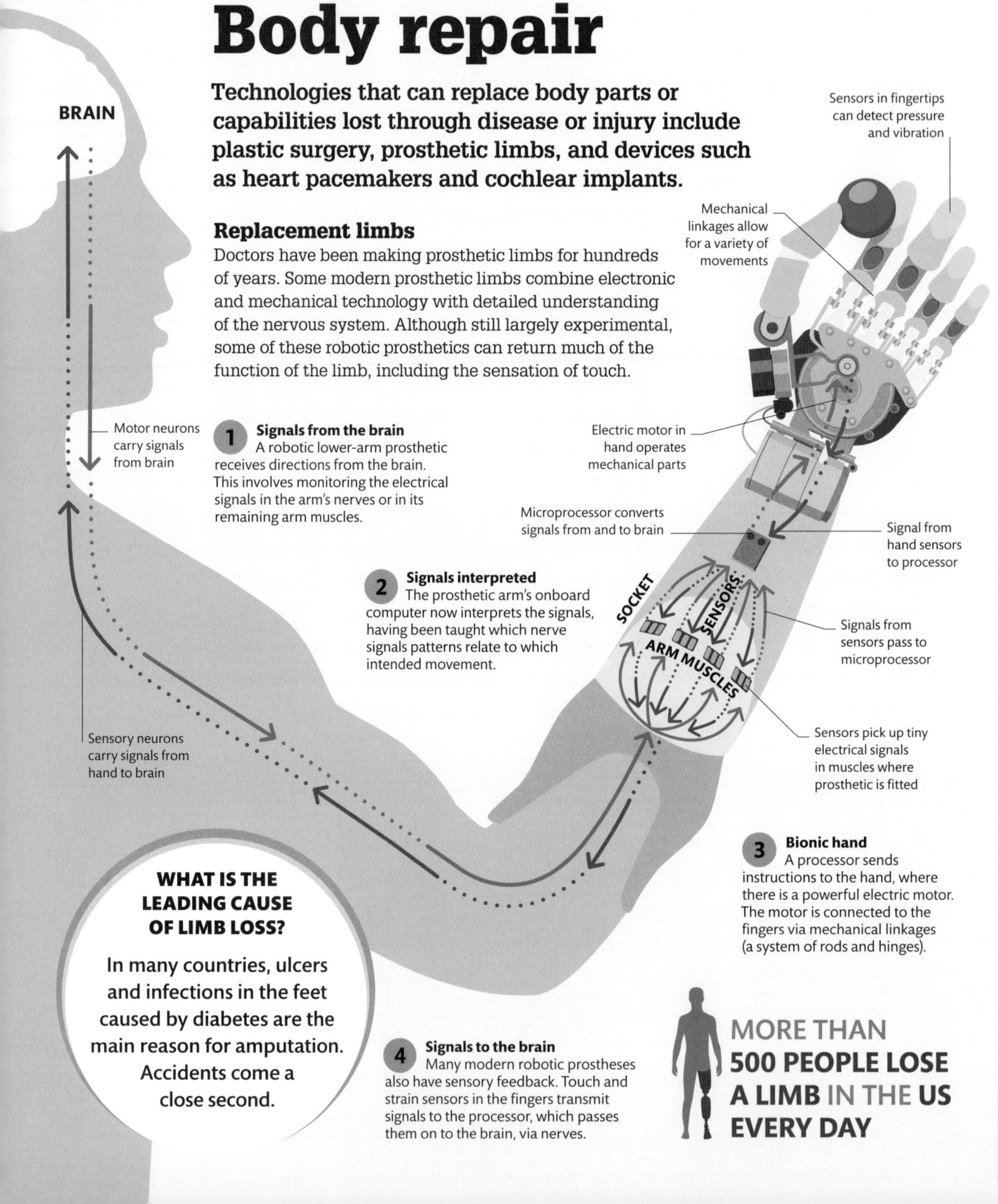

BRAIN
Motor neurons carry signals from brain
Sensory neurons carry signals from hand to brain
Sensors in fingertips can detect pressure and vibration
Mechanical linkages allow for a variety of movements
Electric motor in hand operates mechanical parts
Microprocessor converts signals from and to brain
Signal from hand sensors to processor
SOCKET
SENSORS
ARM MUSCLES
Signals from sensors pass to microprocessor
Sensors pick up tiny electrical signals in muscles where prosthetic is fitted

1 Signals from the brain
A robotic lower-arm prosthetic receives directions from the brain. This involves monitoring the electrical signals in the arm's nerves or in its remaining arm muscles.

2 Signals interpreted
The prosthetic arm's onboard computer now interprets the signals, having been taught which nerve signals patterns relate to which intended movement.

3 Bionic hand
A processor sends instructions to the hand, where there is a powerful electric motor. The motor is connected to the fingers via mechanical linkages (a system of rods and hinges).

4 Signals to the brain
Many modern robotic prostheses also have sensory feedback. Touch and strain sensors in the fingers transmit signals to the processor, which passes them on to the brain, via nerves.

WHAT IS THE LEADING CAUSE OF LIMB LOSS?

In many countries, ulcers and infections in the feet caused by diabetes are the main reason for amputation. Accidents come a close second.

MORE THAN 500 PEOPLE LOSE A LIMB IN THE US EVERY DAY

Artificial senses

Millions of people lack a sense of sight or hearing. For many years, biotechnologists have been working towards technological solutions that can provide or restore those senses. Depending upon the reason for the loss of the sense, artificial retinas may help provide sight, while cochlear implants can provide hearing. There is a long way to go before routine and effective artificial retinas will be commonly offered to people, but cochlear implants are more widely available.

1 Microphone picks up sound
A microphone fitted around the wearer's ear detects sound and converts it into electrical signals. A speech processor inside the microphone analyzes the signals and enhances sounds that it recognizes as speech.

2 Transmitter to receiver
The processed signals pass to a transmitter, which is held to the head by a magnet. The transmitter broadcasts the signal wirelessly to a receiver that has been implanted inside the skull.

3 Converting signals
The receiver converts the signals into a form that the brain will understand, and sends them through fine wires into the cochlea (see p.169), bypassing the middle ear.

4 Signals to brain
The cochlea passes on the electrical impulses to the auditory nerve, which carries it to the auditory cortex—the part of the brain that processes and perceives auditory (hearing) input.

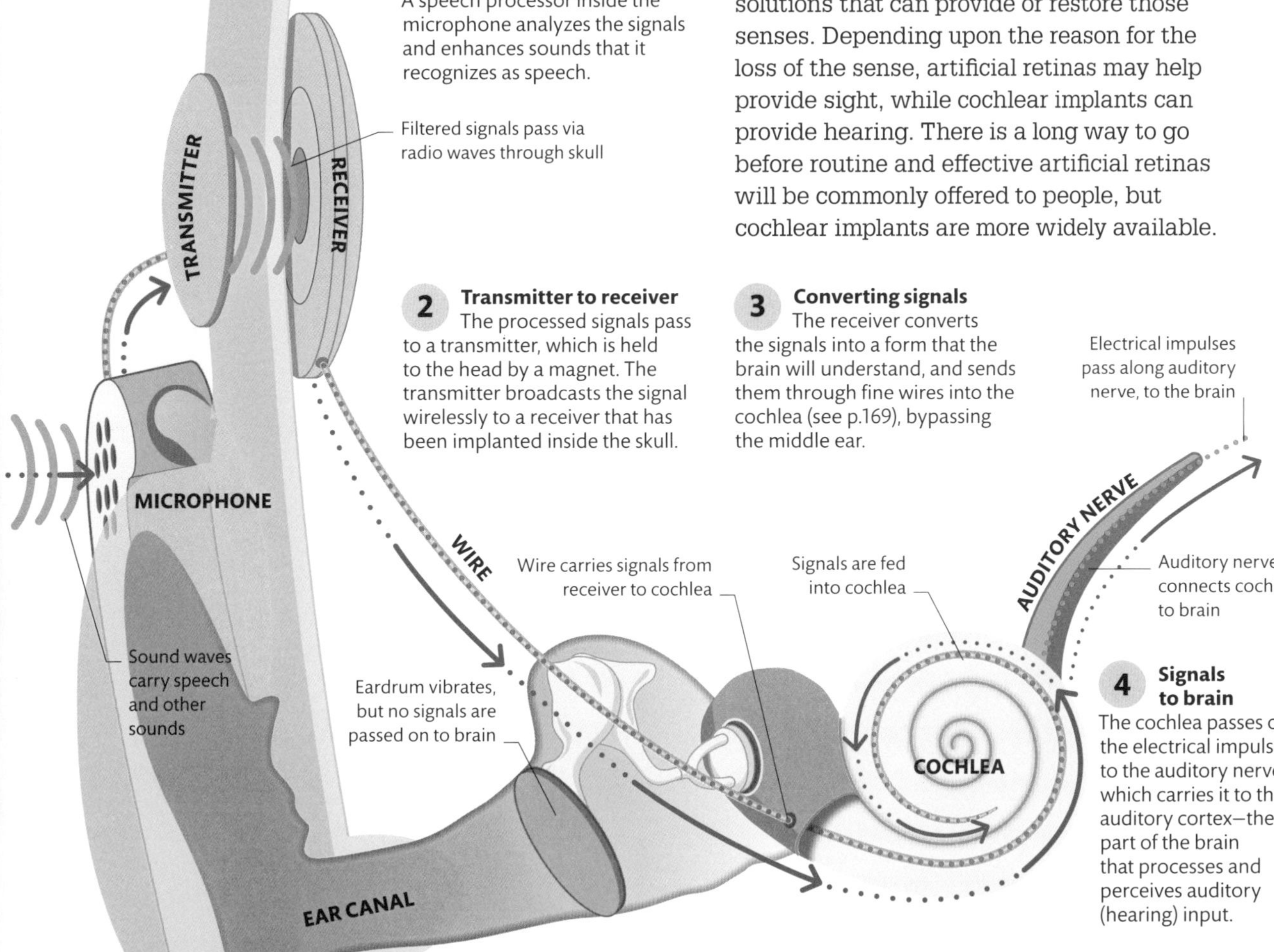
Filtered signals pass via radio waves through skull
TRANSMITTER
RECEIVER
MICROPHONE
WIRE
Wire carries signals from receiver to cochlea
Signals are fed into cochlea
Electrical impulses pass along auditory nerve, to the brain
AUDITORY NERVE
Auditory nerve connects cochlea to brain
Sound waves carry speech and other sounds
Eardrum vibrates, but no signals are passed on to brain
COCHLEA
EAR CANAL

BODY ENHANCEMENT

In the future, it could theoretically be possible to have replacement senses or body parts be fitted even if there were no need. Similarly, engineering and editing the human genome could give future generations new senses, enormous strength, or robustness against disease.

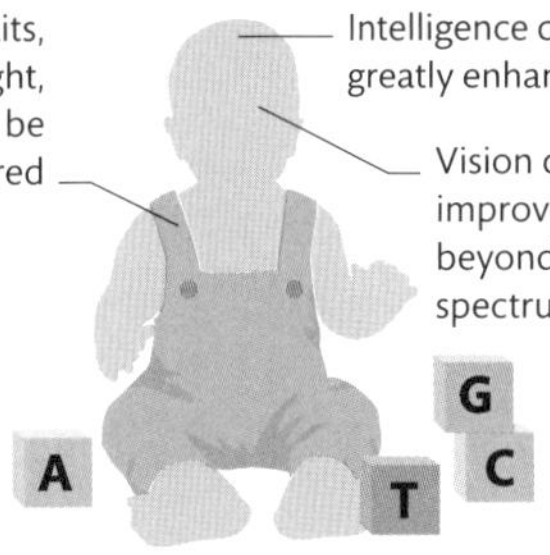
Physical traits, such as height, could be engineered
Intelligence could be greatly enhanced
Vision could be improved, perhaps beyond visual spectrum
A
T
G
C

Synthetic biology

Synthetic biology provides scientists with ways to create genes, proteins, whole chromosomes, and even entire living things that have never existed before. It has a number of exciting possible applications.

LEAD CONTAMINATION
Synthetic genes inserted into bacteria activate in the presence of contaminated water, producing a fluorescent protein.

PATHOGEN DETECTOR
Freeze-dried synthetic cell parts on a strip of paper report the presence of a pathogen by changing color.

VEGAN MEAT
It may be possible to produce meats that have the texture and taste of meat by creating proteins and inserting them into microorganisms.

BREWED MILK
It will one day be possible to create microorganisms with which producers can create milk, without the need for cows.

DRUG COMPOUNDS
It will one day be possible to create new medicinal compounds to order, by creating synthetic genes and inserting them into bacteria.

SPIDER SILK
Proteins that are as strong and durable as spider silk could be produced by microorganisms with synthetic genes inserted into their genomes.

SYNTHETIC PIGMENTS
Researchers are working on synthetic organisms that can produce pigments more sustainably than current pigments, with which to dye textiles in bulk.

CROP PROTECTION
Synthetic insect pheromones, held in a container at the edge of a field, could attract insects away from crops, reducing the need to use insecticides.

Practical applications
The creation of genes and proteins that do not otherwise exist in nature holds great promise for medicine, materials science, environmental monitoring, and many other areas of technology. Synthetic biologists around the world are working on a variety of projects, with some encouraging results.

Proteins made to order

Synthetic biology goes further than genetic engineering (see pp.206–207). Instead of transferring existing genes between species, biologists create new genes, which code for proteins that do not yet exist in nature. Synthetic genes are inserted into organisms such as bacteria, and the bacteria get to work producing the new proteins. It is most often the shape of a protein that gives it its usefulness. Determining a protein's shape from the coded instructions is a hard problem, so this technology is still in its infancy.

XENO-NUCLEIC ACID (XNA)

One area of research is looking to extend the genetic code by using six bases rather than the normal four (see p.37). In living cells, the resulting nucleic acids code for amino acids that cannot be made with the natural code, and therefore for proteins that could never exist in nature.

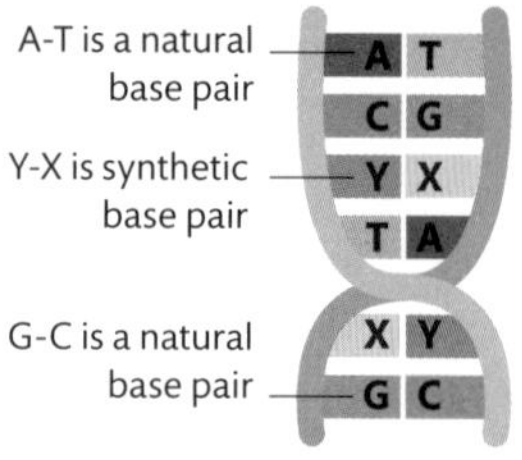
A-T is a natural base pair
Y-X is synthetic base pair
G-C is a natural base pair
A T
C G
Y X
T A
X Y
G C

6 NUCLEOTIDES
3 BASE PAIRS

Synthetic organisms

As the culmination of a project to determine the smallest genome that could sustain a living organism and reproduce, researchers created an organism (called *Mycoplasma laboratorium*) with a synthetic genome. It was a simple bacterium, and its genome was produced in a laboratory, base-pair by base-pair. The genome, based on the DNA of a bacterium that exists naturally, was inserted into cells whose natural DNA had been removed. The cells replicated successfully.

WHY IS NASA EXPLORING SYNTHETIC LIFE?

Creating life forms that do not exist on Earth can help NASA explore biology that may be possible on other planets—so-called exobiology.

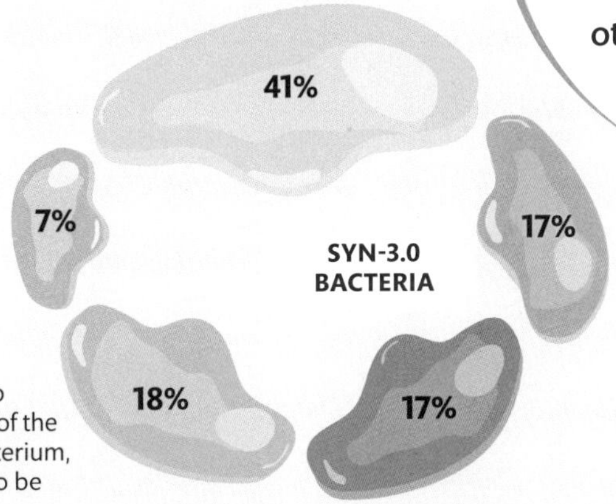
KEY
Gene expression
Preservation of genome
Membrane function and structure
Metabolism
Unknown function
41%
7%
17%
SYN-3.0
BACTERIA
18%
17%

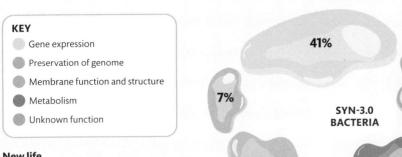

New life
The genome of *Mycoplasma laboratorium* (also known as Syn-3.0) has a total of 473 genes. All of the genetic code was copied from an existing bacterium, and some parts of the genome, while shown to be essential for life, are still of unknown function.

154 OF SHAKESPEARE'S SONNETS HAVE BEEN STORED ON SYNTHETIC DNA

Aptamers

Scientists can create short pieces of synthetic DNA or RNA that will bind to certain DNA sequences in living things. These are aptamers, and they can act as targeted drugs or drug-delivery systems. Some are designed to act as synthetic antibodies or to test for certain diseases. The process of making aptamers involves generating random sequences of RNA or DNA with a gene machine, then seeing which ones fit the target molecule, such as a specific part of a disease-causing pathogen.

Targeted drug delivery
The aim of aptamer designers is for the 3D shape it forms to dock into a pathogen, to deliver a drug molecule in a perfectly targeted way. This targeted approach could also cure diseases such as cancers.

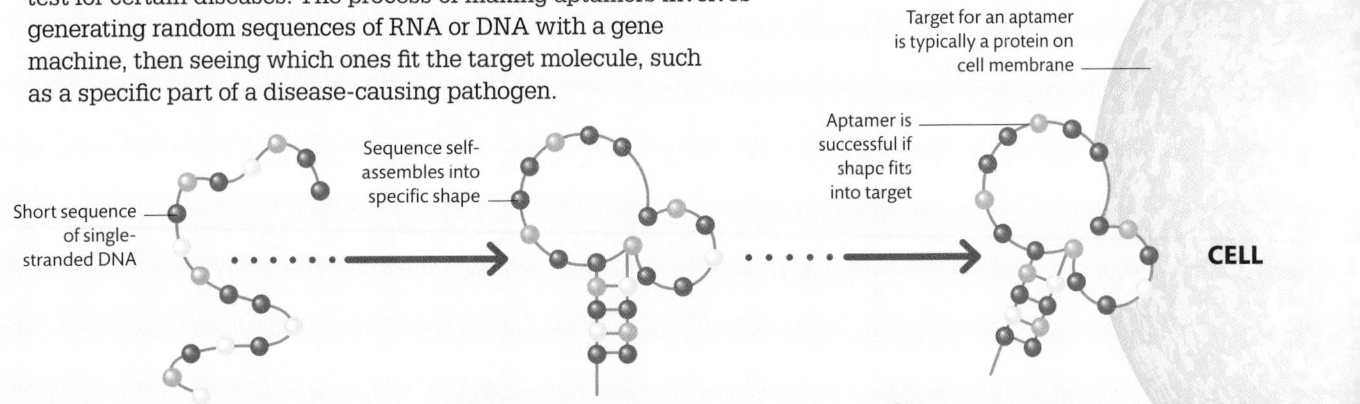
Target for an aptamer is typically a protein on cell membrane
Aptamer is successful if shape fits into target
Sequence self-assembles into specific shape
Short sequence of single-stranded DNA
CELL

1 Sequence
An aptamer is a sequence of single-stranded DNA or RNA. Researchers typically create millions of possible sequences then, through trial and error, find the one that folds to produce the desired shape.

2 3D shape
While the forces between the bases of a double-stranded length of DNA cause it to form a double helix, those same forces cause a single strand to fold into a three-dimensional shape, just as a protein does.

3 Binding to target
Researchers attach a drug molecule to the aptamer they create. When the aptamer enters the body, it will attach to a specific part of the target pathogen or diseased cell but nowhere else.

Index

Page numbers in **bold** refer to main entries

E

F

G

Acknowledgments

DK would like to thank the following people for help in preparing this book: Tom Jackson for help with planning the contents list; Victoria Pyke for additional editing; Helen Peters for compiling the index; Ann Baggaley for proofreading; Senior DTP Designer Harish Aggarwal; and Jackets Editorial Coordinator Priyanka Sharma.

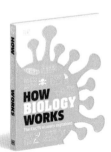

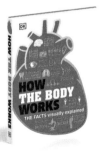

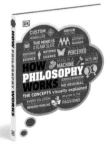

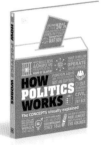

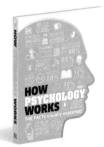

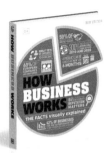

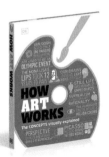

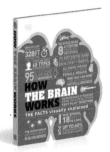

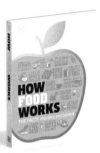

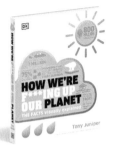

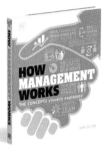

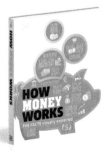

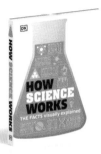

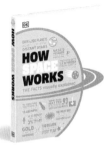

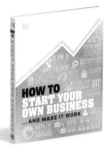

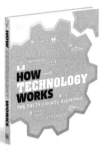